FUNDAMENTALS OF AIR POLLUTION

FUNDAMENTALS OF AIR POLLUTION

SAMUEL J. WILLIAMSON *New York University*

ADDISON–WESLEY PUBLISHING COMPANY
Reading, Massachusetts
Menlo Park, California · London · Amsterdam · Don Mills, Ontario · Sydney

Cover photograph: A. Devaney, Inc., N.Y.

Smoke from many sources drifts toward New York City and the skyscrapers of Manhattan. Because of enacted emission controls such scenes are much less frequent than in 1966 when this photograph was taken. (Courtesy of Mr. A. Proudfit, Sign X Labs, and the U.S. Public Health Service)

To My Parents

PREFACE

This is a book on air pollution. My main concern is with the physical aspects of the phenomenon—the causes and effects. These can be appreciated only if we extract from traditional academic disciplines the essentials which bear on the matter. Thus I approach the subject with an interdisciplinary view.

The effort to write this book was stimulated in part by a growing realization that new lines of communication will have to be developed to meet the challenge posed by the very complexity of the air pollution problem. Especially needed is communication between the scientific and technical disciplines themselves; but no less important is better communication between the disciplines and society. This book is a small contribution to these ends.

Since our focus is on the physical aspects of air pollution, the book is, in this sense, scientific. Our purpose is not to press the polemics of the clean air crusades. Many other books have done that. Rather we hope to draw an objective appraisal of what is known or unknown, what is tested or only theoretical, and what is empirical or speculative. Not only do we need to know what our present level of sophistication can predict, but with what degree of certainty the predictions can be made. Only with these scientific and technical factors in mind can the application of value judgments, inherent in the political decision-making process, be applied to the control of air pollution on a rational basis.

A large economic investment will be required to turn the tide on pollution by the end of the century. In the opinion of some, a change in the lifestyle of the population will be required: more efficient utilization of energy resources, restriction of the convenience and freedom of travel, and reshaping of community patterns of commerce. These suggestions have profound implications, for they foresee a need for a major change in social attitudes. The motivation will come only if the need is clearly demonstrated. And the need can be determined only from an analysis of the medical, scientific, and technical aspects of the current status and future trends.

To achieve the necessary interdisciplinary viewpoint I draw from the fields of meteorology, chemistry, physics, medicine, psychology, and various areas of engineering. Evidently a fair amount of groundwork must be covered in some disciplines to extract their useful contributions. We have attempted to minimize this by taking a very narrow view of the relevant subject and going straight to the heart of the matter. A reader of this book is by no means justified in presuming that he will be instantly converted into an expert in any one of the disciplines. In fact, scientists from any one discipline will undoubtedly be unhappy with the way the concepts of their discipline are treated. But this is the limitation imposed on a book that must cover a lot of territory.

As the title indicates, our emphasis is on the *fundamentals*. This means an *explanatory* as distinguished from a *descriptive* approach. The level of the presentation presumes that the reader has a scientific background at least equivalent to about a year of college-level physics and chemistry.

Although I concentrate on physical reasoning, I have included a mathematical analysis in some sections where this is essential for better understanding, or for comprehension of, the quantitative aspects of the subject. The importance of a quantitative approach cannot be overemphasized, but mathematics itself has been relegated to a place of secondary importance in the pedagogical exposition. Complex statistical features have been eliminated from the explanation of most phenomena; although this has not compromised the accuracy of the physical picture, it has allowed me to avoid explicit use of calculus everywhere but on a few pages of Chapter 7. The introduction of calculus notation in a few equations should not present an insurmountable obstacle to the uninitiated reader who is principally interested in their implications.

Since we shall deal with quantifiable aspects of pollution, we are faced with the question of what is the most useful choice of units in which to express these quantities—the English system, metric system, or a hybrid system. A glance at the literature in the pertinent disciplines reveals a cornucopia of choices. Take the units of energy as an example. Chemists prefer to express an amount of energy in terms of the number of calories, physicists often prefer joules or ergs, engineers use British Thermal Units, spectroscopists like cm^{-1}, and so on. For uniformity and in anticipation of future trends, the Système International will be used as the principal system. The basic units for length, mass, and time are the meter, kilogram, and second (or MKS units). Occasionally only the centimeter-gram-second (CGS) value of a quantity will be given, when the numerical value can be more simply stated. Those readers who cut their teeth on the English system may find it convenient to equate (approximately) the length of one yard with the length of one meter, the length of 60 miles with 100 kilometers, and a weight of 2.2 pounds for the mass of 1 kilogram. Other conversion factors are tabulated in Appendix B. I draw special attention to my use of the *metric ton* for mass (labeled Tonne and abbreviated T), in contradistinction to the short ton for weight (abbreviated *t*), commonly used in the United States, and the long ton formerly used in Great Britain. Of course electric power, even in the United States, is invariably measured in the MKS system, since volts and amperes are MKS quantities. The rate of generation or utilization of one joule of energy each second is equivalent to a power generator or utilization of one watt. Perhaps this association will enable the reader to acquire an intuitive appreciation for the magnitude of one joule of energy (an MKS unit). When a quantity is commonly found in the literature expressed in units different from the MKS units, the second form will often be included in this text as well.

The diverse readership of a book such as this presents special problems for the author. It is unrealistic to assume, for example, that anyone but a chemist or chemical engineer is familiar with the nomenclature of organic chemistry; yet knowledge of some terminology is essential for appreciation of Chapter 10 on Photochemical Smog. Similarly it is unlikely that anyone but a physicist or meteorologist is familiar with the essential features of thermal radiation that play key roles in air pollution meteorology. These subjects and others are briefly treated in appendices, so that the reader can refer to them should he feel the need.

The subject of air pollution is a fabric of cross-woven themes. It is possible to approach the topic from any of several directions. For example, one could discuss each pollutant individually in terms of its sources, methods of emission, how it is affected by chemical reactions in the atmosphere, adverse effects, and ultimate fate. Or by contrast one could deal with pollution from a functional approach, and emphasize the common features shared by a class of pollutants. This book emphasizes the functional approach because in most regions where air pollution is considered a problem, a pollutant coexists with other pollutants and there is a continual evolution of the nature of the atmosphere through chemical reactions and changes in phase of the constituents. Therefore it makes more sense to discuss the evolution and effect of the milieu created by the aggregate rather than concentrate on the individual constituents. This is done when we consider the evolution of sulfurous smog and again for photochemical smog. Another reason for choosing the functional approach is that it provides a logical framework within which the relative contributions of individual constituents can be compared. Thus, for example, we shall focus on the general subject of odors and their characterizations, and, within that context, compare the relative potency of various odorants. In similar fashion we shall take up visibility reduction, respiratory irritation, and climatic influences.

A large part of this book is devoted to meteorology, since atmospheric dynamics is a significant determinant of whether the emission of a given amount of contaminants will cause adverse effects. If the problem of air pollution is to be solved, we must be aware of the constraints imposed by natural processes; meteorology, as it were, establishes the "boundary conditions" of the problem.

Although most of the book concentrates on the physical aspects of air pollution, I have devoted the final section, Chapter 12, to a development of the interface between the technical and political aspects of the air pollution problem. This section deals with concepts which may be applied to the control of pollutant emissions and indicates how social factors may influence the mode by which controls are introduced.

A course which does not have sufficient time to cover this book in detail could most advantageously omit several sections which are devoted to more advanced subjects: Sections 7.2 through 7.8 on Effluent Dispersal from Point and Extended Sources, and Chapter 11 on Aerosols. Organization of the text and cross-references are designed to minimize dislocations for those who do not read these chapters.

Anyone who has taken a science course knows full well that problem-solving is an essential part of the learning process. Problems probe one's depth of understanding. Thus at the end of most chapters I have included a selection of qualitative questions and quantitative problems. Some of these delve into areas not explicitly covered in the text and are meant to stimulate the reader's imagination. Occasionally the quantitative problems will require a knowledge of calculus for their solution, and these are gathered together at the end of the section. I have emphasized variety in choosing the subject matter of the questions, at the expense of not including many similar questions which would give the reader "practice" in solving a given type of problem.

ACKNOWLEDGMENTS

Many of the essential points which this book emphasizes were first brought to my attention during a year-long effort in which a number of residents of California's Ventura County sought to have effective air pollution controls implemented by the Air Pollution Control District. I am particularly indebted to George M. Hidy, D. Eng., Newton Friedman, M.D., Duane Lea, M.A., and Roger Helvey, M. A., for sharing their insights. Special acknowledgment is due to Joseph Kuczek whose advice and expertise in political life provided us with the constant reminder that technical factors are but a part of the problem which we know as air pollution. Subsequently, I have greatly benefited during participation in the evolution of a new interdisciplinary course on Environmental Studies at the Santa Barbara campus of the University of California, and acknowledge informative discussions with Professors John Crowell, Richard Martin, and Peter Mason. I thank Professor William D. Sellers of the University of Arizona, Professor R. Lee Byers of the University of New Hampshire, and Professor Samuel S. Butcher of Bowdoin College for their useful comments on the manuscript. I also acknowledge the valuable suggestions from several colleagues in the Department of Physics at the University of California, Santa Barbara, and thank the entire department and its chairman Professor Vincent Jaccarino for their warm hospitality. And to Mrs. Alice Kladnik goes special praise for her very capable typing of the manuscript. Finally, I acknowledge deep gratitude to my wife Joan B. Williamson for her patience, understanding, and support through a difficult interlude during which work on this book was undertaken.

New York   S. J. W.
September 1972

CONTENTS

CHAPTER 1

INTRODUCTION

Man uses the atmosphere as both a resource and a place for depositing waste. He takes from it oxygen as a necessary ingredient for his industrial activities and indeed his own biological processes. He returns to it a mixture of gases and solids, the by-products of combustion, respiration, and other energy-transferring activities. He has expected Nature to cleanse out his wastes. But we now recognize that Nature cannot do this—not entirely. The natural cycles have their own peculiar rates of, and capacities for, action. It takes time for winds to remove wastes from the atmosphere over a city, time for vegetation to inspire some of the gases, time for soot to settle. The historical development of urbanization and industrialization has produced geographical regions where the natural balance is disturbed. Wastes are emitted into the air at such enormous rates that the natural process of scavenging cannot keep pace. We have all experienced the result: air pollution.

The term "air pollution" has many possible definitions. A useful one is to regard air pollution as the presence in the atmosphere of a substance or substances added directly or indirectly by an act of man, in such amounts as to affect humans, animals, vegetation, or materials adversely. What is classified as a pollutant therefore depends upon a recognition of which substances cause adverse effects. It is an ever-changing definition. Many centuries ago only soot and odorant gases may have been considered pollutants. Now we recognize that pollutants can cause more subtle effects than soiling of clothing or disagreeable odors. Some gaseous pollutants are colorless and odorless, and particles too small to be seen can adversely affect man. Even carbon dioxide may be considered a pollutant.

At the outset it would be well to dispel some confusion concerning the difference between a *pollutant* and a *contaminant*. The current tendency is to regard a contaminant as anything added to the environment that causes a deviation from the geochemical mean composition. Thus contaminants are introduced both by *natural sources*, such as volcanic eruptions and forest fires, and by *anthropogenic sources*, those associated with man's activities. Dust and pollen are generally considered contaminants; but either would properly be classified a pollutant if it had even an indirect anthropogenic origin and were present in sufficient amounts to have an adverse effect. A deviation from the natural composition of the environment need not cause adverse effects, and therefore the label "contaminant" does not always carry the odious connotation associated with the word "pollutant." In many contexts, however, the words are synonymous.

1

By and large, the ways in which pollutants are created and released into the atmosphere are understood from a scientific standpoint. However, why they continue to be released, often to the detriment of human beings, involves considerations which are based upon factors other than those embraced by the domain of science and technology. They include the economic, political, and psychological motivations that traditionally influence human decisions. If solutions are to be found to alleviate the problem of air pollution, they must necessarily recognize the complex nature of the causes. Individual initiative to date has not solved the problem, so governments at various levels have shouldered the responsibility of initiating pollution controls. It is often suggested that substantial advances will be forthcoming only if there is a social or political about-face, from which develops a greater dependence by citizens upon mass transportation, more efficient use of our energy resources, and control over population growth. It is not our purpose to discuss these aspects of the air pollution problem. Certainly many novel social, political, and legal concepts must be introduced to meet the demands that citizens now make for cleaner air. But solutions to the problem of air pollution will not be found unless the *physical* nature of the phenomenon is understood. We must learn where the effort can most effectively be applied.

This book is concerned with the fundamental explanations for the physical causes, evolution, and effects of air pollution. As such, our view of the subject is narrow. Yet it is an essential view, for rational progress cannot be made without an appreciation of the environmental consequences of political and legal decisions. Nor will many of these decisions be motivated without the predictive abilities of the sciences. It may be true that science and technology alone cannot solve the problem, but it is certainly true that the problem cannot be solved *without* science and technology.

Air pollutants are classified according to two categories: *Primary pollutants* are those directly emitted into the air. The most important from the standpoint of the amounts emitted are particulate matter, sulfur oxides, carbon monoxide, hydrocarbons, and nitrogen oxides. In most communities, stationary sources such as power plants, incinerators, and heavy industries contribute the major portion of the first two; and automobiles emit the greater share of the last three. However, these are not the only common pollutants that burden our atmosphere. Under the proper conditions, primary pollutants can undergo chemical reactions within the atmosphere and produce new substances known as *secondary pollutants*. Some reactions require the energy of sunlight to proceed at an appreciable rate, and these are appropriately known as photochemical reactions. Perhaps the best known secondary pollutant is ozone, a product often formed by photochemical reactions in the polluted air over Los Angeles, as well as over other cities with similar primary pollutants and intense sunlight.

The adverse effects of pollution are particularly evident in metropolitan regions where many sources contribute to the atmospheric burden. As a consequence of the mixing of pollutants from different sources, the effect of the emissions from any one is not identifiable. The name *smog* had been given to such community-wide polluted air. The origin of this word is uncertain, but it is known to be a contraction of the words "smoke" and "fog." It may have been first introduced at a public health conference in 1905 by H. A. Des Voeux, a London physician, to describe the frequent condition of the atmosphere which enveloped many British towns; and it was thereafter popularized as a consequence of Des Voeux's report in 1911 to the Manchester Conference of the Smoke Abatement League concerning conditions in Scotland two years earlier, when the combination of smoke, sulfurous gases, and fog was believed to have claimed over a thousand lives in Glasgow and Edinburgh.

It is now recognized that there are two distinct types of smog. One of these obtains its characteristics from the high concentration of sulfur oxides in the air, which may result from the use of sulfur-bearing fuels; this occurs in many cities in Europe and the eastern United States. The conditions are aggravated by dampness and a high concentration of suspended particulate matter in the air. We shall call this *sulfurous smog*. The other smog is typified by the atmospheric conditions over Los Angeles, in which nitrogen oxides and hydrocarbons undergo photochemical reactions and produce ozone and other agents which are chemical oxidizers. This phenomenon depends upon neither smoke nor fog. It is a form of community air pollution known as *photochemical smog*.

The reason why air pollution has become a major problem in some urbanized regions is simple. Emissions of contaminants from anthropogenic sources have continually risen in quantity. However, the volume of air in which they are diluted is determined largely by meteorological factors, and although these display marked variations from day to day, the amount of air circulating over a locality is not on the average increasing. Therefore the average concentrations of pollutants in some regions of the air environment are increasing. We are exposed to more and more pollutants in the ambient air, that air which surrounds us. Common sense indicates that the severity or frequency of adverse effects should also increase. To avoid adverse effects, a balance must therefore be maintained between the amount of contaminants we put into the air and the volume of air available for diluting the substance to a safe level.

To appreciate the enormous volume of air required by some pollution sources to dilute the effluent to inconsequential levels, let us take the automobile as an example. One familiar pollutant it emits is carbon monoxide. An automobile of 1965 vintage emitted about 30 grams of this gas during each kilometer (0.6 miles) of travel. One's first impression might be that this is a small quantity; however, diluting it to a concentration where it is considered to have no adverse effect would require about 2 million liters of air (about 2 million quarts of air). We can appreciate the magnitude of this amount of air when we realize that it takes an entire day for 200 average adults to breathe the same quantity! Great volumes of air are thus needed to dilute pollutants from some sources to safe levels. In many regions of the world, we do not have that much air.

Fortunately, all substances which enter the air do not remain there. Nature has methods by which gaseous and particulate matter may be scavenged from the atmosphere. Although these methods are imperfectly understood, it is known that chemical reactions within the atmosphere, rainfall, sedimentation as a result of gravity, and absorption by vegetation are important mechanisms. However, these natural processes are not always capable of coping with the increasing rate of emission of pollutants, particularly in the urban environment. This is even true on a global scale for certain pollutants. The steady increase in the ambient concentration of carbon dioxide is believed to be an example. The scale of man's influence is now commensurate with Nature's total capacity. No longer may we assume that we can pollute with impunity and that the wind will carry away all our problems.

The toll exacted by the adverse effects of air pollution cannot exactly be gauged, for its effects are often insidious. Some economic effects have been estimated: The annual loss from damage to crops, plants, trees, and materials in the United States alone may amount to as much as $5 billion.[1] And estimates of the health expenses and lost income due to disease range from $2 billion to $6 billion.[1,2] These accountable costs are impressive; and still they do not include a measure for the aesthetic degradation of the environment resulting from numerous assaults on our sensibilities. As an indication, it has been estimated that property values in the United States have been lowered as much as $5 billion as a result of pollution.[1] Yet such estimates fail to include the many health factors that are suspected but not proven to result from exposure over long periods of time to polluted air.

In fact, very little is known about the seriousness of long-term exposure of humans to current levels of community air pollution. In some instances where abnormally severe conditions were sustained for several days, there was an increase in both *mortality* (death rate) and *morbidity* (incidence of health deterioration, however temporary). The often-quoted examples include an episode in Belgium which occurred in the Meuse Valley in 1930, when meteorological conditions led to stagnation of the overlying air and a build-up of pollutant concentrations during a week-long period. A large number of people became ill with respiratory complaints, and 60 died. Other episodes include an incident in Donora, Pennsylvania, in 1948 when almost half of the population of 14,000 became ill and 20 died, and a four-day "killer fog" in London which resulted in 4000 deaths in 1952. But these are rare events, for they occurred only when a population was subjected to a very high level of pollution for several days in succession. Only recently has statistical evidence been obtained for an association between the daily mortality in a large city and the ambient level of pollution, even though the levels are within the realm of the normal concentrations to which the population is exposed.[3] Increases in mortality rates with high pollution levels are particularly evident for older people. Practically unknown, however, are the subtle effects which might result from long or repeated exposure (known as *chronic exposure*) to pollutants which are present in the ambient air at low concentrations. Whether man can adapt to a polluted environment or whether chronic exposure will cause irreparable injury is a question which remains to be answered.

In the next chapter we shall turn our attention to a detailed consideration of some of the adverse effects of air pollution. This will show us the important parameters by which the levels of pollution should be gauged. Perhaps the most significant is the ambient concentration of a pollutant. Hence two factors can be seen to be important in influencing the pollution phenomenon: (1) the amount of pollutant emitted by a source, and (2) the meteorological conditions which affect the pollutant's dispersal. Consequently, the first portion of this book is devoted to an examination of the features of the atmosphere and atmospheric dynamics to gain an appreciation of the conditions under which air pollution is likely to be a problem. This examination also affords us an opportunity to investigate the widespread effects of air pollution, including its possible influence on global and local climatology. Dispersal is also affected by the techniques by which pollutants are released, whether from home fireplaces, a factory smokestack, or automobiles on an expressway. After an evaluation of these factors, we shall be in a position to study the sources of pollutants and trace their evolution as they respond to meteorological conditions and chemical reactions within the atmosphere. We shall focus on the two common types of smog—sulfurous and photochemical—and assess the importance of their adverse effects. Individual contaminants found in these smogs and their effects will also be considered. As we take up the major sources of pollution, we shall also point out how emissions can be reduced and mention proposals for future advances in control methods and devices. Many gaseous pollutants are converted into solid particles through chemical reactions in the atmosphere, so we shall conclude our study of the phenomenon of pollution with an examination of the dynamics of particulate matter and how it is scavenged from the atmosphere by natural processes. In the final chapter we shall explore the concepts which have been applied by governmental bodies to monitor and control air pollution. The discussion will show how value judgments enter the decision-making process and will define the interface between the technical and political sides of the air pollution problem.

A reader who wishes to pursue in more detail subjects covered in this book may consult the list of references under "For Further Reading" included at the end of each chapter. In addition, he is encouraged to consult the three-volume series *Air Pollution*, edited by A. C. Stern (Academic Press, New York, 1968). A more condensed survey can be found in *La Pollution Atmosphérique* by Jean Paul Détrie (Dunod, Paris, 1969). These references are important, because this book is not intended to be an encyclopedia that covers all aspects of air pollution. Many contaminants such as asbestos, fluorides, and ammonia will not be discussed in detail, because they are not commonly experienced in appreciable amounts by the general population. With the basic principles which the reader should gain from this book, he should be in a good position to consult the more inclusive summaries and the most recent professional literature.

NOTES

1. White House Council on Environmental Quality 1970 Annual Report.
2. L. B. Lave and E. P. Seskin, "Air Pollution and Human Health," *Science*, **169**, 723 (1970).
3. T. A. Hodgson, "Short-Term Effects of Air Pollution in New York City," *Environ. Sci. and Tech.*, **4**, 589 (1970).

FOR FURTHER READING

ARTHUR C. STERN, Ed	*Air pollution*, Vols. I, II, and III., 2nd ed., New York: Academic Press, 1968.
ARTHUR C. STERN, Ed	*Air Pollution*, Vols. I and II, 1st ed., New York: Academic Press, 1962.
JEAN PAUL DETRIE	*La Pollution Atmosphérique*, Paris: Dunod, 1969.
A. R. MEETHAM, D. W. BOTTOM, and S. CAYTON	*Atmospheric Pollution*, Oxford: Pergamon Press, 1968.
SEYMOUR M. FARBER and ROGER H. L. WILSON, Eds.	*The Air We Breathe*, Springfield, Ill.: Charles C. Thomas, 1961.
PAUL L. MAGILL, FRANCIS R. HOLDEN, and CHARLES ACKLEY, Eds.	*Air Pollution Handbook*, New York: McGraw-Hill, 1956.

More popular editions

RICHARD SEGAR SCORER	*Air Pollution*, Oxford: Pergamon Press, 1968.
LOUIS J. BATTAN	*The Unclean Sky*, Garden City, N.Y.: Doubleday-Anchor, 1966.
DONALD E. CARR	*The Breath of Life*, New York: W. W. Norton, 1965.

CHAPTER 2

SOME ADVERSE EFFECTS

Air pollution is the focal point of our attention because it affects people adversely. Some effects are direct and invoke physiological response: Eye irritation and respiratory distress are examples. Other effects are indirect, but nonetheless disturbing: Reduced visibility and the soiling of clothing are in this class. We shall examine several of these adverse effects to identify which constituents of polluted air are responsible for them. In so doing, we shall establish a perspective from which the reader should be able to appreciate the relevance of the subjects in the later chapters when we consider in detail the separate factors contributing to conditions of pollution.

We shall first take up the psychology of human response to introduce important concepts which apply to nearly every assessment of the effects of pollution. In gauging the effects of various levels of air pollution, we obtain a measure from which we can determine by how much the emission of contaminants must be reduced to achieve a desired air quality. Not everyone responds in the same way to the presence of a pollutant. Some individuals are more sensitive than others. Thus not everyone in a population will respond equally to an improvement or deterioration in air quality. Whether a significant number of people will benefit from a proposed ordinance for control of a pollutant may have an important bearing on at least two factors: What cost can reasonably be imposed upon industry to implement emission controls, and what cost the public is willing to shoulder. For these reasons, it is important to be able to gauge how individuals and populations will benefit from a proposed reduction in pollutant emissions.

These concepts will then be applied to a study of two adverse effects—objectionable odors and reduced visibility—that are generally the first aspects of air pollution to be noticed by the public. They have also prompted the most numerous and vocal complaints. Air contaminants which are responsible for odors and reduced visibility are well known; and it is advantageous to begin with an analysis of their effects, because they are established in a quantitative sense. By contrast, the effects of pollutants on the human respiratory system are not well understood. Thus when we take up the subject of respiratory diseases in concluding this overview of adverse effects, we will confine our attention to the principal diseases whose symptoms are aggravated by very high levels of pollution. Some physicians hold the opinion that these diseases may be caused by air pollution, because there is statistical evidence that links them with cigarette smoking or chronic exposure to very high concentrations of particulate matter in the air. However, there is no *direct* proof that community air pollution causes respiratory disease. Therefore our present discussion of pulmonary diseases should be regarded as background information, and we shall defer to later chapters on sulfurous smog and photochemical smog a detailed examination of the adverse effects which specific contaminants of community air have upon the respiratory system.

2.1 HUMAN RESPONSE

Man in his everyday activities is subjected to many stimuli. Some of these he detects by a psychological sensation: sight, smell, touch, taste, and hearing. Others cause a physiological response of which he may not consciously be aware. For example, a change in his pulse rate or in the chemical equilibrium in pulmonary tissue can result from stimulation by certain air pollutants. A psychological or physiological response will be observed in many cases only when the stimulus exceeds a certain intensity which is called the *threshold*. If a threshold exists, the presence of a pollutant whose stimulus is below threshold can be considered "safe." The concept of a threshold can also be applied for the response of vegetation or materials when exposed to certain pollutants.

The simplest way to display graphically the concept of a threshold is shown in Fig. 2.1(a). For each level of pollution which causes a given intensity of stimulus, the probability that exposure produces a response is given by the heavy curve. Below threshold, which happens in this example to occur at an intensity of 3 arbitrary units, no response is ever produced. Above threshold there is always a response. If the response is an adverse one, an intensity of less than 3 can be considered harmless for the individual, but not so above that. It should be borne in mind that so far we have said nothing about the magnitude of the adverse response; we are considering only whether or not there is one.

Let us now make more precise what is meant by the "intensity of stimulus" by taking some examples. For odors, the intensity is found to be the ambient concentration of the odorant molecule. For the sensation of pain, it could be the pressure which in an experiment is applied to a portion of skin. If the response is the detection of a distant light on a dark night, the intensity of stimulus could be the apparent brightness of the light. All of these stimuli are immediately present; that is, the intensity of the stimulus does not depend upon the history of past exposure of the subject to the stimulus.

However other types of stimuli are possible. Consider as an example poisoning from an ingested heavy metal, such as lead or mercury. The human body cannot eliminate these metals or their organic compounds if large amounts are ingested during a short interval of time. Concurrent with an increase in the concentration of the contaminant in the blood is a build-up of concentration in several organs of the body. For example lead accumulates in the bone marrow, and mercury in the kidney and liver. In sufficient quantities, both affect the nervous system. The threshold for permanent injury is believed to depend upon the accu-

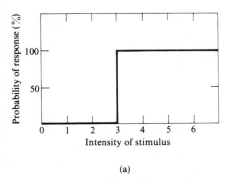
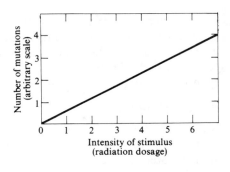

(a)

Fig. 2.1   (a) Idealized response curve showing a well-defined threshold; (b) example of a response which has no threshold. In both cases the units for the intensity of stimulus on the horizontal axis are arbitrary.

mulated amount in the body. The intensity of stimulus is therefore related to the total ingested amount of pollutant which is called the *dosage*. This is a more subtle parameter than the ambient concentration, for it is not necessary to have high ambient concentrations to reach a threshold dosage; a sufficiently long period of exposure can suffice. For some adverse effects it may be that a threshold can be characterized simply by the dosage of a contaminant which an individual has received during his lifetime. But for other cases, where the body continually eliminates some of the contaminant, the time interval over which a dosage is received may also be a factor. We shall see some examples of this in later chapters.

Unfortunately, not all adverse effects have a clearly defined threshold, and it is not possible to define a safe level. One example appears to be structural alteration of human genes which may result from exposure of chromosomes to ionizing radiation. In Fig. 2.1(b) is illustrated the probability of a mutation as it is believed to depend upon the dosage of radiation (in arbitrary units). No matter how low the dosage, there remains a finite probability that damage will occur. With this type of response, there is *no harmless level* of stimulus.

In fact the naturally occurring radioactivity of our environment is believed to cause genetic changes. These may not appear in the characteristics of the next generation, because usually the traits are recessive; that is, both parents must have the affected gene before the offspring displays the new characteristic. However even when recessive, the affected gene is transmitted from generation to generation. So exposure of a large fraction of a population to radiation enhances the probability that both parents possess a similarly altered gene. The resulting changes in the observable characteristics of a population thus depend upon the radiation dosage which each member has received *and* the fraction of the population that has been exposed. A more significant effect may be produced if a low dosage is experienced by many members of a population than if a high dosage is received by only a few.

Thus we see that the threshold for an adverse effect from air pollution can be characterized by: (1) the ambient concentration, (2) the lifetime dosage, (3) the dosage received during a limited time interval, and (4) dosage which is received by a certain fraction of a population. These do not exhaust the possibilities, and we can anticipate that the list will be lengthened as the complexities of human response become better understood.

Response curves

Nature never provides us with an example where the threshold is as clearly defined as in Fig. 2.1(a). Biological systems are too complex to work as repetitively and discriminately as the sharp rise of the curve at the threshold would indicate. The sensitivity of a person may be affected by such factors as the amount of sleep he had the previous night, his state of health, or the simultaneous presence of additional stimuli. Also, the threshold for psychological response to odors, sound, etc., is not uniquely defined because of the customary "noise" in the neural receptors of the body. More common is the "S"-shaped or "sigmoid" response curve illustrated in Fig. 2.2. The individual points in this figure are hypothetical data to emphasize the fact that the response of a particular individual on any given day has a degree of variability. Only after experiments have repeatedly measured the response to many levels of a stimulus will the average response data follow a smooth curve as illustrated.

How can the concept of a threshold be applied to this type of response? The application is somewhat arbitrary. If there is a steeply rising portion of the curve, the threshold is commonly defined by the intensity of the stimulus at the rise, analogous to the situation defined in Fig. 2.1(a). The steepest portion is often found to correspond to a 50% probability of response. This is the case in Fig. 2.2, at an intensity of stimulus equal to 4 units. If the response is to a toxic pollutant, it may be more significant—and provide a greater margin of safety—to define the threshold intensity for a lower probability of adverse response, say the 1% or 5% probability. On the other hand, if the response concerns the visual perception of

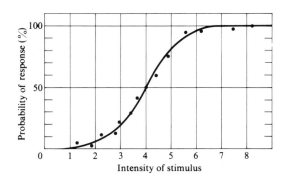

Fig. 2.2 Probability that an observer can detect a stimulus of a given intensity. The smooth curve represents the average response determined from many tests.

the horizon on a smoggy day, a more significant threshold in terms of the observer's confidence in seeing that distance may be one that has a probability of detection at some value significantly above 50%. Observers in tests are found to have no confidence in identifying a dim light on a dark night if on only 50% of their attempts they perceive it.[1] A criterion of 90% probability yields a more meaningful threshold. Therefore, exactly where on a response curve the threshold is chosen depends upon the type of response and the purpose for which the threshold is defined.

The shape of a response curve will vary from individual to individual. Of course it will also depend on the type of stimulus and response.

Response of a population

We have in the preceding considered only the possibility that a stimulus evokes a response from one individual. But an important gauge of the effects of air pollution is how thresholds may vary among members of a group. The group may be all employees of a particular factory, all children in a school, or all citizens of a community, etc. Whatever the group may be, we shall consider its members to form what we have been calling a *population*. A very important question is what percentage of the population is highly sensitive to a particular air pollutant. Presumably the answer to this will in some measure determine the extent of the effort the population will make toward controlling the pollutant.

Figure 2.3(a) shows a hypothetical example of the distribution of threshold concentrations for a given adverse effect. The horizontal axis gives the possible thresholds (denoted by the letter c) which individuals can have; and by reference to the vertical axis the curve indicates what fraction of the population has each threshold. The curve is called the *threshold distribution* of the population and is often found to be what statisticians know as a *normal distribution*, which has a bell shape as illustrated. For this distribution the majority of individuals of a population have thresholds which are close to the mean and the median thresholds. Thus most individuals have a threshold near $c = 4$. Only a few individuals are extremely sensitive or insensitive. The shape of the distribution and its location on the horizontal axis will of course depend upon the definition of the threshold. Different curves will be obtained if the threshold is chosen as the intensity of stimulus for which there is a 50% probability of a response as compared with a choice of 90%.

Distribution curves commonly have shapes other than that taken by a normal distribution. In Fig. 2.3(b) we illustrate a curve skewed toward lower thresholds; so the mean threshold for this population lies considerably above the median. Examples of this type of distribution are found for odor thresholds when members of a population first recognize an odor presented to them. In such cases the median threshold rather than the mean is a better characterization of the level of pollution to which the population as a whole responds, because by comparison the mean threshold is more representative of the insensitive members who have very high thresholds. The particular curve shown in Fig. 2.3(b) is known as a *log-normal distribution*, and its relation to the previous example can be seen by replotting the

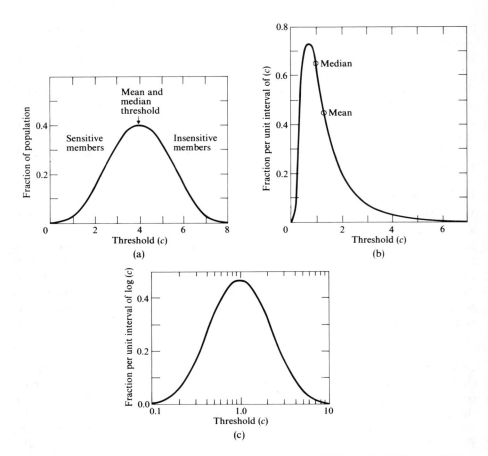

Fig. 2.3 Fraction of a population having a given threshold. (a) Normal distribution; (b) log-normal distribution plotted against the threshold; (c) log-normal distribution plotted against the logarithm of the threshold. For the log-normal curves, the median has been chosen to fall at a threshold of $c = 1.0$.

curve against the logarithm of the threshold concentration, as given in Fig. 2.3(c). This procedure gives us the bell-shaped curve, indicating that a log-normal distribution has the *logarithm* of the threshold normally distributed. We shall later point out examples where the log-normal distribution applies to other phenomena related to air pollution.

A particularly useful way to characterize threshold distributions is by a cumulative distribution. This tells what percentage of the population responds to a given intensity of stimulus. All members whose threshold (c) is less than the given intensity of stimulus will of course respond. Figure 2.4 illustrates a cumulative distribution curve representing the threshold distribution in Fig. 2.3(a). The "S"-shaped curve in the figure has the typical form which characterizes the response of large populations.

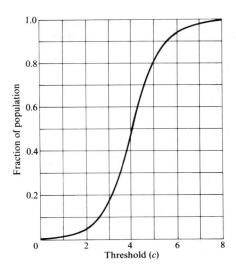

Fig. 2.4 Cumulative distribution. Fraction of a population with thresholds lower than the indicated values.

Suppose that this curve represents an adverse response from members of a population to their exposure to various *concentrations* of an air pollutant. It tells us what fraction of the population exhibits an adverse response when the pollutant is present at the concentration c indicated by the horizontal scale. In this case it is evident that a reduction of the ambient concentration of the pollutant below $c = 4$ which corresponds to the median threshold (where half of the population exhibits a response) will at first make the pollutant undetectable for a substantial fraction of the population; but with successively more severe reductions in concentration a smaller and smaller fraction of the population will benefit from a given percentage reduction. Curves such as Fig. 2.4 can be used to estimate what improvement in air quality is needed to benefit or safeguard a given percentage of a population.[2]

2.2 ODOR

The olfactory organ of man serves many functions: On one hand, it enhances his aesthetic pleasures, but, on the other, it warns him of exposure to certain harmful chemicals. As a defense mechanism it is not perfect, for not all toxic gases have odors. In addition, it can produce strong physiological reactions such as nausea from disagreeable odors, even when they occur at such low concentrations that they cause no injury. In historical times the rapid development of industry and urbanization has subjected man to objectionable odors on a geographic scale from which he could not readily remove himself. He consequently sought recourse by government action. At the end of the twelfth century in England, for example, a Royal letter of complaint was directed to the sheriff of Oxford warning him to correct conditions from which "the air is so corrupted and infected" by the filth in the streets "that an abominable loathing" is "diffused among the masters and scholars," a state of things aggravated by the practice of candlemakers melting tallow before their houses.[3] Elimination of disagreeable odors was long ago recognized as a responsibility of government.

Odors are characterized by two parameters—*quality* and *intensity*. Public appraisal of the quality of an odor is generally hedonic, it is either liked or disliked. However, trained observers can recognize various qualities which can be described in associative terms; for example, spicy, flowery, fruity, resinous, and putrid. An objectionable odor is known as a malodor, but this does not necessarily mean that it is harmful or produces a permanent physiological change. Whether or not an odor is a malodor often depends on a highly subjective appraisal. But the assignment of a quality by a trained observer is unique for a given chemical odorant, with few exceptions.

The olfactory organ which detects odors is a yellow-brown patch of tissue approximately 2.5 cm^2 in area located in the roof of the upper nasal cavity. Receptor cells at the surface are immersed in a covering of aqueous mucus. From each cell up to 1000 hairs protrude into the mucous covering. Molecules of the odorant substance are apparently conveyed by the motion of air to the mucus, where they are trapped. Sniffing increases the chance of impingement by setting up eddying currents of air that increase circulation near the olfactory organ. The olfactory tissue responds rapidly to odorant molecules as the lung draws in large volumes of air. It is known that keenness of perception is affected by the overall physical condition of the observer as well as his skill and experience.

The true mechanism for olfactory stimulus remains unknown. There is considerable evidence that the quality of an odor may depend primarily upon the shape of the odorant molecule, so that molecules having different chemical compositions but similarly shaped would produce the same quality. However, the quality from the simultaneous presence of two different odorants may be quite unlike the qualities of each odorant experienced separately. This fact is the basis for a technique called "odor masking" which is occasionally used by industries to cover up a malodor with a more acceptable one.

For odors, two response curves and their corresponding thresholds are possible. One is the point at which a person detects the presence of an odor, that is, detects a difference from background. This is called the *detection threshold*. The other response is the point when an odor is recognized, defining a corresponding *recognition threshold*. Both thresholds depend upon the ambient concentration of the odorant substance. The thresholds may or may not be sharply defined, depending upon the individual. There is some evidence that determination by trained observers of recognition thresholds as compared with detection thresholds leads to more consistent results.[4] The recognition threshold therefore may have more significance as a basis for air quality standards. For many substances, the ambient concentration for a recognition threshold is approximately ten times higher than for the detection threshold.[5]

Very few studies of odor recognition thresholds have been completed in sufficient detail to provide data for response curves analogous to Fig. 2.2. Usually the stated ambient concentration is the lowest at which the odorant is "surely recognized." It is likely that this criterion corresponds to a probability of response exceeding 90%.

Threshold concentrations for a particular odor are stated as the fraction of odorant molecules contained in the inhaled air. This is equivalent to stating what volume of gaseous odorant molecules must be mixed with a given volume of clean air under the same conditions of temperature and pressure to produce the threshold concentration. It is common to find recognition thresholds at only a few parts of odorant per million parts of air. A more useful designation is the shorter form of "parts per million" or simply "ppm." The extreme sensitivity of the nose is underscored by the fact that the threshold for some odorants may be in the more dilute ranges of several parts per hundred million (pphm) or even parts per billion (ppb).

Occasionally ambient levels of a pollutant may be given in terms of its *mass* concentration; that is, the amount of mass of the pollutant found in a given volume of air. For example, one may find mass concentrations quoted as a certain number

TABLE 2.1

Odor recognition thresholds for a panel of four trained observers. The threshold concentrations are defined as the concentration at which all panelists recognized the odor, and are expressed in parts per million by volume.[*]

Chemical	Odor threshold	Odor quality
Acetic acid	1.0	Sour
Acetone	100.0	Chemical sweet, pungent
Acrolein	0.21	Burnt sweet, pungent
Amine, monomethyl	0.021	Fishy, pungent
Amine, trimethyl	0.00021	Fishy, pungent
Ammonia	46.8	Pungent
Benzyl chloride	0.047	Solventy
Benzyl sulfide	0.0021	Sulfidy
Chlorine	0.314	Bleach, pungent
Ethyl acrylate	0.00047	Hot plastic, earthy
Ethyl mercaptan	0.001	Earthy, sulfidy
Formaldehyde	1.0	Hay/straw-like, pungent
Hydrogen sulfide gas	0.00047	Rotten eggs; nauseating
Methyl mercaptan	0.0021	Sulfidy, pungent
Methyl methacrylate	0.21	Pungent, sulfidy
Phosgene	1.0	Hay-like
Sulfur dioxide	0.47	Sulfidy, pungent
Trichloroethylene	21.4	Solventy

[*]G. Leonardos, D. Kendall, and N. Barnard, *Journal of the Air Pollution Control Association* **19**, 91 1969; courtesy of Arthur D. Little, Inc.

of milligrams of pollutant per cubic meter of air. The conversion between mass concentration and number concentration (which we call simply the "concentration") is straightforward and is given in Appendix B.

Some examples

A summary of recognition thresholds for common air pollutants and familiar chemicals is given in Table 2.1. Some features deserve comment. It can be seen that sulfur-bearing compounds generally have low thresholds. The reason for this is not known. Methyl mercaptan (CH_3SH) and ethyl mercaptan (C_2H_5SH) have thresholds of only a few parts per billion. Their aroma of decayed cabbage is distinctly unpleasant. A variety of similar sulfur-bearing organic compounds are found in crude petroleum in quantities of several percent by weight, and their partial release during refining is often responsible for disagreeable odors in the locality. Another pungent sulfur compound, hydrogen sulfide (H_2S), is released by decaying vegetation and has a characteristic odor of rotten eggs.

Not all odorants with low thresholds produce malodors. Pleasant-smelling artificial musk 2,4,6 trinitro-tert-butylxylene [or $(NO_2)_3C_6(CH_3)_2C_4H_9$] is claimed to have a detection threshold of about 0.007 ppb.[6] The ability of the nose to detect the presence of certain odorants at extremely low concentrations far exceeds the capability of modern chemical techniques. This fact illustrates the difficulty of monitoring and controlling some malodorants.

The familiar odor of natural gas that is served to many homes in the United States is derived from a concentration of only 3 ppm of odorant molecules. These may consist of as many as 20 different organic compounds which occur naturally in the gas, including mercaptans, organic sulfides, disulfides, and polysulfides. In some localities other odorants may be mixed with natural gas to achieve a desired quality and recognition threshold.

The threshold distribution for a few odors has been determined by experiments with selected panels, usually comprising a small population of less than 50 members. As an example, let us consider the cumulative distribution of recognition thresholds for the odorant hydrogen sulfide (H_2S) in Fig. 2.5 which represents a panel of 33 untrained observers.[7] In this panel, the most sensitive individual could recognize H_2S at an ambient concentration as low as 1 ppb, whereas the least sensitive could recognize it only for concentrations exceeding 10 ppb. The factor of 10 that is spanned by the threshold concentrations for the panel illustrates the fact that if we are to characterize population response in a meaningful way, not only must the median or mean threshold of the population be given, but some measure of the spread in thresholds as well.

The vertical scale of Fig. 2.5 is arranged so that a straight line trend of data points in such a graph is indicative of a log-normal distribution. It is apparent that the limited data for this panel is indeed consistent with a log-normal distribution, with a median threshold of about 4 ppb corresponding to the level at which half of the population can recognize the odor. Another noteworthy feature of this curve is the fact that the recognition threshold for these untrained observers is much

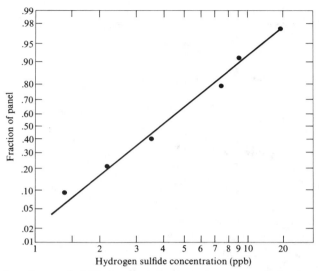

Fig. 2.5 Fraction of a panel of 33 untrained observers that can recognize the odor of hydrogen sulfide at a given concentration. The vertical scale is arranged so that a straight line on this graph represents a log-normal distribution. (Data is from F. V. Wilby, *Journal of the Air Pollution Control Association*, **19**, 96, 1969.)

higher than for the trained observers who provided the data summarized in Table 2.1. A portion of this discrepancy may arise from the much higher sensitivity of trained observers; but a portion may also be due to differing methods of preparing the odorized test atmospheres. This latter factor appears to be an especially important problem with experiments on hydrogen sulfide. The presence of background odors such as SO_2 or gasoline fumes can have an appreciable effect on recognition thresholds.[8] For example, in the presence of diesel engine exhaust, the threshold for H_2S as determined by trained observers was found to increase from 0.47 ppb to about 2 ppb.

It is important to bear in mind the fact that the thresholds listed in Table 2.1 are measured under laboratory conditions with stringently cleaned background air, and therefore many are probably lower than the level at which an average citizen would recognize the odor if he were to experience it in the natural environment.

Intensity of sensation

The olfactory organ has not only high sensitivity and rapid response, but is capable of monitoring a wide range of odor intensities above threshold. One important factor in assessing the effect of odorant air pollutants is knowledge of the relationship between the intensity of the physical stimulation (the ambient concentration of an odorant) and the intensity of the sensation, a psychological response. The only known reliable and accurate instrument for measuring the intensity of odor sensation is, in fact, the human nose.

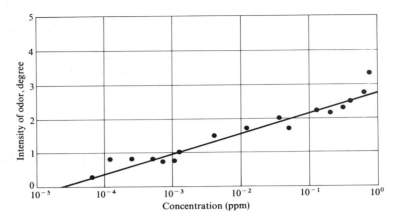

Fig. 2.6 Odor intensity for various concentrations of ethyl mercaptan. (J. F. Byrd and A. H. Phelps, Jr., in *Air Pollution*, 2nd. ed., A. C. Stern, Ed., New York: Academic Press, 1968.)

Unfortunately the appraisal of the intensity of sensation of an odor is highly subjective. It is not proportional to the intensity of the stimulus. Over certain ranges of stimulus it more nearly corresponds to the *logarithm* of the intensity of stimulus. That is, for a geometric increase of the stimulus there is only an arithmetic increase in the perceived sensation. This behavior is summarized by the Fechner law, which has been applied with some success to other perceptions such as hearing, feeling, seeing, and tasting. To apply it to the sense of smell, one commonly introduces an intensity scale of five levels for perceived odor intensity: very faint (1), faint (2), easily noticeable (3), strong (4), and very strong (5). The numbers in parentheses indicate the magnitude which is associated with the appraised intensity. Evidently the assignments are purely subjective; nevertheless, some sense can be derived from the system. In Fig. 2.6 we show the results for the odor intensity of ethyl mercaptan.[9] The straight line describes the Fechner law:

$$\text{(Odor intensity)} = D + 0.5 \log_{10}(\text{Concentration}). \qquad (2.1)$$

In this equation, the constant D is determined from experiment and depends upon the units in which the odorant concentration is expressed. The logarithmic dependence shows that it is necessary for the ambient concentration to increase by almost a factor of 100 to increase the odor intensity of ethyl mercaptan by one intensity level. Investigation of many other odors yields similar results. Often a change in the ambient concentration by only a factor of 10 is sufficient to change the odor intensity by one level.

However, the general applicability of the Fechner law is still a matter of debate, since the response to some odors is found to deviate substantially from Eq. (2.1).[6] For many stimuli such as touch, brightness perception, and sound, the appraised intensity is found to be characterized more accurately by a fractional power law of the form[10]

$$(\text{Odor intensity}) = (\text{Concentration})^n. \qquad (2.2)$$

For some odors, $n = 0.6$ gives fair agreement. The exponent for other types of stimuli has quite different values.

The exact form of the relationship between concentration and odor intensity need not concern us so much as the qualitative features of the dependence. As a general rule, the perceived intensity of an odor is insensitive to changes in the ambient concentration. The implications of this are important. A major effort to control the emissions of an odorant is needed just to achieve a small improvement in the perceived intensity. Figure 2.6 indicates that a hundredfold reduction in ethyl mercaptan emissions would be required to improve an odor from "easily noticeable" to "faint," and another hundredfold to make it "very faint." Reduction of emissions by a factor of 2 or 3 would hardly be appreciated. It has been found that the concentration of a typical odorant would have to be changed by more than 50% before an observer could distinguish *any* difference in intensity. Thus the Fechner law gives us a measure of the size of the problem. When odors are concerned, major efforts are required for improvements.

One feature of the sense of smell greatly compounds the difficulty of assessing the effect which a strong odor has on an individual. This is the phenomenon known as *adaptation* by which the sensitivity of the sensory system is modified due to the continuous presentation of a stimulus at a given level of intensity. Adaptation as used in this sense is distinct from the meaning attributed by the concepts of human evolution. It is found that upon exposure to odors the rate of neural discharge from the olfactory organ may be high initially, but then diminishes with time. During the first minute or two of exposure, there is a rapid decrease in the perceived odor intensity by roughly one level, with a much slower decrease thereafter.[6] People "get used" to an odor. Adaptation may also lead to a change in the quality of an odor. One that was initially pleasant may become disagreeable. Many odorants cause complicated responses because they also affect taste or thermal receptors. Sulfur dioxide, for example, can be "tasted" by some individuals at concentrations below the odor recognition threshold. There is also an apparent decrease in intensity of some odors with increase in the amount of water vapor in the air.

Exposure to malodors does not in itself cause apparent physical harm, although intense odors may lead to nausea, loss of appetite, or sleeplessness. In some cases, detection of a malodor signals the presence of a substance that is toxic, for example hydrogen sulfide. However, it need not have a recognition threshold at levels where it is toxic. The other adverse effects of hydrogen sulfide, such as the discoloration of some paints, tarnishing of copper fixtures and silver, and irritation of the eyes and respiratory system, occur only for concentrations almost a thousandfold higher than the recognition threshold. It is unlikely, except for industrial accidents, that the general public would be exposed to H_2S concentrations which cause damage, even though the odor in industrial areas is often detected. The main adverse effect likely to be experienced is in fact just its odor.

There is no universal relationship yet discovered between odor thresholds of a substance and concentrations which result in physical damage. It is not safe to assume what is true for hydrogen sulfide is true for other odorants. An important example is odorless carbon monoxide, which can be experienced in lethal amounts without causing any forewarning effect on the victim. The CO molecule preferentially takes the place of oxygen in the hemoglobin of red blood cells, thereby decreasing the oxygen supply to various organs of the body. When experienced at sufficiently high concentrations, asphyxia and death may occur. It is ironic that the noted French author Émile Zola, a man famous for his exceptionally acute sense of smell, died as a result of poisoning by carbon monoxide produced by a defective stove as he slept in a tightly sealed bedroom.

2.3 VISIBILITY

We turn in this section to the effect which air pollutants have on visibility. The reduction of visibility resulting from the turbidity of polluted air has been experienced by nearly everyone. The inability to see great distances is considered to be an adverse effect and has caused numerous complaints to air pollution control authorities. The distance one can see is a parameter by which nearly everyone at one time or another gauges the severity of air pollution.

An inventory of community air pollutants would reveal the presence of diverse gases, liquid droplets, and solid particles. These contaminants affect the transmission of light, and thereby the visibility. Perhaps one of the more disagreeable effects is the brownish coloration of the atmosphere effected by photochemical smog. This arises in many cases from the absorption of light of short wavelengths by gaseous nitrogen dioxide (NO_2). In fact, it is precisely the absorption of light by this gas that initiates the atmospheric reactions which produce the smog, as we shall explain in Chapter 10.

However, absorption of light is not the primary mechanism by which visibility is reduced. Much more important is the *scattering* of light by particles and droplets. By scattering we mean deflecting the direction of travel of light. Visibility is reduced because the polluted atmosphere between an observer and a distant object scatters light which comes from the sun and other parts of the sky; some of this scattered light enters the eyes of the observer. The intervening atmosphere thus acquires a luminance—the greater the intervening distance, the greater the amount of light scattered toward the observer. The observer tries to perceive the light reflected from the distant object and distinguishes this object by its contrast with its surroundings and this additional scattered light. When the contrast is reduced because of more scattered light, the visibility is reduced. Thus visibility is affected by particles and droplets in the air, and the duration of poor visibility is determined by how long they remain.

The length of time required for a particle or droplet to settle to the ground depends upon its size, and it is important to appreciate how this comes about. A convenient unit for indicating sizes is the *micron*, a length which is one-millionth of a meter (10^{-6} meters). When we refer to the size of a particle, we refer to a measure of its average radius. Many particles have irregular shapes, so the average radius does not give us complete information about its appearance. However, this simplification will suffice for our purposes.

It is found that particles whose radius exceeds about 10 microns usually settle to the ground within several hours of their emission into the air. This is because the force of gravity is sufficient to overcome the buffeting which they suffer from the surrounding air molecules. One can see this by noting that the force of gravity on a particle is proportional to its mass and therefore to the cube of its radius; however, the frictional force from air which opposes its downward motion is proportional to only the first power of the radius. Thus larger particles will fall faster due to the relatively more important influence of gravity. (The details of how this comes about will be examined in Chapter 11.) On the other hand, particles which are smaller than 10 microns do not fall out so rapidly. They have a greater surface area in comparison to their mass, and therefore the frictional effect of the air molecules is greater. As a consequence of the fairly long period of residence that small particles enjoy in the atmosphere, they are given a special name: *aerosols*. Some aerosols are primary pollutants; the fine ash from coal-burning electrical power generating stations is an example. Other aerosols are secondary pollutants formed by reactions between primary pollutants and components of the atmosphere. The development of droplets of sulfuric acid in sulfurous smog illustrates this case. Aerosols can remain in the air for many days if the particles are sufficiently fine.

Light scattering

The essential features of how aerosols scatter light are complicated, because particles of different size have different effects. Some of the important details are given in Appendix C. At this point, we shall just summarize the results of theory and experimental observations. An important parameter which characterizes the pattern of light scattered by a particle is the ratio of the radius of the particle to the wavelength of light. Visible light which affects the retina of the eye has wavelengths from about 0.40 to 0.70 micron. Thus we are concerned with how light of such wavelengths is affected by particles of various sizes.

Large particles, those whose size exceeds several microns, will scatter light by three processes which are fundamental to optics, as illustrated in Fig. 2.7. A portion of the light incident on the particle will be *reflected* off the surface, and another portion will be *diffracted* around the edges. Depending upon the composition of the particle, some light may be *refracted*, pass through the interior, and exit while again being refracted. These three processes cause light to be deflected from its original direction of travel (called the "forward direction") and therefore are responsible for light scattering. We are all familiar with instances in which we have seen large particles in the air by means of the light they reflect; and we have witnessed the phenomenon of a rainbow, which is light refracted by water droplets. But diffracted light is much less familiar, even though the amount of light diffracted around the edges of a large particle is equal to the amount reflected and refracted. Perhaps some of the incident light of certain wavelengths

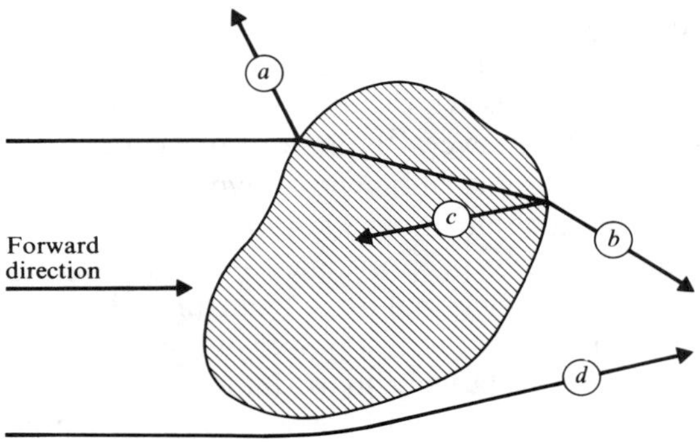

Fig. 2.7 Light which is incident on a large particle may be (*a*) reflected, (*b*) refracted twice on passing clear through, (*c*) refracted then internally reflected, or (*d*) diffracted around the edge.

will be absorbed within the particle, and consequently the light which is reflected or refracted may have a different color than the incident light. (Black smoke is an example of the case in which most of the incident light is absorbed.) If there is no absorption, the scattered light has essentially the same color as the incident light. Most of the light scattered by large particles has its direction only slightly altered from the forward direction, and thus continues to travel in nearly the same direction as it did originally. As a result, these particles cause the air near the direction of the sun to appear white, nearly the same color as the sun.

But aerosols most effective in scattering are smaller ones, whose radius is comparable to the wavelength of light. That is, a greater proportion of the incident light will be scattered well away from the forward direction when the wavelength of the light is about equal to the size of the particle. Concepts from geometrical optics such as reflection and refraction cannot be applied to describe the scattering process, because the wave nature of light must be explicitly considered. Detailed calculations show that most of the scattered light is deflected by more than 1 degree, but less than 45 degrees, from the forward direction. This pattern is distinctly different from the predominantly forward scattering from large particles. But the scattered light from intermediate size particles, like that from large ones, is essentially the same color as the incident light. As a result of scattering from intermediate size particles, the sky takes on a hazy appearance if the atmosphere is sufficiently polluted with them. This effect is particularly pronounced near the horizon, where one's view encompasses air with a greater concentration of aerosols and where the number of particles in the line of sight is greater.

Light scattering from large and intermediate size aerosols is responsible for the haze often found along seacoasts, over deserts, and within inland valleys. Burning refuse releases great numbers of these particles and can dramatically increase the effectiveness with which light is scattered by the atmosphere.

The characteristics of scattering are quite different if the size of the particle is much smaller than the wavelength of light. Aerosols of a size less than about 0.1 micron scatter light of a particular wavelength equally well in the forward and backward directions, and nearly as much intensity is scattered to the sides. This is in sharp contrast with the forward scattering from larger particles. Another contrast is the fact that small particles scatter light of short wavelengths more effectively than long wavelengths. That is, a larger fraction of the incident blue light than of incident red light will be scattered in all directions. This effect is responsible for the red color of sunsets, which results because the blue component of sunlight has been almost completely scattered out of the beam before it reaches us. The reddish hue of the sun is more pronounced with a greater concentration of small aerosols in the atmosphere.

The theory of light scattering by small particles was first developed by Lord Rayleigh in 1871, and the process now bears his name. However, the notion that the brightness of the daytime sky is due to the scattering of sunlight by particles suspended in the air is reported to have been already formulated in the early eleventh century by Alhazen of Basra, an Arabic physicist who carried out much of his work in Cairo.[11] Rayleigh added the suggestion that individual gas molecules could also scatter light and that this was actually responsible for the blue of the sky. We now know that this is not strictly correct; instead, Rayleigh scattering causing the blue color is due not to scattering from individual molecules but to scattering from groups of molecules in regions of the atmosphere where the concentration of molecules is greater than the average momentarily, as indicated schematically in Fig. 2.8. Thus, fluctuations in the density of air caused by the random motion of the gas molecules, increasing the concentration in some regions

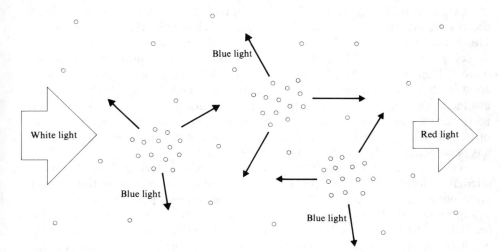

Fig. 2.8 Rayleigh scattering from regions of the atmosphere where the air density is momentarily greater than average. The circles represent air molecules.

and decreasing it in others, give rise to scattering which accounts for the brightness of the sky, even if the atmosphere were to contain no particles. Because the fluctuations in the density of air are appreciable only over small volumes of atmosphere whose dimensions are much less than the wavelength of visible light, the scattering has all the characteristics of Rayleigh scattering from small particles and accounts for a portion of the blue color which we see.

Contrast reduction

Now we are in a position to examine the important aspects of how light scattering reduces visibility. The basis for our discussion is the fact that man discerns objects by their contrast with the surroundings. The contrast may be in the color or brightness of an object compared with its background. This contrast is reduced when extraneous light is scattered toward the observer by particles in the intervening distance.

How light scattering produces a reduction in contrast can be formulated in mathematical terms; we shall derive a relationship between the light scattering properties of a polluted atmosphere and the resulting reduction of visibility to predict how far a person can see when community air carries a given burden of particulate matter. But first we must sharpen our thinking by introducing an important term: the *visual range*. This is commonly defined as the greatest distance at which an observer can distinguish a contrast between an object and its background. Thus the visual range indicates how far we can see. It is unfortunate that "visibility" rather than "visual range" is often found in the literature describing the same concept. The nontechnical meaning of visibility is usually equated with contrast or the clarity with which objects stand out from their surroundings, with no reference to the distance at which the objects are perceived.

Let us consider as well another concept which is important because it indicates the extent of the geographical area in which the visual range is restricted. This is the *prevailing visibility*, the greatest visual range which is attained or surpassed around at least half of the horizon, but not necessarily in continuous sectors. Roughly speaking, the prevailing visibility tells us how far out we have good visibility around half the horizon. It is a useful concept because it is a distance that can be determined by visual observations of the surrounding landmarks as viewed from one location, such as an airport control tower. As a result of its importance to aviation, a considerable body of data on prevailing visibility has been collected for locations around the world.

An accurate formula for the visual range was first advanced by H. Koschmieder in 1924.[12,13] His derivation is based upon many assumptions about atmospheric conditions and the psychology of human perception; but three assumptions are fundamental. The first is that the object which we view (or the target) is black and therefore reflects no light, and that it is perceived against a white background. This condition approximates the common situation of a dark building seen against a light horizon sky. We will let the symbol $C(x)$ represent the contrast as perceived by an observer at any distance

x from the target. The contrast $C(x)$ is defined as the relative difference between the apparent luminance of the target and the background. The apparent luminance of an object is determined by the rate at which light energy is incident on the eye from the direction of the object. This in turn is directly proportional to the rate at which light energy is incident on a unit area of a hypothetical surface perpendicular to the direction of travel of the light, a quantity called the *light intensity*. It will be most convenient to deal with the intensity of light which travels from a region of the target to the observer and from a region of the background to the observer. Therefore we shall let $I_1(x)$ be the intensity of light directed toward the observer from the direction of the target, at a distance x from the target; and let $I_2(x)$ be the intensity from the direction of the background.

The contrast between the background and the target is then the relative difference in intensities. At the location of the target ($x = 0$), it is simply

$$C(0) = \frac{I_2(0) - I_1(0)}{I_2(0)}. \tag{2.3}$$

A perfectly black target absorbs all of the light which is incident on it, reflecting none. So therefore we must have $I_1(0) = 0$. The contrast at the target is thus $C(0) = +1$, independent of the background intensity.

As the distance x separating the observer and target increases, $I_1(x)$ will be altered by two factors: (1) absorption of light by constituents of the atmosphere, and (2) addition of extraneous light which is scattered along the line of sight toward the observer. We can calculate these effects by referring to Fig. 2.9. We introduce Koschmieder's second hypothesis to simplify the derivation: The atmosphere is assumed to be homogeneously polluted so that scattering and absorption of radiation is the same everywhere. Thus we can consider the change in $I_1(x)$ as light travels through a short distance of the atmosphere toward the observer, along the line of sight an *arbitrary distance x* from the target.

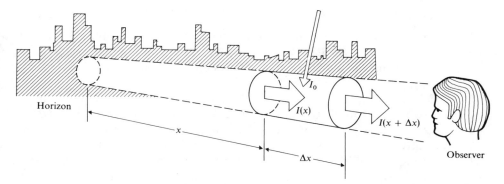

Fig. 2.9   The intensity of light as it passes from x to $x + \Delta x$ is reduced by loss through absorption and scattering but augmented by the fraction of the side-incident light I_0 which is scattered toward the observer.

For convenience, we shall introduce a notation which is conventionally used to denote the value of a small change in any quantity. This is the "delta" symbol, and when it precedes a quantity (for example Δx) it means that the pair of symbols represents the *change* in x, not the actual value of x. So when the beam of light from the direction of the target goes from a position x to a slightly greater distance $x + \Delta x$, the intensity of the beam will be affected by absorption and scattering in the intervening distance Δx; and in general the intensity at $x + \Delta x$, denoted $I_1(x + \Delta x)$, will differ slightly from the value at x, denoted $I_1(x)$. The smaller we assume the magnitude of Δx to be, the smaller the difference. To a good approximation, we can assume that a certain fraction of $I_1(x)$ is lost to absorption and another fraction to scattering as the beam travels the distance Δx, and we can neglect the fact that I_1 actually varies slightly within that distance.

The fraction lost will be proportional to Δx, because it depends upon the number of particles which absorb or scatter the light. We can introduce two constants k and b, as yet of unknown value, and say that the intensity $I_1(x + \Delta x)$ will be less than $I_1(x)$ by the fraction $k\Delta x$ of $I_1(x)$ which is absorbed by aerosols and gases and by the fraction $b\Delta x$ of $I_1(x)$ which is scattered away from the line of sight. Both k and b are constants that are assumed to be independent of the position x. However, while traversing the distance Δx, $I_1(x)$ will be augmented by a fraction $b'\Delta x$ of light intensity I_0 which is incident from other portions of the sky onto the line of sight between observer and target where it is scattered toward the observer. The constant b' is not equal to b in general. The net gain in the intensity is therefore

$$I_1(x + \Delta x) - I_1(x) = [b'I_0 - (b + k)I_1]\,\Delta x, \qquad (2.4)$$
$$\Delta I_1(x) = [b'I_0 - (b + k)I_1]\,\Delta x.$$

In going from the first equation to the second, we have introduced a new symbol for the change in I_1 produced by a change Δx, and following our convention for the "delta" notation, we have let

$$\Delta I_1(x) = I_1(x + \Delta x) - I_1(x).$$

The fact that scattering and absorption of light depend upon the wavelength of the light need not concern us at this time; the current imprecision of measurements of the visual range in polluted atmospheres does not justify including such details.

A tacit assumption is contained in the derivation of Eq. (2.4). We have assumed that turbulent motion of the air ensures that particles are in random positions, moving with random velocities, and are separated on the average by sufficiently large distances that they move independently from one another. The scattering of light by each particle is therefore independent of the process by which light is scattered from each of the others. Then the fraction of light intensity which is absorbed or scattered is proportional to the number of particles in the path of the light; that is, it is proportional to the length Δx of the path.

If the observer now directs his attention to the background, he will see $I_2(x)$

as it is similarly affected by absorption, scattering away from the line of sight, and by introduction of extraneous light. However, it is essential to note that for a homogeneous atmosphere every position x is a horizon for another properly located observer (excluding pathological cases of intervening hills!). Thus $I_2(x)$ must be independent of x, so $\Delta I_2(x) = 0$. Equation (2.4) adapted for this case can be written consequently as

$$\Delta I_2(x) = 0 = b'I_0 - (b + k)\, I_2. \tag{2.5}$$

The reason for writing this condition is that it allows us to replace the unknown quantity I_0 in Eq. (2.4) with the more useful quantity $I_2(x)$. Thus in view of Eq. (2.5), we can rewrite Eq. (2.4) as

$$\Delta I_1(x) = -(b + k)(I_1 - I_2)\, \Delta x. \tag{2.6}$$

This gives the change in $I_1(x)$ with increase Δx in distance. But since $I_2(x)$ does not change with distance, we can subtract the quantity $\Delta I_2(x)$, a quantity equal to zero, from the left side of Eq. (2.6). This yields an equation for the change in *contrast* with distance, provided that we also divide both sides of the equation by $I_2(x)$. Thus we have

$$\Delta C(x) = -(b + k)\, C(x)\, \Delta x. \tag{2.7}$$

The negative sign indicates that the contrast *decreases* with *increase* in distance. Readers who are familiar with calculus will immediately recognize this equation. The smaller we assume Δx to be, the more accurately the equation represents the true situation. The solution of this equation has the form of an *exponential function.*

The exponential function

Equation (2.7) is an example of an equation of a well-known type which can be represented by the more general form

$$\Delta y = -ay\, \Delta x, \tag{2.8}$$

where x represents an independent variable and y a dependent variable. The parameter a is a constant. This equation will frequently appear in this book, so it is worthwhile to ensure that its implications and solution are clearly understood. Equation (2.8) says that the incremental change Δy in a quantity y is proportional not only to the change Δx in the independent variable x but also to the value of the quantity itself $y(x)$. We seek a solution of this equation which will tell us how y depends upon x. Another way to look at this equation is to divide both sides by y:

$$\frac{\Delta y}{y} = -a\Delta x.$$

This means that the fractional change in y depends only upon the constant a and the change in x given by Δx. For a succession of equal increments Δx, y will change each time by the same fractional amount. Equation (2.8) also describes the number of unstable nuclei in a sample of radioactive material: The number of radioactive decays Δy is proportional to the number of undecayed nuclei y and the time interval Δx during which they are observed. Thus if a is a positive quantity, y decreases with increasing time. An identical equation (with $-a$ a positive quantity) describes the growth of a large population, if y is the number of individuals and Δx is an interval of time.

Equation (2.8) has an exact solution if Δx is assumed to be infinitesimally small, as is customarily done in calculus. The solution is given by the *exponential function*, and y is said to depend exponentially on x:

$$\frac{y(x)}{y(0)} = e^{-ax}. \tag{2.9}$$

The symbol e represents a number, $e = 2.718\ldots$, and derives its value from the definition of the exponential function in calculus. It should be noted that the exponent $(-ax)$ means the same thing in this formula as it does in algebra; namely, that e is raised to the power $(-ax)$. Thus for example, $e^0 = 1$, so that $y(x) = y(0)$ at $x = 0$ according to Eq. (2.9). The exponential function is occasionally written in a different notation for convenience:

$$\frac{y(x)}{y(0)} = \exp(-ax). \tag{2.10}$$

The right-hand side here is identical to the quantity in the right-hand side in Eq. (2.9). The form of the exponential function (for positive a) is illustrated in Fig. 2.10 by the decreasing curve.

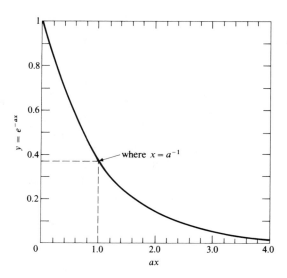

Fig. 2.10   Decreasing exponential.

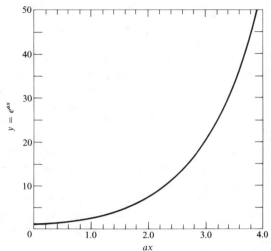

Fig. 2.11   Increasing exponential.

The parameter a^{-1} gives a natural scale of length which indicates how far x must be increased to produce a marked decrease in y. To put it more precisely, for the curve shown in Fig. 2.10, when x increases by an amount a^{-1}, the value of y diminishes to e^{-1} of its original value. Thus for $x = a^{-1}$, $y(a^{-1})$ is only 37% of its initial value $y(0)$. In the example of radioactive nuclei cited above, a^{-1} is called the "lifetime" of the radioactive sample, for it tells how long we must wait until a large majority (63%) of the radioactive nuclei have decayed.

For our example of population growth, $(-a)$ must be positive to reflect the fact that y increases with an increase in x. In this case, the exponential function appears as in Fig. 2.11, showing a steady increase with time x. The rules of algebra for exponents tell us that the function illustrated in Fig. 2.11 is simply the inverse (or reciprocal) of the function which is shown in Fig. 2.10.

Some predictions

A comparison of Eq. (2.7) for the reduction of contrast with distance and the general form Eq. (2.8) immediately indicates that the contrast should decrease with distance according to an exponential function.[14] Thus Koschmieder's theory predicts that $C(x)$ decreases according to

$$C(x) = C(0)\ e^{-(b + k)x}. \tag{2.11}$$

The constant $(b + k)^{-1}$ provides a length scale which indicates how rapid the decrease will be. The quantity $(b + k)$ is called the *attenuation coefficient* or *extinction coefficient*. It is the sum of a *scattering coefficient* b and an *absorption coefficient* k.

Equation (2.11) provides a means by which visual range can be related to the attentuation coefficient. The visual range is the maximum value of x for which the contrast $C(x)$ may be perceived by an observer. If x is any greater, the target cannot be seen. Thus a quantitative definition of visual range is predicated upon a numerical determination of the contrast threshold for the human eye. Considerable efforts have been invested in measurements of contrast thresholds for individuals and representative groups of people. The results of H. R. Blackwell[1] are particularly noteworthy since they involved the analysis of almost half a million responses from a panel of 19 trained observers under controlled conditions. His response curves for individuals and for his population of 19 were found to be similar to those which we discussed in Section 2.2. The results indicate that a contrast threshold can most precisely be determined if it is defined as the contrast which presented to an observer results in a 50% probability of its being perceived. With this definition, the threshold $\epsilon = C(x)$ is found to be $\epsilon = 0.02$. The value of ϵ is insensitive to the background intensity I_2 provided that the visual angle subtended by the target exceeds about 1°, and daylight illumination prevails.[13] This value for the threshold was in fact previously used by Koschmieder as his third assumption.

With $C(x) = \epsilon$, we have by Eq. (2.11) a mathematical condition for the greatest distance at which we can see the target. This distance is given a special name, the *meteorological range*, and is denoted by the symbol L_v. Figure 2.12 illustrates the definition of this term. By taking the natural logarithm of both sides of Eq. (2.11) (taking the logarithm to the base e), we can obtain an explicit expression for the

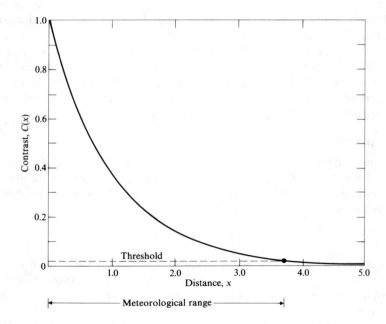

Fig. 2.12 Reduction in the contrast of a black object on the horizon with increasing distance between the object and an observer. The "meteorological range" is defined as the distance at which the contrast is reduced to the threshold value $C(x) = 0.02$.

meteorological range:

$$L_v = \frac{-ln\,(\epsilon)}{(b + k)} = \frac{3.9}{(b + k)}. \tag{2.12}$$

Since the contrast threshold ϵ appears logarithmically in this equation, the definition of the meteorological range is relatively insensitive to the exact value of ϵ. Doubling ϵ decreases the meteorological range by less than 20%. This is fortunate, for the relationship between the thresholds determined by controlled experiments and the thresholds for casual observers in a polluted city is unknown. In fact, the controlled experiments suggest that $\epsilon = 0.04$ may be more appropriate than $\epsilon = 0.02$. The reason for this is the fact that the 50% probability criterion previously mentioned left the observers with practically no confidence that they had really detected a contrast; whereas, a threshold of $\epsilon = 0.04$ is found to correspond to a 90% probability of detecting a contrast, for which the trained observers felt considerably more confident in their identification. We shall find that the inaccuracy of current data makes moot any debate as to whether $\epsilon = 0.02$ or $\epsilon = 0.04$ is more appropriate.

Although the meteorological range is defined in terms of the attenuation coefficient $(b + k)$ in Eq. (2.12), Koschmieder's derivation emphasizes that it is *not through the attenuation of light from the target* that the contrast is degraded, for the target, being black, reflects no visible light. In fact *degradation results from extraneous light* which impinges on aerosols in the line of sight between target and observer, with the subsequent scattering of a portion of this light toward the observer. Numerical estimates which are based upon Eq. (2.12) should of course employ commensurate units for L_v and the attenuation coefficient; for example, to have L_v expressed in meters, we should express $(b + k)$ in (meters)$^{-1}$.

It is important to note that an effect not envisioned in Koschmieder's derivation is atmospheric "shimmer" or "twinkling." Particularly on sunny days, the path taken by a light beam is bent as it passes through air in which the density varies from place to place in irregular fashion. These density fluctuations result from the eddy motion of air in response to local heating by solar radiation, and occur on a much larger scale than the density fluctuations responsible for Rayleigh scattering by the atmosphere. As a consequence of their thermal origin, these fluctuations are known colloquially as heat waves. Such processes will obliterate fine details and distort the appearance of an object, perhaps leading to reduced visibility and a shorter visual range. The visual range may also be influenced by extraneous light sources within the field of vision, in the same way that the contrast threshold is known to depend upon such factors.[1] Thus Eq. (2.12) is not expected to be valid at sunrise and sunset when the target is near the direction of the sun.

Some comparisons

Now we shall face the important question as to how accurately Eq. (2.12) represents the distance we can see through a polluted atmosphere. Experiments by Horvath, et al.[15] have been designed to answer this question by comparing

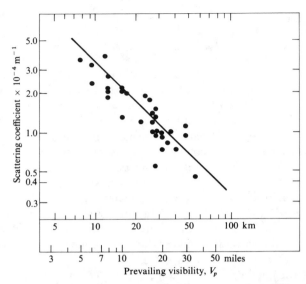

Fig. 2.13   Reduction of prevailing visibility with increasing scattering coefficient. (Reprinted with permission from H. Horvath and K. E. Noll, *Atmospheric Environment*, **3**, 543, 1969, Pergamon Press)

the prevailing visibility (denoted by the symbol V_p) with the scattering coefficient b for ambient air as measured by a scientific instrument called a nepthelometer. Figure 2.13 shows the observed reduction of prevailing visibility with increase in the scattering coefficient for a series of measurements from the top of a four-story building in Seattle, Washington. The straight line represents Eq. (2.12), assuming that the attenuation coefficient arises from scattering alone, and assuming that the meteorological range L_v is essentially the same as the prevailing visibility V_p. This experiment and other similar ones indicate that Eq. (2.12) adequately describes the prevailing visibility to within the imprecision of the experimental data, generally less than ±50%. Attempts to fit the data by an empirical formula of the form

$$V_p = \frac{A}{(b + k)} \tag{2.13}$$

resulted in values of k which were insignificantly different from zero. Thus we reach the important conclusion that light absorption is not a major factor influencing visibility.[13]

Furthermore the values of A determined by the most precise measurements are found to differ from 3.9 by less than 20%. The experimental values are generally on the low side and would be in better agreement if the threshold contrast ϵ had been chosen to have a value slightly larger than 0.02. Thus we conclude that the data support the notion that light scattering alone is the dominant cause of reduced visibility through polluted air.

The scattering theory discussed in Appendix C suggests that the relationship between the scattering coefficient b and the composition and size of aerosols may be complex. Thus Eq. (2.13) does not provide us with a clear indication of how V_p relates to the *amount* of aerosol in the air. It would be useful to find a simple empirical relationship between the scattering coefficient and some easily measured parameter of the aerosol burden of ambient air, such as the concentration of particles, or possibly the total mass of particles that occupies a cubic meter of air (called the *mass concentration*).[16,17] In fact it is possible to find a simple relation: Studies in a number of cities have revealed the existence of an inverse dependence between *prevailing visibility* and the *mass concentration*, of the form:[18]

$$V_p = \frac{K}{M},\qquad(2.14)$$

where K is a constant and M is the mass concentration. An example of the experimental results is given in Fig. 2.14; the line shows the best fit to the data. The constant K for different locations (Berkeley, California; St. Louis, Missouri;

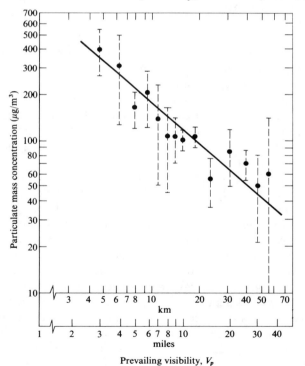

Fig. 2.14 Reduction of prevailing visibility with increasing aerosol mass concentration. The vertical bars represent the estimated uncertainty in the measurements. (Reprinted with permission from K. E. Noll, P. K. Mueller, and M. Imada, *Atmospheric Environment*, **2**, 465, 1969, Pergamon Press)

and Seattle, Washington) and different measuring techniques shows fairly wide variations; the data from these and other measurements give an average value for K amounting to 1800 km-μg/m^3 (or 1.8 g/m^2), but the scatter in the data is such that values of K between 900 km-μg/m^3 and 3600 km-μg/m^3 are possible. (We choose to use this unusual system of units for K so that the prevailing visibility expressed in kilometers can be obtained simply by dividing K by the mass concentration, as commonly expressed in micrograms per cubic meter.) Thus for a particulate burden of $M = 100\ \mu$g/m^3, the prevailing visibility is only 18 km (about 11 mi) on the average. These are convenient numbers to remember. In heavily polluted cities, the mass concentration often can amount to 300 μg/m^3, and the prevailing visibility is correspondingly down to 6 km (3.5 mi).

How much of the variation in the constant K between different locations arises from differences in the measuring techniques and how much is due to physical or chemical differences in the aerosols is presently unknown. The relationship in Eq. (2.14) cannot therefore be assumed applicable to geographical regions where the types of aerosol are grossly different from those where the studies were made. However, it is useful as a first approximation to indicate what prevailing visibility is expected from a given burden of aerosol, with an accuracy of about a factor of 2.

To gain an appreciation of the numerical significance of parameters which are used to characterize the aerosol burden of the atmosphere, let us take an over-simplified example: an aerosol which consists of identical particles each with a radius of 0.05 micron and composition density of 2 g/cm^3. Each particle therefore has a mass of only 10^{-15} g. Then an aerosol mass concentration of 100 μg/m^3 would correspond to a concentration of about 10^{11} particles per cubic meter (or 10^5 particles per cubic centimeter). In some cities this would be regarded as relatively clean air. According to Eq. (2.14) the prevailing visibility for this mass concentration would be about 18 km.

Of course it is a gross oversimplification to assume that aerosols consist of particles of a uniform size. In fact aerosols are found to include particles from the size of macromolecules (0.001 micron) to visible dust (10 microns). It is remarkable that the overall effect of light scattering by this variety of particles can often be summarized by the simple relation in Eq. (2.14) which depends solely on the aerosol total mass concentration. This fact indicates that the aerosols in urban localities have common characteristics, such as a similar size distribution of particles. Some reasons for this will be examined in Chapter 11.

The observed inverse dependence of prevailing visibility and mass concentration tells us that a reduction of the aerosol content of ambient air by a factor of 2 would double the visual range. This direct relationship is in marked contrast with the relatively insensitive logarithmic response of a perceived odor intensity and the ambient concentration of the odorant. For a reduction of pollutant concentrations, an improvement in visual range from reductions in aerosol mass concentration ought to be more readily perceived than an improvement in the intensity of an odor from the same percentage reduction in odorant concentration.

Humid air

The preceding discussion applies only to the prevailing visibility in fairly dry air. This is because the formation of water droplets on aerosol particles may com- complicate the situation if the relative humidity exceeds about 65%. We shall take up the subject of humidity in some detail in Chapter 3. For the moment we shall rely on the common notion that the relative humidity is a measure of the concentration of water vapor in the air compared with the concentration when the air is saturated. The ratio of the ambient concentration of water vapor to the concentration at saturation, expressed as a percentage, is the relative humidity.

When air is nearly saturated, the high concentration of water vapor molecules encourages their condensation on available solid surfaces such as are provided by aerosols. The increased size of a particle enclosed in a water droplet influences its light scattering properties. The ability of particles to nucleate droplets depends upon their composition. Some particles such as salt are hygroscopic; that is, they are soluble in water and absorb it even at a fairly low relative humidity. Salt, for example, is continually released from the sea by bursting bubbles, and it can then be entrained in the wind and circulated over the globe. It is an aerosol commonly found in many regions, even several thousand kilometers inland from the seacoast. Salt is effective in nucleating droplets for relative humidities in excess of 66%. And when the relative humidity exceeds 75%, the salt is found to dissolve in the droplet which it forms.[19] The reduction of visibility due to condensation on salt nuclei is of course most frequently observed in coastal areas. An illustration of

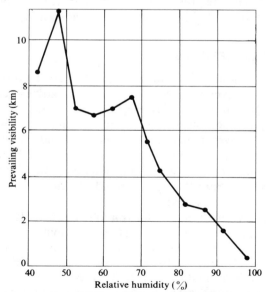

Fig. 2.15 Decrease in prevailing visibility with increasing relative humidity for measurements at Los Angeles International Airport. (After N. Neiburger and M. G. Wurtele, *Chemical Reviews*, **44**, 321, 1949.)

this is shown in Fig. 2.15 for the average prevailing visibility at the Los Angeles International Airport, adjacent to the coast of the Pacific Ocean. A marked reduction is apparent for relative humidities in excess of 66%. The presence of sea salt aerosols explains why, over continental regions, air with a given humidity, is more likely to have reduced visibility if it is of maritime, rather than continental, origin.

Another hygroscopic substance which occurs in polluted air is sulfur trioxide (SO_3). This gas is produced during the burning of sulfur-bearing fossil fuels such as coal and oil, although much more of the sulfur is released as sulfur dioxide (SO_2). As we shall find in Chapter 8, the gaseous SO_2 may be oxidized *in the atmosphere* to form SO_3, thus augmenting the concentration of the hygroscopic component. The trioxide unites with atmospheric water vapor to form a solution of sulfuric acid (H_2SO_4). These droplets of acid grow in size as they collide with, and incorporate more water molecules. This agglomeration leads to a high concentration of small particles which are very effective as light scatterers. The exhaust from a coal-burning power generating station may exhibit a characteristic dense plume which on close observation appears to have a bluish tinge as a result of light scattering from the small particles of condensed vapor. Experiments indicate that SO_3 is a cloud-forming agent at all relative humidities above about 34%, but is especially effective above 75%.[20]

Thus for relative humidities above about 65%, the composition of an aerosol and the water content of the atmosphere affect the visibility. The simple relationship in Eq. (2.4) between the prevailing visibility and the mass concentration of an aerosol can be expected to hold only for lower relative humidities.

2.4 RESPIRATORY EFFECTS

Malodors and poor visibility in polluted air are aesthetic factors, and their effect on humans is usually psychological. However, other adverse effects may cause physiological change and perhaps disease. Interest in the potential pathological effects is generally focused on the respiratory system, and it is that subject we shall now take up. However, it is well to keep in mind that air pollutants can affect man in other ways. Some contaminants such as lead-bearing aerosols, after settling to the ground or onto vegetation, may enter the food chain and subsequently be introduced into the body through the digestive system. Pesticides can also be introduced in this way.

The respiratory system of man is illustrated in Fig. 2.16. It is often convenient to divide it into an upper system, including the nasal cavity, pharynx, and trachea, and a lower system of the air passages below the trachea, including the bronchi and lungs. The bifurcation at the lower end of the trachea from which the right and left bronchi lead is the first of a series that produces a tree-like arrangement of successively smaller bronchi. The trachea and bronchi are kept open by transverse rings of cartilage, spaced at intervals along the length of the air passages. The narrowest bronchi, about 1–3 mm in diameter, occur after eight to 13

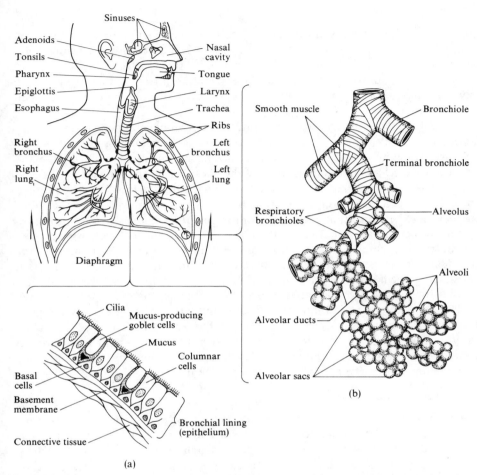

Fig. 2.16 (a) Air passages of the respiratory system of man. (b) Details of the terminal respiratory units of the lung.

generations. They lead to even finer passages called bronchioles, which are distinguished from the bronchi in that they have no cartilage. Branching continues for another 10–15 generations, ending in terminal bronchioles, approximately 0.6 mm in diameter. Bronchioles and terminal bronchioles are encircled by muscle which during spasms, as in asthma, can block off the air passages.

The terminal bronchioles divide into three or more respiratory bronchioles, which are distinguished by the fact that they have occasional outpouchings called alveoli. In turn, the respiratory bronchioles branch into several passages known as alveolar ducts, whose walls contain five or six alcoves or sacs formed by groups of alveoli. The alveoli are the functional unit of the lung. It is across their thin membrane that oxygen diffuses from the air in the lung to the pulmonary capillaries; and carbon dioxide diffuses in the opposite direction. Although each alveolus is small, with a diameter of about 0.2 mm, efficient transfer of oxygen and carbon dioxide in respiration is ensured by their great number. It is estimated that as many as 400 million alveoli are contained in a healthy adult lung, for a total surface area of about 50 m^2, an area more than 25 times the area of the skin. Air can pass from one alveolar duct to another through pores between the alveoli. Thus, should a passage in one duct become blocked, air can be supplied from another.

The lung plays a passive role in respiration, depending upon muscular activity of other parts of the body for expansion and contraction. During inspiration, the diaphragm below the lung contracts and lowers, and the ribs are elevated by the chest muscles. The increasing volume of the chest causes the lung to expand and establish a low pressure in the rarified air in the lung. In this way air is forced into the nose and respiratory system by the higher atmospheric pressure of ambient air. An important parameter for gauging the response of the respiratory system to certain air pollutants is the *airway resistance*, which, when multiplied by the rate at which air flows through the upper respiratory system, gives the difference in pressure between the ambient air and the air in the lung. High resistance indicates that there is an obstruction or constriction of the air passages, so that for a given pressure difference air flow is slower than in a normal system. In later chapters we shall indicate which pollutants can invoke a physiological response which increases airway resistance. The amount of inspired air, called the *tidal volume*, is about 500 cm^3 for the average adult. About two-thirds to three-fourths of the tidal volume occupies the spaces of the lung and bronchi, the balance remaining in the upper respiratory system.

Defense mechanisms

The respiratory system has several defense mechanisms to protect it and counter-act the effect of exposure to irritants. Normally, inspired air is warmed and moistened as it makes its turbulent passage through the nasal cavity and trachea. By the time it reaches the bronchi, it is completely saturated with water vapor thus keeping the pulmonary membranes from drying out. Gaseous air pollutants of appreciable solubility, such as sulfur dioxide, are mainly absorbed by the moist lining of the upper respiratory tract and have little chance of entering the lung. For this reason, the irritating effects of SO$_2$ are localized when experienced in the relatively low concentrations normally found in community air. Insoluble gases will of course enter the lung in the same concentration as they occur in the ambient air. We shall consider the effects of gaseous pollutants on lung structure and function in Chapters 8 and 10 when we examine the phenomena and consequences of sulfurous and photochemical smogs. For the balance of this chapter, we shall therefore devote our attention to the effects of aerosols and larger particles.

The upper respiratory system effectively filters larger particles from the air stream during inspiration. Hairs in the nasal passage form the first line of defense. Particles which succeed in avoiding entrapment may be caught by the second line, the mucous covering over the lining of the nasal cavity and trachea. This is effective in stopping large particles because of the phenomenon of *impaction.* Such particles, because of their large mass, have a relatively high momentum when moving in an air stream and cannot readily follow the turbulent pattern of the air flow because of their inertia. Thus at sharp bends in the passages of the nasal cavity and trachea, large particles tend to impact on the wall forming the outside of the curve, where they can be entrapped in the mucus. Practically all particles whose radius exceeds 5 microns are thereby filtered from the air before it leaves the trachea.

Some particles may also be caught by fine hair-like cilia which line the walls of the upper respiratory system, the bronchi, and bronchioles. A continual wave-like motion of the cilia has the important effect of moving the mucus and entrapped particles to the pharynx where they can be eliminated by swallowing or expectoration. Irritants such as tobacco smoke stimulate ciliary activity and improve mucus transport, but very high concentrations of some pollutants can have the opposite effect. The reasons for these responses are not understood.

Particles smaller than about 3 microns radius can escape the defense mechanisms of the upper respiratory system and enter the lung. A small fraction of these may impact on the bronchiolar membranes and become trapped; but the air flow in the lung is not as rapid and the inertia of the particles is less, so impaction is not the most important mechanism by which particles are retained. More important is *sedimentation,* the downward motion of particles in response to gravity. The slow movement of air in the lung, together with the large horizontal surface area of membrane, permit an appreciable fraction of the larger particles to settle onto the lower surfaces of the bronchioles and alveoli. The fraction of inspired particles which are retained in the upper and lower respiratory tracts is illustrated in Fig. 2.17.[21] The marked dependence upon the size of the particles is evident. It is highly significant that there is a peak in the curve representing the proportion retained in the lung, and that particles are most effectively retained if they have an equivalent size of about 0.5 micron radius. The curve falls off toward larger sizes, because such particles do not enter the lung; and there is a decrease for smaller sizes, because their rate of sedimentation is slower and they do not have sufficient time to reach the lung membrane.

But the curve shows an upward trend again for particles smaller than 0.25 micron radius.[22] The reason for this is intimately associated with the molecular nature of air. A small particle is more strongly influenced by the continual bombardment of the surrounding air molecules than by gravity. The reason that small particles respond differently to the surroundings is because the rate at which they are struck by air molecules is correspondingly lower; and due to the random nature of molecular motion, fluctuations in the rate at which they are struck are consequently more important. Fluctuations in the rate at which air molecules collide on opposite sides of very small particles result in momentary imbalances in the rate at which momentum is conveyed to the particle. The small mass of the particle permits it to respond to a momentary imbalance with an

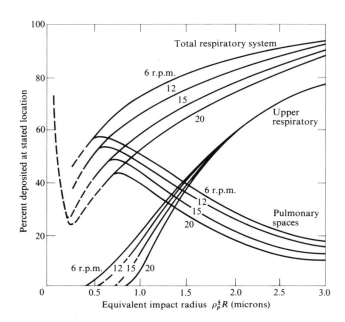

Fig. 2.17 Retention of particles in the respiratory system. The horizontal scale encorporates the particle density ρ_P, because impaction, sedimentation, and Brownian motion depend on the particle mass. For $\rho_P = 1$ g/cm^3, the horizontal scale would give the particle radius in microns. Curves are shown for various respiratory rates, indicated by the number of respirations per minute. (From T. F. Hatch, in *The Air We Breathe*, S. M. Farber, Ed., 1961. Courtesy of Charles C. Thomas, Publisher, Springfield, Illinois).

appreciable velocity in the direction of the net momentum it receives. Thus it moves about in an irregular fashion, both upward and horizontally, as well as downward. This random motion is called *Brownian motion*, and is named after the botanist Robert Brown who in 1827 first investigated a similar wriggling motion of small inert particles in fluids. During a given time interval, smaller particles travel farther by their Brownian motion than larger ones; thus in the lung the smaller particles have a greater chance to strike the bronchiolar or alveolar membrane. A result is that a substantial fraction of particles smaller than about 0.1 micron are also retained in the lung. It is interesting that Brownian motion of particles which are the size of macromolecules (about 0.001 micron) is so vigorous that they reportedly impact on the walls of the nasal cavity and trachea before reaching the lung.[23] But this feature is not included in the curves of Fig. 2.17.

We conclude that particles in two size ranges are most effectively retained in the lung: those whose radius is about 0.5 micron, and smaller ones whose radius is about 0.05 micron. It is sobering to recall that aerosols of the latter class do not scatter light effectively and consequently their presence in the air is difficult to detect by optical techniques. For particles of intermediate size, about 0.25 micron, neither Brownian motion nor sedimentation is very effective, and only 20–40% of the inspired particles are retained, the rest being exhaled.

Alveolar membranes have no mucous covering or cilia. When irritants become lodged on their surface a different defense mechanism comes into play. The irritated tissue releases cells which attach to the foreign body and build up an engulfing mass called a *macrophage*. The mobile cells are known as *phagocytes*, and the process is called *phagocytosis*. If the particle is soluble, it may be removed through the alveolar wall and taken into a lymph node or the blood stream. If not, the macrophage may remain in place. Some dusts such as pure carbon, tin, and iron are attached to the wall without further affect and are therefore known as "benign" dusts. But other particles, because of composition or shape, may be sufficiently irritating that scar tissue and infection may develop. Whether the particle is benign or pathogenic, eventually the macrophage may be released from the wall. The free body can then be eliminated by exhaling, or it may be caught by the cilia in the bronchioles and moved up the ciliary escalator to be eliminated by swallowing or expectoration. Phagocytosis serves also to maintain sterility in the lung, for bacteria can be killed by phagocytes and subsequently eliminated from the body. Thus we see that phagocytosis is a central mechanism of pulmonary defense.

Diseases

In the opinion of some physicians, certain chronic diseases such as chronic bronchitis, emphysema, and lung cancer may be caused by polluted air. There is little doubt that certain gases and aerosols when experienced at very high concentrations and for sufficiently long periods can cause these diseases. For example, strong statistical correlations have been found between excessive smoking and the incidence of emphysema and lung cancer. There is also a strong association between certain occupations, such as coal mining and asbestos operations, and a high incidence of chronic lung diseases. However, the concentrations of pollutants experienced by cigarette smokers or workers in such occupations are a factor of 10 or 100 greater than normally found in the polluted ambient air which the general population experiences. It remains an open question as to whether chronic exposure to these much lower levels can also lead to chronic diseases, although there is strong evidence that this probably does occur for bronchitis and lung cancer. We shall examine the associations between specific community pollutants and lung diseases in more detail in subsequent chapters, after we have laid the groundwork for understanding the circumstances under which large populations are exposed to pollutants. Now we shall turn our attention to the general characteristics of the diseases.

Bronchitis is a vaguely used term for an inflammation of all or part of the bronchial tree. In response to the irritation of the respiratory tract, mucus production is increased by the goblet cells in the bronchial wall, and a cough reflex is initiated to assist in eliminating the secretions. Airway resistance increases as the mucous layer grows in thickness. *Acute bronchitis* is a short-lasting disease, caused by one or more irritants such as a virus, fumes of chemical agents, and dust. Infection and fever is frequently, but not necessarily, present. The disease runs its course within a week; however, if untreated or if the irritants persist, acute bronchitis can develop into *chronic bronchitis*. This is a long-standing inflammation of the bronchial lining, resulting in permanent destruction of cells and loss of cilia. Mucus glands are commonly enlarged and are excessively productive. Scar tissue narrows the air passages and produces rigidity and distortion. The condition is characterized by a persistent cough and production of sputum. Infection and swelling of the terminal bronchi in chronic bronchitis can cause a collapse of the walls with a result similar to pulmonary emphysema. In a significant number of cases, chronic bronchitis probably develops into emphysema. There is strong statistical evidence that the incidence of chronic bronchitis depends upon the average levels of sulfurous smog; and this is a subject which we shall take up in Chapter 8.

Pulmonary emphysema (from the Greek, to inflate or blow into) is a condition of the lung in which a thinning and destruction of the alveolar walls results in an enlargement of the air sacs. The lung becomes hyperinflated due to dilatation and coalescence of the alveoli. The disease is progressive; there is no known cure. It is characterized by shortness of breath, due to the incapacitation of the lung. When the disease is well developed, any physical exertion is followed by great difficulty in breathing; simultaneously the heart is placed under stress as it attempts to increase blood circulation through the viable portions of the lung to meet the oxygen demands of the body. Because the available alveolar surface area for producing phagocytes is markedly reduced, foreign bodies cannot be eliminated efficiently, with the result that the lung is susceptible to further injury. Death finally occurs either from heart failure or from an acute episode of infection. Emphysema is the fastest growing cause of death in the United States. The toll has increased 17-fold since 1950 to an annual rate of 20,252 deaths in 1966, and it is popularly suspected that air pollution plays a causal role. Evidence is accumulating that susceptibility to emphysema may be enhanced by hereditary factors.[24]

Lung cancer (bronchogenic carcinoma) is a condition in which cells of the linings of the air passages of the lung undergo a change that leads to their uncontrolled growth. Typically it develops in the epithelial lining of the bronchi or bronchioles. The mechanism by which it is caused is unknown. The resulting malignant tumors block air circulation and destroy pulmonary structure, thereby reducing the ability of the lung to function. Unless growth is arrested or the tumors removed at an early stage, the disease evolves to a condition where metastases occur, and the disease spreads through the lymphatic system to other parts of the body, resulting eventually in death. Other carcinomas may originate in the alveoli, but this is rare.

Some components of cigarette smoke are known to be carcinogenic (cancer producing) when applied in high concentration to the skin of animals. These include the polycyclic aromatic hydrocarbons benzo(a)pyrene (or benzo-3,4 pyrene, abbreviated as "BaP") and benz(a)anthracene. Whether they cause lung cancer in humans has not been proven. Perhaps multiple causative factors are important.

Carcinogens are also known to be in polluted community air. The tar emitted upon burning soft coal in a stove, or gasoline in an automobile engine, contains several carcinogenic polycyclic compounds.[25] Trace metals such as vanadium which are found in urban air are also carcinogenic when experienced at high concentrations. Asbestos, nickel, chromium, and soot are additional examples.[26] Thus the presence of carcinogens in polluted air is well established.

It is important to keep in mind however that carcinogens occur at much lower concentrations in ambient air than in cigarette smoke. In community air, BaP for example is found at concentrations of several parts per billion, whereas it may be more than ten times more concentrated in cigarette smoke. The question remains as to whether carcinogens occur at harmful levels. It is difficult to show a causal relationship for carcinogens for several reasons. One factor is a long period of latency usually amounting to many months between exposure and development of a recognizable tumor. Another is the high mobility of populations, in which members continually move from one city to another, thus accumulating a history of exposure to a wide variety of atmospheric contaminants under various conditions.

Occupational diseases

Primarily we are concerned in this book with air pollution and its effect on the general population. However, several occupational diseases are caused by inspiration of dust-laden air and may in fact be extreme examples of afflictions of the lung of urban man. The most common disease of world-wide distribution is *pneumoconiosis* (from the Greek words meaning "lung" and "dust"), a generic term for a group of diseases of the lung caused by inhalation of specific dusts. The different types include silicosis, asbestosis, and anthracosis. These diseases are most commonly found to afflict workers in the mineral industries.

Silicosis can result from exposure to high concentrations of dust-bearing free silica (SiO_2), such as quartz and sand, which are used in foundries, sandblasting, and pottery manufacturing. The particles when lodged in the lung pass through the alveolar walls and collect in the lymphatic system. Damage to the alveolar membrane is caused by production of silicotic nodules and a fibrous structure in the lung wall. The condition may be irreversible when developed beyond a certain stage. Emphysema can result from the alveolar walls losing their elasticity and breaking down, thereby forming large cavities perhaps 3–4 mm in diameter.

The other forms of pneumoconiosis are also characterized by fibrosis of the lung. Asbestos—a combination of silicates of magnesium and iron—is also a potent lung irritant. This may be due partially to the shape of the material, which even in the most finely divided state maintains its needle-like form. The resulting fibrotic development in the lung is known as asbestosis. As we mentioned, asbestos is also carcinogenic.

Perhaps the most widespread pneumoconiosis is anthracosis, commonly known as coal miner's "black lung." In some mines as many as one-third of the workers suffer from the disease.[27] Autopsies reveal that the lung in the advanced stage of the disease does indeed have a black color, although the cause of the color is not due simply to the dust which is retained. Physiological changes such as fibrosis contribute to the effect. Some 10–40% of the dust may be free silica. The observation that the lungs of city dwellers are blackened is probable evidence that some of the inhaled solid aerosols of urban air are retained in the lung. However, again we emphasize that a clear distinction should be maintained between occupational exposures to such things as coal dust and exposure to soot in community air pollution. No disease has as yet been causally related to community exposures to particulate matter alone. The concentration of aerosols in urban air is much smaller than the concentrations found in mines and foundries, so the exposures are markedly different.

Synergy

When dealing with the effects of air pollution, we must recognize that complex factors are at work. A common example is the phenomenon known as *synergy*. A synergistic effect is one in which the reactivity or potency of a chemical is affected by the presence of a second agent. The adverse effects that result when they are experienced simultaneously differ from the sum of the consequent effects when experienced separately. For example, it is known that the effect of sulfur dioxide on the lung is considerably more severe when aerosols are present simultaneously. To understand this synergistic effect, recall that SO_2 itself is absorbed by the moist lining of the upper respiratory system, so it is regarded as a mild irritant of only that portion of the air tract. This is no longer true when aerosols are present. In ambient air, sulfur dioxide can be adsorbed (coated) onto the surface of the particles and be carried deep into the lung if the particles are the appropriate size. There, through sedimentation or Brownian motion, the particles with SO_2 can lodge on the pulmonary membrane. So situated, SO_2 becomes an an irritant of the bronchioles and alveoli.

Aerosols can have still another synergistic effect. Many particles are a loosely knit and porous assembly of molecules, having a large surface area on which gases such as SO_2 can be readily adsorbed in great amounts. When entrapped on a pulmonary membrane, the particle applies the irritant over a very localized region. Thus the resulting concentration of absorbed gas molecules is, in effect, much greater than that which would result from a uniform distribution of the same amount of gas over the entire pulmonary surface. The irritant concentrated at the particle's site might exceed the threshold for injury, whereas evenly distributed it would not.

2.5 EPIDEMIOLOGY

Epidemiology is the branch of medicine concerned with the conditions associated with the occurrence of widespread diseases. When dealing with the effects of air pollution, epidemiologists have endeavored to correlate environmental factors with the occurrence of disease, discomfort, or physiologic change in body functions. Frequently such studies have sought to find associations between levels of specific air pollutants and the rate of mortality or morbidity within an exposed population. Although an association between a specific pollutant or combination of pollutants and an accompanying or subsequent effect does not prove the existence of a causal relationship, the association may be so obvious as to provide sufficient basis for policy decisions by individuals or governments. As epidemiology has played a central role in assessing the effect of air pollution, we shall in this section consider some of its strengths and weaknesses.

The challenge for epidemiology is in gathering and processing the appropriate statistical information in a manner that will separate out effects arising from causes other than air pollution.

For example, to obtain a true measure of the correlation between the levels of air pollutants in a community and the incidence of lung cancer, the effects of smoking must be isolated. The correlation between smoking and lung cancer is strong enough that it may obscure in some cases a statistical relationship between community air pollution and lung cancer. On the other hand, if a true measure of the effects of community air pollution is to be obtained, the smokers of a population cannot be neglected. Due to synergistic effects, air pollution may in fact have a more serious effect upon smokers than nonsmokers. A dramatic illustration is provided by the results of a study of the incidence of occupational lung cancer among a group of asbestos workers.[28] Of a population of 87 nonsmoking workers, during the period from 1963 to 1967 there were no deaths from lung cancer; however, of 283 workers who did smoke, there were reported 24 deaths from lung cancer, compared with only 3 that had been expected on the basis of their smoking habits or occupations alone. Thus multiple causative factors can be very important.

One difficulty in obtaining reliable epidemiologic data for the effect of pollutants in community air is the heterogeneous character of urban populations. The residents of a community at the time of a given study have a varied background with respect to previous exposure to contaminants, partly as a result of moving from one place to another, but also partly because the nature of the pollutants in some communities has changed during the past twenty years, as new industries are introduced and automobile traffic increases. Another difficulty facing epidemiologic studies is lack of reliable measurements of ambient levels of pollution, a consequence of an insufficient number of monitoring stations throughout most of our metropolitan regions. Thus the reported air pollution levels for a city may in fact represent conditions for only one location, and this is perhaps not indicative of either the average levels or peak levels for other parts of the city. It has therefore been difficult to obtain reliable assessments for the long-term health effects of specific pollutants.

By comparison, clinical studies of *air pollution toxicology* in laboratory experiments offer better control over the conditions of exposure for human or animal subjects; and in the case of animal experiments, physiological change in the respiratory system can be studied systematically by autopsy for different degrees of exposure to a given pollutant. But animal studies of toxicology have a limited duration, at most a few years or so, corresponding to the lifetime of the subject. This restricts the possibility of studying the development of chronic lung diseases which might follow exposure to low pollution levels over much longer durations, as experienced by inhabitants of metropolitan areas. Thus to obtain positive results, the animal subjects often must be exposed to much higher concentrations of pollutants than are normally experienced in community air. Such studies can permit the establishment of mechanisms by which physiological change occurs, but whether these mechanisms are operative at much lower concentrations cannot be established by extrapolation for so complex a system as living tissue. In addition, there is no assurance that the sensitivity of humans for a given pollutant is comparable to the sensitivity of animals. The lower respiratory system in rats and mice, for example, differs from that in man.

Toxicologic studies compared with epidemiologic are also limited in another way: the gross statistical insensitivity imposed by the small number of animals normally tested. Thus experiments involving low levels of pollution and 50 animals would have little chance from a statistical basis alone of detecting diseases that might be produced at a rate of only 100 individuals per 100,000 similarly exposed population. And this is a statistically significant rate for many types of epidemiologic studies.

Epidemiologic comparisons of urban and rural populations consistently show a higher incidence of chronic lung disease for the former. Mortality rates from lung disease are also higher. This is true even when the effects of smoking are removed from the statistics. The remaining higher morbidity and mortality is

TABLE 2.2

Lung cancer mortality studies. Number of deaths
from lung cancer per 100,000 population[*]

Smokers, standardized for age and smoking			Nonsmokers			Study[29]
Urban	Rural	Urban/ Rural	Urban	Rural	Urban/ Rural	
101	80	1.26	36	11	3.27	California men (Buell, 1967)
52	39	1.33	15	0		American men (Hammond, 1958)
189	85	2.23	50	22	2.27	England and Wales (Stocks, 1957)
			38	10	3.80	Northern Ireland (Dean, 1966)
149	69	2.15	23	29	0.79	England; no adjustment for smoking (Golledge, 1964)
100	50	2.00	16	5	3.20	American men (Haenszel, 1962)

[*]From L. B. Lave and E. P. Seskin, *Science* **169**, 723 (1970). Copyright © 1970 by the American Association for the Advancement of Science. Reprinted by permission.

known as the *urban factor*. Several examples of the urban factor for various studies
of mortality from lung cancer are given in Table 2.2. For nonsmokers, the mortality rate from lung cancer is more than 2.3 times higher in an urban environment
than rural. For smokers, the difference is generally not as pronounced, ranging
from 1.26 to 2.23; but the mortality in both urban and rural environments is
considerably greater. Although statistical evidence such as this is compelling, it
is not *a priori* necessary to conclude that the cause of the urban factor is air
pollution, for there are social and economic differences between urban and rural
populations, including the social stress from crowded conditions, noise level; also
diet, customary amounts of exercise, etc.[30]

One study by Ishikawa *et al.* may have avoided many of these complications
and provided evidence that air pollution may cause or contribute to emphysema.[31]
A comparison was made of autopsy lung material from residents of two cities:
Winnipeg, Manitoba, and St. Louis, Missouri. The Canadian city has a relatively
low level of air pollution, whereas the American city characteristically has high
levels of industrial contaminants. Emphysema was found to be seven times more
common in St. Louis for ages 20–49 and twice as common for ages over 60.
A more detailed comparison is given in Fig. 2.18. Smoking was significantly
associated with the disease, but could not be isolated as a factor. The incidence of
severe emphysema was found to be four times as high among cigarette smokers in
St. Louis as among a comparable group of smokers in Winnipeg. These results are

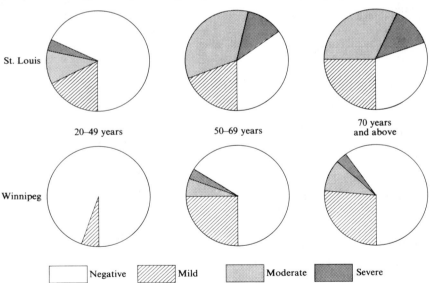

Fig. 2.18 Prevalence of emphysema, as found in a 1960–66 post-mortem examination of the
lungs of 300 residents of St. Louis, Missouri, and an equal number from Winnipeg, Canada. The
subjects were well matched by sex, occupation, socio-economic status, length of residence,
smoking habits, and age at death. (Data from S. Ishikawa, D. H. Bowden, V. Fisher, and J. P.
Wyatt, *Arch. Environ. Health*, **18**, 660, 1969)

highly suggestive, but yet do not prove a causal link between air pollution and emphysema. A question remains as to whether the markedly different climatic conditions of the two cities may be a contributing factor.

As a result of difficulties such as this in making definitive statements from epidemiologic studies on different populations, there is considerable interest in the temporal variation and effects of air pollution on the same population. The simultaneous double phenomena of abnormally polluted air and excess morbidity and mortality may be so pronounced as to seem to imply a causal relationship; indeed, in the pollution episodes which occurred in Donora (1948) and London (1952), there can be little doubt about this.

TABLE 2.3

Categories of respiratory and heart diseases included in a study
of mortality and air pollution levels in New York City *

Respiratory diseases	Heart diseases
Tuberculosis of respiratory system	Arteriosclerotic heart disease, including
Malignant neoplasm of respiratory system	coronary disease
Asthma	Hypertensive heart disease
Influenza	Rheumatic fever and chronic rheumatic
Pneumonia	heart disease
Bronchitis	Other diseases of the heart, arteries, veins
Pneumonia of newborn	Certain types of nephritis and nephrosis
	of the kidneys

*T. A. Hodgson, *Environmental Science and Technology*, **4**, 589, 1970.

However, the more subtle episodes that occur several times a year in many regions of the world require more care in their analysis because, as we shall find, the weather patterns that are largely responsible for the episode are frequently accompanied by characteristic temperature changes, and this in itself can give rise to an increase in respiratory illness. Abrupt temperature changes put a strain on the body and nervous system, while prolonged warm spells contribute to exhaustion. Furthermore, cold decreases mucus transport, thereby reducing the efficiency for removing airborne materials from the lung.

Thomas Hodgson, Jr., has found that mortality in New York City from respiratory and heart disease is quite significantly related to environmental conditions, especially air pollution (as indicated by the mass concentration of aerosols) and temperature (in terms of degree-days).[32] Some 73% of the variation in daily mortality from the heart and respiratory diseases listed in Table 2.3 are explained by concurrent variations of air pollution levels and temperature, with pollution showing a more significant correlation. The expected increase in average daily mortality for a moderate increase in particulate concentrations is 20–30 deaths, almost 20% of the average daily total number of deaths attributable to heart and respiratory diseases. Studies of this nature are providing evidence that air pollution is continually exacting a toll, and that an increase of mortality and morbidity is not limited to just the spectacular but rare episodes.

A comment

We have indicated in the previous discussion that the pollutant thresholds for causing chronic diseases are not known. Whether dosage, peak levels, or average levels provide the most relevant parameter has not been determined. Complications arise from effects such as synergy. Adaptation of the lung is another factor.[33] That adaptation may be important has been shown by experiments in which animals were exposed to high concentrations of gaseous ozone (O_3); it was found that previous exposure to somewhat lower levels of O_3 permitted the animals to survive a dosage that would otherwise have been fatal.[34] It is also possible that complex temporal relationships are important in influencing the severity of effects, since some of the defense mechanisms such as ciliary activity do not reach maximum effectiveness until several hours after irritation is first experienced. Uncovering the precise causes of chronic pulmonary diseases is a challenge still facing medical science.

One factor is certain: The severity of adverse effects depends upon the *ambient concentration* of pollutants. In two simple cases, reduction of visibility and perception of odors, the concentration is the most important parameter. In others, dosage may also be important. All adverse effects, however, can be avoided if it is possible to keep pollution concentrations sufficiently low. It is essential that we realize what this means: Meteorological factors are an integral part of the air pollution syndrome, for they determine how much air is available for diluting contaminants. If we are to understand how local air pollution problems arise, we must consider not only the source of the pollution but meteorological conditions as well. In the next five chapters we shall consider the role played by atmospheric dynamics and their effects on the dispersal of pollutants. Then we shall be in a position to examine the interplay between the emission of pollutants and their dilution. The net result of these two processes determines whether the adverse effects we have discussed in this chapter will occur.

SUMMARY

The olfactory organ has high sensitivity in detecting and recognizing odors; but the intensity of sensation (the "strength" of an odor) is fairly insensitive to the ambient concentration, as indicated by the logarithmic dependence on concentration given by the Fechner law. Thus the rate at which odorants are released into the atmosphere has to be reduced substantially to yield just a small reduction in the strength of an odor. An assessment of the benefits to a population from improved controls over the emissions of odorants is complicated by the phenomenon of adaptation and by the fact that members of a population have a distribution of recognition thresholds. Some other adverse effects of air pollution, however, can be more simply characterized. Prevailing visibility, for example, is found to be inversely related to the mass concentration of particulate matter in the air. Thus a decrease in the aerosol mass concentration by a given factor would increase the prevailing visibility by the same factor. A substantial reduction in primary and secondary aerosol pollutants would be expected to produce a correspondingly substantial improvement in visibility. Adverse effects on the respiratory system are harder to quantify. Exposure to sufficiently high concentrations of certain gaseous or particulate contaminants can cause chronic lung diseases such as bronchitis, emphysema, and lung cancer. There is strong statistical evidence that bronchitis may be caused by exposure to certain pollutants in community air and there is suggestive evidence for lung cancer but there is no direct proof. Some diseases may have multiple causes, and it will be a difficult task to sort out the important factors. Aerosols can act in synergism with gaseous contaminants to affect the lower respiratory system, the size of the particles determining how far they can penetrate into the respiratory tract.

NOTES

1. H. R. Blackwell, "Contrast Thresholds of the Human Eye," *J. Opt. Soc. Am.*, **36**, 624 (1946).

2. R. I. Larsen, "Relating Air Pollution Effects to Concentration and Control," *J. Air. Poll. Control Assoc.*, **20**, 214 (1970).

3. H. Rashdall, *Universities of Europe in the Middle Ages*, Vol. III, Oxford: Oxford University Press, 1936, p. 80.

4. G. Leonardos, D. A. Kendall, and N. Barnard, "Odor Threshold Determinations of 53 Odorant Chemicals," *J. Air Poll. Control Assoc.*, **19**, 91 (1969).

5. H. Stone and G. Pryor, "Some Properties of the Olfactory System of Man," *Perception and Psychophysics*, **2**, 516 (1967).

6. E. Kaiser, "Odor and Its Measurement," in *Air Pollution*, Vol. I, 1st ed., A. C. Stern, Ed., New York: Academic Press, 1962.

7. F. V. Wilby, "Variation in Recognition Odor Threshold of a Panel," *J. Air Poll. Cont. Assoc.*, **19**, 96 (1969).

8. G. Leonardos, D. A. Kendall, and E. R. Rubin, "The Role of Background Odor in Affecting the Perception of Odorant Chemicals," presented at 63rd Annual Meeting of the Air Pollution Control Assoc., St. Louis, Missouri, June 14–18, 1970.

9. J. F. Byrd and A. H. Phelps, "Odor and Its Measurement," in *Air Pollution*, 2nd ed., A. C. Stern, Ed., New York: Academic Press, 1968, p. 308.

10. S. S. Stevens, "Neural Events and the Psychophysical Law," *Science*, **170**, 1043 (1970); *Science*, **133**, 80 (1961).

11. M. Kerker, *The Scattering of Light and Other Electromagnetic Radiation*, New York: Academic Press, 1969.

12. H. Koschmieder, "Theorie der Horizontalen Sichtweite," *Beitr. Physik. Freien Atmosphäre*, **12**, 33 and 171 (1924).

13. W. E. K. Middleton, *Vision Through the Atmosphere*, Toronto: University of Toronto Press, 1963.

14. S. Q. Duntley, "The Reduction of Apparent Contrast by the Atmosphere," *J. Opt. Soc. Am.*, **38**, 179 (1948); "The Visibility of Distant Objects," *J. Opt. Soc. Am.*, **38**, 237 (1948).

15. H. Horvath and K. E. Noll, "The Relationship Between Atmospheric Light Scattering Coefficient and Visibility," *Atmos. Environ.*, **3**, 543 (1969).

16. W. Fett, *Beitr. Physik. Atmosphäre*, **40**, 262 (1967).

17. R. J. Charlson, H. Horvath, and R. F. Pueschel, "The Direct Measurement of Atmospheric Light Scattering Coefficient for Studies of Visibility and Pollution," *Atmos. Environ.*, **1**, 469 (1967).

18. K. E. Noll, P. K. Mueller, and M. Imada, "Visibility and Aerosol Concentration in Urban Air," *Atmos. Environ.*, **2**, 465 (1968).

19. S. Twomey, in *Atmospheric Chemistry and Sulfur Compounds*, Geophysical Monograph No. 3, Washington, D.C.: Geophys. Union, 1959.

20. E. Robinson, "Effects of Air Pollution on Visibility," in *Air Pollution*, Vol. I, 1st ed., A. C. Stern, Ed., New York: Academic Press, 1962, p. 220.

21. T. F. Hatch, "Dust and Disease," in *The Air We Breathe*, S. M. Farber and R. H. L. Wilson, Eds. with J. R. Goldsmith and N. Pace, Springfield, Ill.: Charles C. Thomas, 1961.

22. B. Altshuler, L. Yarmus, E. D. Palmes, and N. Nelson, "Aerosol Deposition in the Human Respiratory Tract, I. Experimental Procedures and Total Deposition," *Arch. Ind. Health*, **15**, 293 (1957).

23. A. C. Chamberlain and E. D. Dyson, *Brit. J. Radiology*, **29**, 317 (1956).

24. J. Lieberman, "Hetero- and Homo-zygous Alpha₁-Antitrypsin Deficiency in Pulmonary Emphysema," *New England J. Medicine*, **281**, 279 (1969); S. Eriksson, "Studies in Alpha₁-Antitrypsin Deficiency," *Acta Med. Scandinav.* **117** (Supp. 432) 1 (1965).

25. W. Agnew, "Automotive Air Pollution Research," *Proc. Roy. Soc.*, **A307**, 153 (1968).

26. W. C. Hueper, "Carcinogens in the Human Environment," *Arch. Path.*, **71**, 237 and 355 (1961).

27. G. K. Tokuhata, P. Dessaver, E. P. Pendergrass, T. Hartman, E. Digon, and W. Miller, "Pneumoconiosis among Anthracite Coal Miners in Pennsylvania," *Am. J. Public Health*, **60**, 441 (1970).

28. I. J. Selikoff, E. C. Hammond, and J. Chung, "Asbestos Exposure, Smoking, and Neoplasia," *J. Am. Med. Assn.*, **204**, 106 (1968).

29. P. Buell, J. E. Dunn, Jr., and L. Breslow, *Cancer*, **20**, 2139 (1967); E. C. Hammond and D. Horn, *J. Amer. Med. Ass.*, **166**, 1294 (1958); P. Stocks, "British Empire Cancer Campaign," supplement to "Cancer in North Wales and Liverpool Region," part 2 (Summerfield and Day, London, 1957); G. Dean, *Brit. Med. J.*, **1**, 1506 (1966); A. H. Golledge and A. J. Wicken, *Med. Officer*, **112**, 273 (1964); W. Haenszel, D. B. Loveland, and M. G. Sirken, *J. Nat. Cancer Inst.*, **28**, 947 (1962).

30. D. D. Reid, "Air Pollution as a Cause of Chronic Bronchitis," *Proc. Roy. Soc. Med.*, **57**, 965 (1964).

31. S. Ishikawa, D. H. Bowden, V. Fisher, and J. P. Wyatt, "'The Emphysema Profile' in two Midwestern Cities in North America," *Arch. Environ. Health*, **18**, 660 (1969).

32. T. A. Hodgson, Jr., "Short-Term Effects of Air Pollution on Mortality in New York City," *Environ. Sci. and Tech.*, **4**, 589 (1970); and *Environ. Sci. and Tech.*, **5**, 548 (1971).

33. E. J. Fairchild, II, "Tolerance Mechanisms," *Arch. Environ. Health*, **14**, 111 (1967); P. E. Morrow, "Adaptations of the Respiratory Tract to Air Pollutants," *Arch. Environ. Health*, **14**, 127 (1967).

34. H. E. Stokinger and D. L. Coffin, "Biological Effects of Air Pollutants," in *Air Pollution*, Vol. I, 2nd ed., A. C. Stern, Ed., New York: Academic Press, 1968, p. 446.

FOR FURTHER READING

J. F. CORSO   *The Experimental Psychology of Sensory Behavior*, New York: Holt, Rinehart and Winston, 1967.

S. H. BARTLEY   "Adaptation to the Environment," in *Handbook of Physiology*, J. Field *et al.*, eds.; American Physiology Society, Baltimore: Williams and Wilkins, 1964, Chap. 7.

M. EISENBUD   *Environmental Radioactivity*, New York: McGraw-Hill, 1963.

C. I. BLISS   *Statistics in Biology*, New York: McGraw-Hill, 1967 (Vol. I), 1970 (Vol. II).

W. E. K. MIDDLETON   "Vision Through the Atmosphere," *Handbuch der Physik* XLVIII, Geophysik II, Berlin: Springer-Verlag, 1957, p. 254. Also *Vision Through the Atmosphere*, Toronto: University of Toronto Press, 1963.

D. V. BATES and   *Respiratory Function in Disease: An Introduction to an Integrated*
R. V. CHRISTIE   *Study of the Lung*, Philadelphia: Saunders, 1964.

B. ALTSHULER   "Behavior of Airborne Particles in the Respiratory Tract," Ciba Foundation Symposium on Circulatory and Respiratory Mass Transport, G.E.W. Wolstenholme and J. Knight. Eds., London: J. and A. Churchill, 1969, p. 215.

QUESTIONS

1. What fraction of the population of a city do you believe should be safeguarded from an odor that is generally considered offensive? How might your answer be influenced by the cost of controlling emissions of the odorant?

2. Describe the quality of odors which you have smelled today. How would you rank their relative intensities? What do you consider to be an easily noticeable odor? A faint odor?

3. What are the implications when human perception of odors is described by the Fechner law? What advantages are offered by a logarithmic response to the intensity of a stimulus? What disadvantages?

4. What is the difference between light reflection, diffraction, scattering, and absorption?

5. How much would you be willing to pay annually for more costly goods and services if by so doing you could be sure that 95% of the time the visual range would be at least 15 km (about 9 miles)? For 30 km?

6. What factors might limit the accuracy of data such as are displayed in Fig. 2.14?

7. Sometimes the visual range is considerably less than is given by formulas of the type of Eq. (2.14). One instance is when the sun is close to the horizon near the line of sight of the observer. Why might the observer's visual range be especially short?

8. What defense mechanisms serve to protect the respiratory system?

9. How might aerosols and gaseous irritants have a synergistic effect on the lung?

10. What mechanisms cause inhaled aerosols to contact membranes of the upper and lower respiratory tract? What size particles are most affected by each mechanism and where are they retained in the tract?

PROBLEMS

1. Sulfur dioxide is emitted from an exhaust stack of an oil refinery at a concentration of 1500 ppm. By what minimum factor must the exhaust plume be diluted by air in order that the odor is unrecognizable?

2. Suppose that the volume of exhaust gas in a plume were emitted at a rate of $10 \ m^3$ per second, and it had an SO_2 concentration of 1000 ppm. At what rate must air be mixed with the plume to dilute the SO_2 to the recognition threshold within ten seconds of emission?

3. A bottle contains $50 \ cm^3$ of liquid ammonia. If the bottle is left open, and the ammonia completely evaporates and mixes with the air in a closed room, could the odor of ammonia be recognized? Assume that the volume of ammonia expands by a factor of 800 when it evaporates at atmospheric pressure. The room has a floor area of $90 \ m^2$ and a height of 4 m.

4. A pulp mill emits hydrogen sulfide (H_2S) at a rate of 100 kg per day. What mass of air must be mixed with the H_2S to make the odor just barely recognizable? Assume that air has a molecular weight of 29 g/mole.

5. Near a petroleum refinery, ethyl mercaptan is found at an ambient concentration of 2 ppm, and the intensity of the odor is generally regarded as easily noticeable. By what factor must emissions be suppressed to reduce the intensity to a "very faint" level, assuming that meteorological conditions do not vary?

6. If an aerosol consists of identical particles of density $2 \ g/cm^3$ and radius of 1 micron, what concentration of particles would correspond to a mass concentration of $100 \ \mu g/m^3$? Compare your answer with the example given in the text following Eq. (2.14).

7. An air pollution advisory committee recommends that the ambient concentration of particulate matter does not exceed $75 \ \mu g/m^3$. How far could people see if the air had a uniformly distributed burden of this concentration? What would you estimate as the uncertainty in your prediction?

8. Suppose that a burning municipal dump emits 7 tons of particulate matter each

hour. If half of this remains in the air and mixes to a uniform concentration, what volume of air would be needed to dilute an hour's emissions to the point where the meteorological range could be maintained at 20 km?

9. On an abnormally polluted day, the ambient concentration of particulates is found to be 1 mg/m^3. What is the meteorological range as defined on the basis of a 50% probability of contrast detection versus a 90% probability?

10. Suppose that a target object on the horizon were not black but grey, with a contrast $C(0) = +0.5$. How would this change the condition for the meteorological range?

11. The density of particles in a smokestack plume is often estimated from its opacity, a measure of how much light is absorbed or scattered as it traverses the plume. If the extinction coefficient is $k = 1.5$ m^{-1}, what fraction of light intensity is absorbed as sunlight passes transversely through a plume of thickness $d = 1.2$ m?

12. An average adult takes 15 breaths a minute, inhaling 500 cm^3 of air each time. If the air he inhales contains 100 μg/m^3 of particulate matter, what is the total daily amount of particulates retained in the lung if all particles have a radius of 0.5 micron and a density of 1 g/cm^3? A radius of 0.25 micron and a density of 1 g/cm^3?

13. Suppose that the threshold for an adverse reaction from members of a population is normally distributed according to the formula $A \exp\left[-(c - c_0)^2/2B\right]$. What do the constants A, c_0, and B represent? Find an expression for A in terms of c_0, B, and the number of people in the population.

CHAPTER 3

THE ATMOSPHERE

Ordinarily our atmosphere is a hospitable environment in which to live. The dynamic processes which move air from one place to another encourage the mixing of its gaseous components, thereby avoiding a permanent build-up of the concentration of a pollutant over a given locality. Global circulation ensures a uniformity of composition of the major gaseous constituents. This is all the more significant when it is appreciated that the atmosphere participates in many processes that cycle elements and compounds throughout the biosphere, wherever biological activity is found. The atmosphere's relationship to the oxygen–carbon dioxide cycle is well known, but the atmosphere also figures in the nitrogen cycle and the sulfur cycle (involving SO_2 and H_2S). Despite the uneven distribution of the sources of these gases over various regions of the earth, the continual motion of the air eventually establishes a uniform distribution, from polar regions to equatorial, with a few notable exceptions. In addition it serves to transport water and water vapor from one portion of the world to another. It is with respect to this background of the natural atmosphere that we must assess the importance of contaminants from anthropogenic sources.

Some air pollution is solely a local phenomenon; explanations for its cause are readily found to depend upon the types of sources, the rate of emission of pollutants, and the meteorological conditions in the locality. However, sustained industrial growth during the past century has been accompanied by an increase of affected area, which in some instances now encompasses a million square kilometers. Some effects of air pollution are even world-wide in scope and have affected the global balance between the *sources* of atmospheric constituents and their *sinks* (whatever removes gases or particulate matter from the air).

To provide a perspective for the content of subsequent chapters, we shall devote this one to an examination of the chemical and physical properties of the natural atmosphere. We shall emphasize especially the factors which influence or are affected by air pollution. Water vapor and carbon dioxide are prime examples. Water and water vapor are the most important constituents of the atmosphere with regard to determining its effect on solar and terrestrial radiation. Not only do they have a global effect in influencing the mean surface temperature of the earth, but under the proper circumstances their presence in sufficient concentration can produce local meteorological conditions which are conducive for the accumulation of pollutants. Carbon dioxide also influences global surface temperatures, but it is of secondary importance. Nevertheless, it is the focal point of current interest, for anthropogenic sources appear to be causing a steady increase in its ambient concentration. Most of this chapter will concentrate on a brief review of the gas laws and the physical aspects of the atmosphere as they are influenced by gravity. We shall develop the molecular model for how a gas exerts a pressure and how this pressure and the density of air decrease with altitude. In subsequent chapters, we shall see how this variation is instrumental in affecting the vertical mixing of pollutants.

3.1 CONSTITUENTS

Our globe is blanketed by a mixture of many gases. The concentration of some of them at sea level is uniform and essentially unvarying; consequently they are sometimes called *permanent gases*. The best known are also the major constituents: nitrogen and oxygen. Over three-quarters of all the molecules in the atmosphere are nitrogen, about one-fifth are oxygen. No change has been detected in their ambient concentrations. For example, measurements over the past 50 years indicate that the ground-level concentration of oxygen has varied by less

TABLE 3.1

Composition of the clean atmosphere near sea level

Permanent gases			
Constituent	Chemical formula	Percent by volume	Parts per million by volume
Nitrogen	N_2	78.084	
Oxygen	O_2	20.946	
Argon	Ar	0.934	
Neon	Ne		18.2
Helium	He		5.2
Krypton	Kr		1.1
Hydrogen	H_2		0.5
Nitrous oxide	N_2O		0.3
Xenon	Xe		0.09
Variable gases			
Water vapor	H_2O	0–7	
Carbon dioxide	CO_2	0.032	
Methane	CH_4		1.5
Carbon monoxide	CO		0.1
Ozone	O_3		0.02
Ammonia	NH_3		0.01
Nitrogen dioxide	NO_2		0.001
Sulfur dioxide	SO_2		0.0002
Hydrogen sulfide	H_2S		0.0002

than the 0.03% uncertainty of the measurements.[1] Air contains many other gases, notably the inert elements argon, neon, and helium. Some of them are found in only trace amounts, at concentrations of a few parts per million or less. Table 3.1 lists the more common constituents of "clean" air and their normal ambient concentrations.

It is notable that several gases may exhibit marked temporal or spatial variations in concentration. These are called *variable gases*. The best known are water vapor and carbon dioxide. The atmospheric concentration of water vapor may vary from 0–7% by volume, depending upon the past history of the air and its temperature. Carbon dioxide shows seasonal variations of concentration in response to the growth cycle of vegetation in each hemisphere of the world, but generally the relative changes amount to only a few percent. In addition, the average concentration of CO_2 is found to have a steady relative increase of about 0.2% per year. Many variable gases are pollutants and occur in high concentrations in urban air; but gases such as ammonia, nitrogen dioxide, ozone, and hydrogen sulfide also have natural origins and are found world-wide. The concentrations listed in Table 3.1 are "background levels" found in areas of the world that are remote from anthropogenic sources.

Water

Water is a unique component, since in the atmosphere it can exist simultaneously in solid, liquid, and gaseous phase. The three phases are often present in close association in certain clouds; the transformation between one phase and another is a significant factor in weather processes.

Water molecules continually move in both directions across the interfaces separating the different phases. They can escape from the liquid to the gas, because the molecules in the liquid phase continually exchange energy as they vibrate and strike each other. On occasion a molecule at the surface of the liquid may be struck so hard by its neighbors that it overcomes their mutual attraction and enters the air, as illustrated by A in Fig. 3.1. The energy required to liberate a

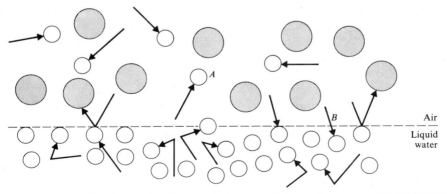

Fig. 3.1 Water molecules continually cross the boundary between the gaseous and liquid phase. Shaded circles represent other air molecules.

molecule is thereby lost by the liquid, so the energy and temperature of the remaining liquid is slightly decreased. The minimal energy which is needed to evaporate a molecule from a liquid surface is called the *latent heat of vaporization*. A similar process takes place at the surface of ice, where the loss of molecules to the gas phase is known as sublimation and the corresponding latent heat is the *latent heat of sublimation*.

There is a reverse process as well. Water molecules in the air over a surface of liquid water are also moving about. They continually strike other air molecules and, if sufficiently close, the surface of the liquid water. Some of those which strike the liquid, such as *B* in Fig. 3.1, are captured by attractive forces and remain in close association in the liquid; this is known as condensation. As they are captured, they relinquish their latent heat and cause a rise in the temperature of the liquid. We shall later find that the shape of the water molecule and its electrical properties give rise to very strong forces of attraction when molecules are close to one another in the liquid (or solid) form. For this reason, the latent heat is exceptionally large. In fact, the latent heat of vaporization for a large number of molecules, 2.43×10^6 joules per kilogram (580 calories per gram at 30°C), is greater than that of almost all other common liquids. The large latent heat of water, and the large energy which is absorbed on evaporation or released on condensation, makes water an effective medium by which energy can be transmitted from one region of the world to another.

Let us consider again the molecular picture of air. Molecules of water vapor, as well as those of the other atmospheric constituents such as nitrogen and oxygen, continually strike and rebound off any solid surface, for example a window-pane, or the ground. They therefore exert a pressure on the surface. We can imagine the total pressure as the sum of the pressures due to collisions from the water molecules alone, the nitrogen molecules alone, the oxygen molecules alone, etc. Each separate contributing pressure is known as the *partial pressure* of the particular atmospheric constituent. The partial pressure of water vapor is also known as its *vapor pressure*. Evidently the vapor pressure of water will increase if the ambient concentration of water vapor increases, assuming that other factors remain constant. We shall later point out that the vapor pressure is in fact proportional to the concentration of water molecules. The concentration of water vapor is called the *absolute humidity*.

Vapor pressure is a useful parameter but it does not tell us everything that we frequently need to know. Let us consider again what happens at the interface between air and water in the liquid (or solid) form. When conditions are such that there is no net movement of water molecules across the interface, the vapor is said to be *saturated*, and the pressure exerted by the water vapor alone is called the *saturation vapor pressure*. Although there is no *net* motion across the interface for the condition of saturation, there is a continual random motion as molecules move in either direction. It has been estimated that during each second, 2 to 3 kilograms of water are moving in every direction across a square

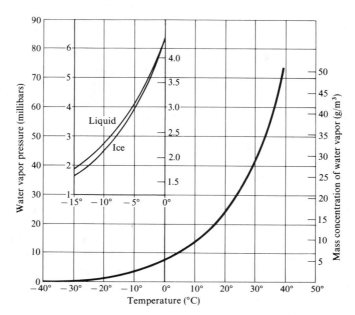

Fig. 3.2 Saturation vapor pressure and mass concentration of water vapor as a function of temperature. The inset compares the saturation vapor pressure over ice with that over water below 0°C. The pressure unit of 1 millibar (1 mb) corresponds to 10^2 newtons/m^2 (10^3 dynes/cm^2).

meter of open water surface.[2] At saturation, there is a dynamic, rather than a static, equilibrium. Should the vapor pressure exceed its saturation value, there will be a net flow of molecules into the liquid; that is, condensation takes place. Should the vapor pressure be less than the saturation value, evaporation occurs. As a result of the motion of water molecules between air and ice, similar processes take place in response to imbalances between the actual vapor pressure and the saturation vapor pressure of ice, but in this case, movement of molecules in *either* direction across the surface of the solid is known as sublimation.

The saturation vapor pressures of water and of ice increase with temperature, since the kinetic energy of molecular motion within the liquid or solid increases with temperature and thereby the chance that molecules can overcome their mutual attraction within the liquid is enhanced. The dependence of the saturation vapor pressures upon temperature is shown for water and ice in Fig. 3.2. It is interesting that at an ambient temperature of 30°C (about 86°F) the curves indicate the saturation vapor pressure of water to be about 4% of the total atmospheric pressure at sea level. Water vapor in warm, saturated air can therefore be the third most plentiful constituent of the atmosphere.

Air is not usually saturated with water vapor. The fraction of the saturated vapor pressure represented by the actual vapor pressure is called the *relative humidity*, a parameter usually expressed as a percentage. The relative humidity shows how closely atmospheric conditions approximate to saturation. By contrast, we recall that the absolute humidity is a measure of the concentration of water vapor. Thus if we were to vary the temperature of an enclosed volume of air containing water vapor, we would change its relative humidity through changing the saturation vapor pressure, even though the absolute humidity remains unchanged. For example, a decrease in temperature would cause a sharp drop in the saturation vapor pressure as illustrated in Fig. 3.2; but as we shall find in Section 3.5 the pressure of water vapor in a container of fixed volume decreases much less rapidly, being proportional to the absolute temperature. Thus the ratio of the actual vapor pressure to the saturation vapor pressure will increase with decrease in temperature, and the relative humidity will rise.

Thus if unsaturated air is cooled, and the gain or loss of water vapor is negligible, its relative humidity increases until at some low temperature 100% relative humidity will be achieved. Condensation will occur as the temperature is lowered further. The temperature at which the air reaches saturation is called the *dew point*. As the name suggests, morning dew is often a result of nighttime cooling of moist air below its dew point. The process of condensation resulting from the cooling of air at high altitudes is responsible for the clouds. These are conglomerations of condensed water droplets whose size, perhaps 7 microns in radius, is too small to permit them to fall at any appreciable speed.

Carbon dioxide

Of the other gases occurring in appreciable quantity in the atmosphere, only carbon dioxide has a freezing temperature sufficiently high ($-56°C$) for it to be naturally produced on earth. However, neither the liquid nor solid form is stable, for the saturated vapor pressures of both forms are much greater than the ambient vapor pressure of CO_2 which corresponds to the existing ambient concentration of about 0.032%. Carbon dioxide does not therefore affect atmospheric dynamics by changes of phase, because it exists to an appreciable extent only as a gas. However, carbon dioxide is an essential constituent of the atmosphere owing to its role in life processes.

Solar energy enters the biological cycle with the action of chlorophyll-bearing organisms. Carbon dioxide is an essential ingredient for the process of photosynthesis in which this energy is used to decompose water and combine it with CO_2 to build organic molecules. The exact mechanism by which photosynthesis occurs is complicated, but the essential results of the process can be summarized in simplified form by the reaction

$$n\,CO_2 + n\,H_2O + E \rightarrow (CH_2O)_n + n\,O_2. \qquad (3.1)$$

The symbol E represents solar energy which is absorbed, and n is an integer ($n = 1, 2, 3$, etc.). The end products of the reaction are molecular oxygen, which is released to the atmosphere, and various carbohydrates, whose different formulas correspond to the different values of n. Carbohydrates are the common fuels stored in plants. One such fuel is a sugar, sucrose ($C_6H_{12}O_6$) indicated by

$n = 6$. Experimental studies of photosynthesis using oxygen tracer isotopes have revealed that the molecular oxygen released to the atmosphere is derived from the decomposed water; therefore the oxygen component of the carbohydrate is supplied by atmospheric carbon dioxide.

The other side of the biological cycle involves the extraction of O_2 from the air, and release of CO_2. This in part is performed by multicellular animals which take in O_2 to oxidize carbohydrate foodstuffs and release energy for physical activity and life-supporting functions. The oxidation and subsequent release of CO_2 is known as *respiration*. The process can be summarized by Eq. (3.1) written in the reverse direction.

Other processes which involve atmospheric O_2 and CO_2 are the oxidation of decayed organic materials on forest floors, which may in fact increase the local concentration of CO_2 by a factor of 2 or 3 above the average ambient level. By contrast, other areas of lush plant growth may absorb so much CO_2 that the local concentration is reduced by a factor of more than 2. Oxidation of iron from the ferrous to ferric state may also account for a major utilization of atmospheric oxygen.

It is evident that the processes of the carbon dioxide and oxygen cycles occur across the globe. Local depletion or enrichment of a gas in one area is quickly eliminated by mixing within the atmosphere and by circulation around the world. This is an important role in the natural cycles which is played by large-scale circulation.

3.2 GRAVITY

Our atmosphere is confined near the surface of the earth by gravity. The gravitational force has a paramount role in establishing conditions which permit the vertical motion of air, and therefore a firm understanding of this phenomenon is essential.

Experience with the physical world has revealed a constancy in the behavior of matter which was first formulated by Sir Isaac Newton in 1687 as the law of universal gravitation. He observed that every body in the universe attracts every other body with a force given by

$$F = G\frac{m_1 m_2}{r^2}, \qquad (3.2)$$

where m_1 and m_2 are the masses of the interacting bodies, r is the distance from one body to the other which it is attracting, and G is a universal constant $G = 6.67 \times 10^{-11}$ newton-m^2/kg^2. Thus the gravitational force decreases as the square of the distance separating the bodies. The implications of gravitational attraction are more than just the everyday experience that a released object falls toward the surface of the earth. Surprising, perhaps, is the indirect result that a volume of air will rise when warmed to a temperature in excess of its surroundings. How gravity causes warmer air to rise will be taken up in Chapter 5.

Newton also recognized three other consistent behaviors of matter which were stated together with the law of universal gravitation in his *Principia Mathematica*. His *first law of motion* observes that a body at rest remains at rest, and a body in motion remains in straight line motion with unchanging speed unless acted upon by a force. This property of matter is called *inertia*.

We shall later discuss several devices for controlling pollution which depend upon the inertia of large particles to cause their removal from a stream of exhaust gas. We shall also find that because of inertia, winds appear to be deflected as air moves over the globe, and the trajectories do not appear to be straight lines but are curves. For now, however, let us continue with our summary of the laws of motion.

The *second law of motion* states that a force can be defined by the acceleration *a* which results from the force alone acting on a mass *m*:

$$F = ma. \tag{3.3}$$

The force and acceleration have the same direction. However, this equation is perhaps more often interpreted in another way: The left-hand side can represent the total force or net force on the object when several forces are acting simultaneously [the gravitational force of Eq. (3.2) could be one such force]. Then the acceleration of an object such as a quantity of air is determined by the total force acting on it.

From Eqs. (3.2) and (3.3), we conclude that the force of gravitation from the earth causes a "downward" acceleration of every unsupported body, an acceleration which is directed toward the center of the earth [assuming that the earth is precisely spherical]. The value of the acceleration predicted by Eqs. (3.2) and (3.3) is independent of the mass of the affected body and therefore is indicated by a special symbol g^*:

$$g^* = \frac{GM_e}{(R_e + z)^2}, \tag{3.4}$$

where M_e and R_e are the mass and radius of the earth, and z is the altitude of the affected body.

Newton's *third law of motion* recognizes that for every force exerted on a body A by a body B there is an equal and oppositely directed force exerted on body B by body A. The symmetrical appearance of m_1 and m_2 in Eq. (3.2) accounts for this effect in the case of gravitational attraction. Thus, as an object is accelerating toward the earth, the earth is accelerating toward the object, but at a rate inversely proportional to the ratio of the masses, as predicted by Eq. (3.3). The large mass of the earth (6.0×10^{24} kg) compared with that of familiar objects ensures that the earth's acceleration is hardly noticeable.

For Newton's three laws to be valid, the acceleration a must be measured with respect to an *inertial reference system*. We emphasize this because later we shall examine important consequences of using noninertial systems when studying the motion of air over the earth. For our purposes, we can define an inertial reference system as one whose position and orientation is fixed with respect to the positions of the stars, neither accelerating in any direction nor rotating.

Now let us consider the gravitational force experienced by an object which is at an altitude z above the surface of the earth. If z is much smaller than the radius of the earth R_e, the downward acceleration due to the gravitational force can be approximated by expressing Eq. (3.4) as a formula containing a series of terms in ascending powers of the quantity (z/R_e); if we keep only the first two terms, since they are the largest, we have

$$g^* = g_0^*(1 - 2z/R_e). \qquad (3.5)$$

In this expression, g_0^* is the acceleration due to gravity at the surface of the earth. The second term, $2z/R_e$, is small compared with unity near the surface and so the acceleration due to gravity which we experience is essentially independent of altitude. For altitudes as high as $z = 400$ km, where g^* is only 87% of the value g_0^*, our neglect of the third and remaining terms of the series for the exact value of g^* results in an error of less than 1%. Since the earth's mean radius is $R_e = 6.4 \times 10^3$ km, Eq. (3.4) indicates that the acceleration due to gravity for altitudes up to 100 km can be considered constant, with a resulting inaccuracy of less than 3%. Therefore, for considerations of weather and large-scale air motion, which occur at altitudes much lower than this, the variation of g^* with altitude may generally be neglected.

3.3 INERTIAL EFFECTS ON A ROTATING GLOBE

When we regard the motion of air over the face of the globe, as we do when viewing a weather map, we are not usually conscious of the fact that the globe itself is rotating. Only the diurnal progression of the sun across the heavens may recall to mind that the earth does indeed rotate. Usually we are not concerned with the rotation, for the major forces earth exerts on us are from gravitation, which attracts us, and from the solidity of its surface, which supports us. In our everyday experiences, such as when we put a book on a desk or drive an automobile down a steep hill, we are well aware of these effects and can quite safely forget about possible additional effects of the earth's rotation.

However, when we consider motions on a larger scale, such as the global circulation of air, we must be more careful. Then it is important to keep in mind the fact that our points of reference—the features on the globe—are a rotating reference system. It rotates with respect to an inertial reference system fixed in relation to the stars. Since the earth rotates, each point on the globe describes a circular trajectory. The velocity vector of each point when measured from an inertial reference system continually changes its direction; therefore each point is accelerating. Accelerating reference systems such as the earth's surface are not inertial. Therefore we are obliged to determine how Newton's laws of motion are affected when they are applied to a description of vertical or horizontal motion of air as it moves across the globe.

We shall find there are two effects. The first is experienced by every object, depending only upon its *location* on the globe, and will be the subject of our investigation in this chapter. The second influences only objects which are moving with respect to the earth's surface, and depends upon both the *location* and the *velocity* in relation to the earth's surface. The results of this effect on the horizontal patterns of air motion are surprising and pronounced and will be examined in detail in Chapter 5 when we deal with global circulation.

Let us first consider what it means to be on the rotating reference system which we know as the earth. The period for a complete rotation about the polar axis is known as the sidereal day and is 23 hr 56 min 4.09 sec. A complete rotation corresponds to an angular rotation by 360 degrees or 2π radians. Thus the angular velocity of rotation (sometimes called the angular frequency), denoted by Ω, is $\Omega = 7.29 \times 10^{-5}$ rad/sec. Suppose that an observer, whom we shall call A, is standing on the surface of the earth at a latitude θ, as illustrated in Fig. 3.3(a). His distance from the polar axis is

$$r = R_e \cos \theta,$$

where R_e is the radius of the earth. As the globe rotates, A will execute circular motion, for he rotates with the earth. To accomplish a movement through the distance $2\pi r$ during one sidereal day would require that A have a speed of $r\Omega$ with respect to our inertial system which is fixed in relation to the stars. He would have a maximum speed of 1.7×10^3 km/hr if he were standing at the equator ($\theta = 0$) and he would have no speed if he were at the pole. At any instant, his velocity is tangential to the circle of his trajectory as shown in the figure. His acceleration for circular motion is directed inward toward the center of the circle about which

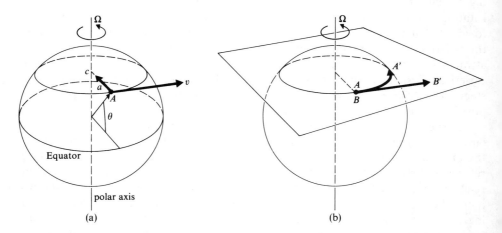

(a)   (b)

Fig. 3.3   (a) Observer A standing on the earth at a latitude θ moves in a circular orbit as the earth rotates. The acceleration of this observer A is directed toward the polar axis of rotation. (b) As A accelerates inward, moving to A', a second observer B, imagined to be unaffected by gravity, continues on in a straight line to B'.

he revolves, for this is the direction in which the velocity always changes. The magnitude of the acceleration appropriate for circular motion is $a_{ce} = r\Omega^2$. Thus any stationary observer on the earth is continually accelerating inward, toward the closest point on the polar axis, and the magnitude of the acceleration is

$$a_{ce} = R_e \Omega^2 \cos \theta. \tag{3.6}$$

The significance of this acceleration can be appreciated from a consequence we shall now describe. Suppose that a second observer B is at some instant stationary on the earth beside A. However, B is hypothetically unaffected by any force, including gravity. Because of his inertia, B will continue his motion in a straight line, as illustrated in Fig. 3.3(b). However A, because he is accelerating inward, will see B as appearing to accelerate outward away from him with an acceleration of magnitude a_{ce}. Earth-bound A, if he did not realize he were rotating, would therefore conclude from Newton's second law, Eq. (3.3), that the apparent acceleration of B was caused by a force F_{ce} of magnitude ma_{ce}:

$$F_{ce} = mR_e \Omega^2 \cos \theta, \tag{3.7}$$

where m is the mass of B. The force F_{ce} is called the *centrifugal force*. It exists only because the behavior of B is measured with respect to a rotating reference system. It does not exist in an inertial reference system and therefore is not a real force in that sense. The apparent force or centrifugal force in the rotating system is called an *inertial force*, because it merely expresses the inertial tendency of any object to continue its present motion in a straight line. The centrifugal force appears to act in addition to other forces such as gravity.

Any object which rotates with the earth thus experiences when viewed from the earth both the gravitational force $F_g^* = mg^*$ directed toward the center of the earth and the centrifugal force $F_{ce} = ma_{ce}$ directed away from the polar axis.

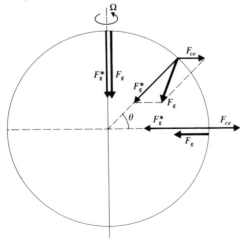

Fig. 3.4   Force of gravity F_g^* and centrifugal force F_{ce} (not drawn to scale) at three different latitudes on the surface of a spherical earth. The sum of the two forces is F_g.

The vector sum of these two forces, illustrated in Fig. (3.4), does not point toward the center of the earth except at the poles and equator. If the earth had a precisely spherical shape, F_{ce} would have the effect of accelerating toward the equator any object that was on the earth's surface. In fact, however, the centrifugal force has had a measurable effect on the shape of the earth. Since the earth is not a rigid body but is plastic, it has deformed during its billions of years of existence so that the total force F_g is everywhere perpendicular to the surface of the crust, as illustrated in Fig. 3.5. Thus the polar radius is now about 0.34% less than the equatorial radius. The ratio $F_g/m = g$ is the observed acceleration of a mass m in the earth's gravity. The percentage variation of g with latitude is only slightly larger than the percentage variation of the earth's radius, since $g = 9.832$ m/sec^2 at the poles and 9.780 m/sec^2 at the equator.[3] We can therefore assume for a good approximation that the combined effect of gravity and centrifugal forces exerts a force whose magnitude is uniform everywhere on the globe. What deserves emphasis is the fact that the total force is everywhere in the vertical direction. Consequently, it cannot by itself cause the horizontal motion of the air which we know as the winds.

3.4 PRESSURE

In previous sections, we have referred to the notion of gas pressure as we commonly use it in everyday terms. Now we will examine the concept in more detail, for in order to discuss important physical characteristics of the atmosphere we shall need to develop the relationship between pressure and density of a gas and show how these parameters of atmospheric gases vary with altitude.

Whenever a gas is in contact with a surface, it exerts a force on that surface. The way this comes about is apparent when we view the dynamics of the molecules which form the gas. Air occupies the space it does because the molecules are continually moving; they move in a straight line until they strike other molecules, whereupon they rebound and move off in another direction. The

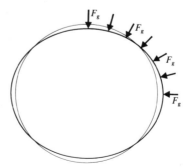

Fig. 3.5   Surface of the earth has deformed as a result of the centrifugal force. The deformation in this figure is exaggerated.

average effect is to keep air molecules well separated. At sea level the average distance between air molecules is about 2×10^{-8} m, and a given molecule travels 10^{-7} m before striking another (a distance called the *mean free path*). From thermodynamics and statistical mechanics, we know that the average kinetic energy of a gas molecule is proportional to the absolute temperature (the temperature which is indicated by the Kelvin scale). The force on a surface is caused by gas molecules continually striking it and rebounding. Momentum is thereby transferred from the gas molecules to the surface; the amount gained by the surface and whatever supports it is equal to the amount lost by the molecules during their collisions. The *force* experienced by the surface is the rate at which momentum is transferred to it, and in most practical situations involving large surface areas, innumerable gas molecules strike with such rapidity that it appears that there is a steady force. The *pressure* is defined as the force which is exerted on a unit area of surface. In the MKS system, pressure is measured in the units of newtons/m^2 or kg/m-sec^2. (In the English system it is in lb/in^2.)

We will now develop an expression for gas pressure; this is known as the *ideal gas law*. By "ideal" we mean that we can neglect the fact that molecules of the gas collide with each other; the properties of the gas are caused by the motion of each molecule through space, as though the others did not exist. The pressure of such a gas can be derived if we know the rate at which molecules strike a surface and the momentum each delivers. Unfortunately, this derivation cannot strictly be achieved without complicated statistics. As this would take us too far afield, we will instead illustrate the physics by a simple model.

We first need to find the rate at which molecules strike an area A of a flat surface for a situation as depicted in Fig. 3.6. We assume that all molecules

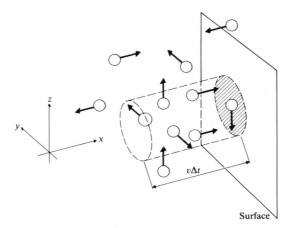

Fig. 3.6   Simplified model of a gas in which molecules travel in only six directions, corresponding to motion in the positive or negative direction along the coordinate axes x, y, or z. The x-axis is arranged to be perpendicular to the surface which experiences the pressure of this gas.

of the gas have the same speed v (in reality, of course, they have a range of speeds). It is apparent from the figure that only molecules within a distance $v\Delta t$ of the surface could hope to strike during an interval of time Δt. Then all of the molecules within a cylindrical volume of the gas, of area A and height v could strike the area A per unit time, provided that they were directed toward it. If n is the concentration or number of gas molecules per cubic meter, this would mean that nvA is the rate at which they strike. But not all of the molecules are headed toward the area; there are six principal directions in which each molecule could be traveling, so only 1/6 of them are directed toward the surface. Thus the rate at which they strike the area A is $nvA/6$. (Our argument using six principal directions is admittedly oversimplified, but it yields the correct result.) A molecule which is headed along the principal direction normal to the surface strikes and rebounds in the opposite direction, imparting a momentum $2mv$. Thus the rate at which momentum is absorbed by the surface is

$$\text{Force} = (2mv)(nvA/6) = nmv^2A/3.$$

The average force per unit area, the pressure, is thus seen to be proportional to both the concentration of molecules n and the average kinetic energy of each molecule $mv^2/2$. Since the latter is proportional to the absolute temperature T, the pressure P must be given by

$$P = nR'T. \tag{3.8}$$

The parameter R' is a constant which our derivation shows is independent of the type of molecule exerting the pressure. The derivation indicates that the pressure is equal in all directions at any point within a gas at equilibrium.

Equation (3.8) is known as the *perfect gas law* or the *ideal gas law*. Historically, it was first determined by combining two empirical laws, one relating P and V at constant T known as Boyle's law (1660), and one for V and T at constant P called Charles's law (1802). The constant R' is called the *universal gas constant*. If n is expressed as the number of moles of gas per cubic meter,

$$R' = 8.314 \text{ J/deg-mole}.$$

Because Eq. (3.8) indicates that P depends upon the total concentration of molecules, the total pressure of a mixture of several different gases is the sum of the pressures of the individual gases. Thus the pressure of any one component is independent of the pressures of the other components, a feature first enunciated by John Dalton in 1790.

It is often convenient to express Eq. (3.8) in terms of the mass density of air rather than the number density. Table 3.1 provides the necessary data for computation of the molecular weight of dry air. One mole of dry air (which at $T = 0°C$ and a pressure of 1 atm occupies 22.4×10^{-3} m^3 or 22.4 liters) has a molecular weight of $M_a = 28.97$ g. This weight is remarkably independent of altitude. Only for altitudes above 80 km does the molecular weight change, and then it decreases due to dissociation of O_2 into atomic oxygen by incoming solar radiation. The ideal gas law which is valid at lower altitudes can therefore be written as

$$P = \rho R_a T, \tag{3.9}$$

where, for a density ρ expressed in kg/m^3, the gas constant for air has the value $R_a = R'/M_a = 287.0$ J/kg-deg.

3.5 DENSITY AND PRESSURE PROFILES

A vertical column of air will exert a force on any horizontal surface upon which it rests, for the surface must support the overlying gas. This force equals the weight of the air within the overlying column. The pressure and, by Eq. (3.9), the density in an isothermal (uniform temperature) atmosphere will decrease with altitude because the weight of the overlying air decreases. How a quantity depends upon altitude is called its *profile*. The pressure profile can be calculated with reference to Fig. 3.7. In equilibrium, the molecules at any altitude z must through their mutual collisions support the weight of the mass of air above. Therefore the pressure at altitude z, indicated by $P(z)$, is greater than the pressure at a higher altitude $z + \Delta z$ by an amount which is equal to the weight of the air over each unit horizontal area in the interval:

$$P(z) - P(z + \Delta z) = \rho g \Delta z.$$

Thus writing $\Delta P(z) = P(z + \Delta z) - P(z)$, we have

$$\Delta P(z) = -\rho g \Delta z. \tag{3.10}$$

This is known as the *hydrostatic condition*, for it applies in cases of equilibrium for the pressure profile of any fluid, not just air. The negative sign appears, since

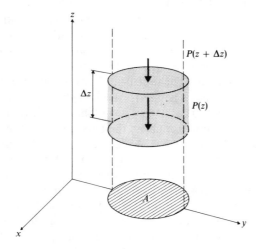

Fig. 3.7   A column of air of cross-sectional area A.

the required pressure *decreases* with *increase* in altitude. By applying the ideal gas law, Eq. (3.9), we may eliminate the gas density as a variable in Eq. (3.10). The equation that P must satisfy is then

$$\Delta P(z) = -\frac{M_a g}{R'T} P \Delta z. \tag{3.11}$$

If the atmosphere is isothermal, so that T is independent of z, this equation has the standard form of Eq. (2.8); hence the solution is the exponential function:

$$P(z) = P_0 e^{-M_a gz/R'T}, \tag{3.12}$$

where P_0 is the pressure at sea level ($z = 0$). Since the ideal gas law indicates that gas density is proportional to pressure at constant T, the gas density must also exhibit a corresponding exponential decrease with altitude:

$$\rho = \rho_0 e^{-M_a gz/R'T}, \tag{3.13}$$

where ρ_0 is the density at sea level. An exponential decrease of pressure and density with altitude is characteristic of compressible fluids such as air.

Water, on the other hand, is nearly incompressible since it undergoes only a slight volume decrease upon isothermal application of pressure. Doubling the pressure on a confined volume of water causes a relative decrease in volume of only 5×10^{-5}. As a result, the density of water is nearly independent of depth, and the pressure shows a nearly linear increase with depth.

The factor $R'T/M_a g$ appearing in the exponent of Eq. (3.13) has the dimensions of a length and is called the *scale height*. It determines the rate at which the density and pressure decrease with increasing altitude for an atmosphere which is approximately isothermal. For an average temperature of 288°K at sea level, the scale height is 8 km; therefore at an altitude of 8 km, the pressure and density would be reduced to e^{-1} or 37% of the pressure and density respectively at sea level, if the temperature did not change with altitude. Every additional increase in altitude by 8 km would see a similar reduction. A comparison of this exponential behavior, given by Eq. (3.13), with the observed profile is shown in Fig. 3.8. From this it is evident that Eq. (3.13) is only qualitatively in accord with the data. The discrepancy is due to the fact that in deriving Eq. (3.13) we neglected dynamic aspects of air circulation; the actual density profile at any instant may be quite different from both the exponential approximation and the long-term average, which is indicated in Fig. 3.8 by the solid curve. Furthermore, the temperature of the atmosphere is not uniform, but decreases with altitude up to about 20 km, where it is often found to be only 220°K. At altitudes above 20 km, the temperature is observed to increase with altitude as a result of the absorption of solar radiation by ozone in the upper atmosphere and the subsequent warming of the surrounding air; at much higher altitudes the temperature decreases again. The temperature profile of the atmosphere will be discussed in

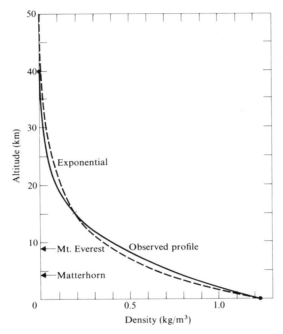

Fig. 3.8   Density profiles of the atmosphere.

Chapter 4 when we consider the processes that influence it, and it is summarized in Fig. 4.7.

The observed decrease of air density with altitude shown in Fig. 3.8 has important implications for biological processes, as any mountain climber knows. At the highest permanently inhabited village in the Peruvian Andes, located at an altitude of 5.3 km, the air density is about half of the sea level density. In adapting to these conditions the inhabitants have developed unusually large lung capacities.

Because of the exponential decrease of density with altitude, most of the atmospheric mass is beneath an altitude of 33 km, about three times the altitude of Mt. Everest. Thus the atmosphere, which is so essential for sustaining life, forms but a thin covering over the globe.

Equilibrium conditions in the atmosphere are so nearly established that the sea-level pressure P_0 rarely varies by more than $\pm 3\%$ about its mean.[3] The mean pressure P_0 is $1.013 \times 10^5 \mathrm{N/m^2} (1.013 \times 10^6 \mathrm{dyn/cm^2})$. For convenience, meteorologists use a unit of pressure called the "bar," abbreviated b, which is defined as $10^5 \mathrm{N/m^2}$. This is chosen, because it is nearly equal to the mean sea level pressure. Atmospheric pressures are invariably expressed in millibars, abbreviated mb. Thus sea level pressure is generally stated to lie between 980 mb and 1040 mb.

The observed average pressure together with the value of acceleration due to gravity and the surface area of the earth provide sufficient information for a calculation of the total mass of the atmosphere. The calculated value of 5×10^{18} kg is less than 10^{-6} of the mass of the earth.

Equation (3.13) has an important physical interpretation in the statistics of molecular behavior, with implications bearing upon the evolution of the atmosphere. If numerator and denominator of the exponent are divided by the number of molecules comprising a mole of gas $N = 6.022 \times 10^{23}$ (Avogadro's number), the resulting expression for the gas density is

$$\rho = \rho_0 e^{-mgz/k_B T}, \tag{3.14}$$

where m is the mass per molecule and k_B is the Boltzmann constant, $k_B = R'/N$, named for the nineteenth-century physicist Ludwig Boltzmann. In the theory of equilibrium statistics, the exponential factor multiplying ρ_0 is known as the Boltzmann factor and is proportional to the probability that a molecule in a gas at temperature T subject to a downward force mg will be found to have the potential energy mgz; that is, to be at an altitude z. The fraction of the total number of air molecules expected at the altitude z is obtained by multiplication of the Boltzmann factor by a suitable constant; in this case multiplication by ρ_0 gives the observed value of ρ at all altitudes, including $z = 0$. The properties of the exponential factor reflect the statistical tendency for molecules to be found with the lowest possible energy. To minimize their potential energy, they tend to concentrate near $z = 0$. This tendency to fall is countered by those collisions between molecules from which at least one of the molecules can rebound upward. The average kinetic energy of the molecules, proportional to the temperature, is a rough measure of how high the rebounding molecule can penetrate before being decelerated to a momentary standstill by gravity. Of course, with a typical *mean free path* between collisions of 10^{-5} m at sea level, it is unlikely that a given molecule will rise undeflected until the initial kinetic energy is completely converted into potential energy. However, if it strikes another molecule, the second will propagate the disturbance upward to a third, and so on. Equation (3.14) portrays the most probable behavior of a large number of molecules in an equilibrium condition.

Gas mixing

If Eq. (3.14) is correct in detail, heavier molecules should be found concentrated closer to sea level than lighter ones; that is, nitrogen should constitute a smaller fraction of air than oxygen at high altitudes. In fact, such a variation is observed, but only for altitudes exceeding 120 km. A uniform mixture is observed at much lower altitudes, because the vigorous motion of air serves to mix all of the component gases, much as milk and cream may be homogenized by vigorous shaking. Such nonequilibrium circumstances invalidate the application of the Boltzmann factor. However, above 120 km, vertical movement of the air is practically non-existent, and the different distributions suggested by Eq. (3.14) for gases of differing molecular mass are more apparent. Lighter molecules such as hydrogen and helium are found in much higher relative concentration than at sea level.

In Section (3.2) it was pointed out that the force of gravity decreases with altitude in accordance with the law of universal gravitation. Therefore molecules at high altitudes which have a sufficiently high kinetic energy may actually overcome the earth's gravitational attraction and escape. Above 700 km the number density of molecules is less than 10^{12} m^{-3}(10^6 cm^{-3}), so the mean free path between molecular collisions is correspondingly greater than at sea level where the number density is 2.5×10^{25} m^{-3}.[4] The mean free path in this upper region is comparable to the diameter of the earth. Should a molecule at or above this level be directed upward with sufficient kinetic energy, it has a high probability of escaping from earth without a collision. The earth continually loses such molecules. This region of the upper atmosphere from which molecules can escape is called the *exosphere*.

Since Eq. (3.14) indicates that lighter molecules are more likely to be found in the upper atmosphere than heavier, it is the lighter molecules that predominantly escape from earth. Such a feature of molecular statistics has had an important influence on the evolution of the atmosphere.

Evolution

It is believed that the original composition of the atmosphere was similar in atomic components to the composition of the sun, consisting chiefly of a mixture of hydrogen and helium with small amounts of methane (CH_4), ammonia (NH_3), water, carbon dioxide, and nitrogen. Hydrogen and helium because of their small molecular mass could escape relatively quickly from earth, in a matter of perhaps 10^5 years. Decomposition of water and ammonia by high-energy solar radiation encouraged formation of molecular nitrogen and oxygen, the latter being able to survive in the atmosphere without violent recombination with hydrogen once the bulk of the hydrogen gas had escaped. Vigorous outgassing of the earth's interior, vented through the surface via volcanic eruption, added steam, carbon dioxide, nitrogen, and small amounts of sulfur-bearing gases, all of which are still the principal emissions from currently active volcanoes. Solar radiation and electrical discharges in the atmosphere (lightning) provided energy for the creation of organic molecules. These eventually found an hospitable home once the surface of the earth had cooled sufficiently for water to condense.

It is generally believed that oxygen first became an abundant and permanent constituent of the atmosphere with the advent of primitive plant life. This probably dates back 2 billion years, as evidenced by the earliest appearance of oxidized continental red bed ores rich in the oxidized ferric ion, but this date is in dispute. Formation of ozone in the upper atmosphere, as we shall discuss in the following chapters, provided a shield to protect the fragile organic molecules from high-energy solar radiation. As water vapor condensed in response to terrestrial cooling, the growing oceans absorbed most of the atmospheric carbon dioxide, leaving nitrogen as the predominant gas. The present atmosphere therefore is very much a product of the effects of solar radiation and the advent of photosynthesis. Necessarily then, the division of atmospheric gases into the two classifications of "permanent" and "variable" gases is somewhat arbitrary, for the atmosphere has been continually evolving. Gases are permanent only in the sense that their concentration has shown little change during the span of technical measurements on this planet.

SUMMARY

The atmosphere is a mixture of gases and particles. Almost 99% of the molecules are gaseous nitrogen and oxygen. Because of gravity the atmosphere is confined close to the surface of the earth, forming a thin blanket about the globe. The net sum of gravitational and centrifugal forces is a force which is solely in the vertical direction near the earth. Because of this force, the density of air decreases approximately exponentially with altitude; this variation is determined by a balance between the thermal agitation of the molecules, which tends to keep them separated, and the gravitational attraction of the earth, which tends to draw them together to the surface. All but 1% of the atmospheric gases are within 33 km of the surface, so the thickness of the atmosphere represents only 0.5% of the radius of the planet. Some constituents such as carbon dioxide and oxygen are continually cycled through the atmosphere as they are utilized by plants and animals and effect chemical changes in mineral deposits. Water is also cycled as it evaporates and condenses in response to changing atmospheric conditions. Of the predominant gases in the atmosphere, water vapor displays the most marked variations in concentration from place to place. On the other hand, the relative concentrations of the permanent gases are remarkably uniform in the lower atmosphere, a result of vigorous mixing. This is no longer true at altitudes above about 120 km where there is little vertical air circulation.

NOTES

1. L. Machta and E. Hughes, "Atmospheric Oxygen from 1967 to 1970," *Science*, **168**, 1582 (1970).
2. K. C. Hickman, as quoted by R. E. Munn, *Descriptive Micrometeorology*, New York: Academic Press, 1966, p. 92.
3. R. G. Fleagle and J. A. Businger, *An Introduction to Atmospheric Physics*, New York: Academic Press, 1963.
4. S. I. Rasool, "Structure of Planetary Atmospheres," *AIAA Journal*, **1**, 6 (1963).

FOR FURTHER READING

R. A. CRAIG   *The Upper Atmosphere; Meteorology and Physics*, New York: Academic Press, 1965.

R. J. FLEAGLE and   *An Introduction to Atmospheric Physcis*, International Geophysics
J. A. BUSINGER   Series, New York: Academic Press, 1963.
SCIENTIFIC AMERICAN   "The Biosphere and Atmospheric Cycles," Sept. 1970.

QUESTIONS

1. On the basis of the molecular nature of a gas, explain why the pressure of an enclosed gas increases with increase in its temperature.

2. If the temperature of the air over a portion of the globe is increased due to solar heating, does the air become less dense or more dense at ground level? Explain your answer.

3. The vapor pressure of water increases with temperature as illustrated in Fig. 3.2. The pressure of an enclosed gas also increases with temperature, as given by Eq. (3.9). If a dish of water is placed in the bottom of an air-tight container, will the humidity of the air in the container increase if the temperature increases, indicating a loss of water from the dish? Why or why not?

4. If the temperature of the atmosphere decreases with altitude, will the density of air decrease more or less rapidly than it does when the temperature is uniform?

5. Why is argon so much more abundant in the earth's atmosphere than helium and neon (which are lighter) and krypton (which is heavier)?

6. The ambient concentration of carbon dioxide in the Northern Hemisphere exhibits seasonal variations. Would you expect that the concentration would be higher in summer or winter? Why?

7. It is popularly believed that the hydrocarbon content of urban air is a result mainly of emissions from automobiles. This is not true if nonreactive hydrocarbons are also considered. What molecule occurs naturally as the predominant hydrocarbon constituent of the atmosphere and at what concentration is it found?

PROBLEMS

1. The absolute temperature scale is convenient for describing the thermodynamic state of matter, but it is perhaps not so convenient for everyday use. Derive a formula which relates the Fahrenheit, Celsius (Centigrade), and Kelvin temperature scales. The following information is sufficient for this purpose: at a pressure of one atmosphere, the melting point of ice is at about $273°K$ or $0°C$ or $32°F$, and the boiling point of water is at $373°K$ or $100°C$ or $212°F$.

2. Calculate the average molecular weight of dry air from the data in Table 3.1. How many components of the atmosphere must you include to obtain an answer which is accurate to 0.1%?

3. If a water droplet of 10 micron radius were to evaporate completely at sea level and at an ambient temperature of $20°C$, what volume of initially dry air could it saturate?

4. What is the partial pressure of O_2 at an altitude of 6000 m compared with the pressure at sea level? Assume that the temperature is uniform and is $10°C$.

5. The average density of air at sea level is 1.2 kg/m^3, and the average pressure is 1.01 × 10^5 N/m^2 (1.01 × 10^6 dyn/cm^2). If the density were *independent* of altitude, what would be the altitude of the top of the atmosphere, where the pressure is zero?

6. Suppose that the earth is spherical with radius R_e = 6.4 × 10^3 km. What is the difference between the acceleration resulting from the total gravitational and centrifugal forces experienced by a molecule of air at sea level at the North Pole and that resulting from the same forces a molecule experiences at the equator? If the atmospheric pressure at sea level were to be uniform over the globe, and assuming the temperature of the air is uniform, how much more air would be found over a point at the equator than over the poles? You may express your answer as a ratio.

7. Suppose that the mean altitude at which the pressure is 700 mb at 10°N latitude is 3500 m and that the mean temperature is 280°K for the atmosphere over this latitude. What is the mean temperature of the air over the North Pole if the 700 mb altitude is at 3300 m?

8. The mean free path of an air molecule is inversely proportional to the ambient concentration of air molecules. If the mean free path is 7 × 10^{-8} m at sea level, what is it at 10 km? At 30 km?

9. Find a first-order correction to our simplified derivation of the density profile of the atmosphere by permitting the temperature in Eq. (3.11) to decrease with altitude. A good approximation for the lower atmosphere can be obtained by assuming that $T = T_0 - \Gamma z$, where T_0 = 288°K, Γ = 6.5 × 10^{-3} degrees Kelvin per meter, and z is the altitude expressed in meters. The hydrostatic equation can be solved in a convenient way by assuming that $T^{-1} = (1 + \Gamma z/T_0)T_0^{-1}$. For this approximation, what is the density of air at $z = 10^4$ m relative to the sea-level density, as compared with the density for an atmosphere with a uniform temperature of $T_0 = 288°K$?

ENERGY BALANCE OF THE EARTH

Everyone is aware that weather and the circulation of air involve a movement of sensible heat from one place to another. However, it is much less evident that the absorption or loss of heat by portions of the atmosphere gives rise to air motion in the first place. Frequently the quality of ventilation in a community is determined by just this factor. The modes of energy transfer and their effectiveness directly influence meteorological conditions and the consequent accumulation of air pollutants.

Of the three mechanisms by which energy can be transported—*conduction, convection*, and *radiation*—the first two require a transporting medium such as a liquid or gas, whereas the third permits propagation through a vacuum. All three operate within the earth's atmosphere, although convection and radiation are by far the more important. *Conduction* is the diffusion of energy as sensible heat, transported through a medium via the interactions of neighboring atoms with no net movement of the atoms from their mean positions. *Convection*, on the other hand, is the conveyance of energy, either sensible heat or latent heat, by a movement of the molecules of the medium from one place to another. In meteorology, the term is specialized to apply only to movement in the vertical direction; a second term, *advection* is used for movement in the horizontal direction. (In addition, convection and advection are generalized to signify the transport of *any* atmospheric characteristic such as humidity and temperature, through the movement of air in either the vertical or horizontal direction, respectively.) *Radiation* is the propagation of energy in the form of electric and magnetic fields of an electromagnetic wave such as visible light; radiant energy can thus be conveyed through a vacuum as well as a transparent media, such as air.

The sun is the main source of energy for atmospheric circulation. It emits energy almost entirely by radiation, a portion of which is intercepted by the earth and atmosphere. By comparison, conduction of energy from the earth's molten core to its surface appears to contribute much less than 1% of the energy which the earth absorbs from solar radiation. Were the earth to absorb solar radiation without losing energy itself, its temperature would continually increase. In fact the mean temperature of the globe is fairly constant; and thus the earth (including its atmosphere) must lose energy at the same rate as it is absorbed. The only way the earth can rid itself of energy is by radiating into the vacuum of outer space. The balance between incoming and outgoing radiant energy is called

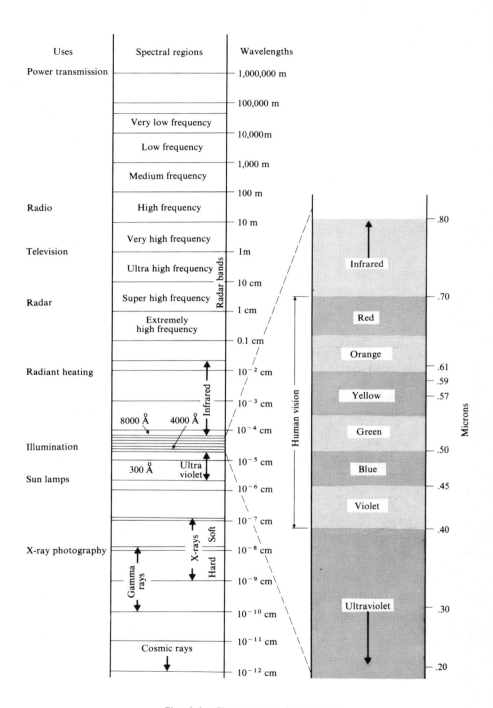

Fig. 4.1   Electromagnetic spectrum.

the *energy balance* or *heat balance* of the earth and is responsible for maintaining an environment which is hospitable for life as we know it. Any major disruption of the balance could have serious consequences, such as global warming or cooling trends. The atmosphere plays a role, for it determines how much of the incoming solar radiation reaches the earth's surface and how much of the terrestrial radiation escapes, thereby influencing the mean surface temperature and determining climatology and meteorology around the world.

This chapter on energy balance is devoted to an examination of the role which the atmosphere plays in the energy balance of the earth. From this we learn that minor constituents of the atmosphere—water, water vapor, and carbon dioxide—have predominant importance in absorbing terrestrial radiation, whereas the major constituents—nitrogen and oxygen—have practically no effect.

Electromagnetic spectrum

The sun emits electromagnetic waves with wavelengths from hundreds of meters to less than 10^{-10} meters. But about 99% of the solar energy is concentrated in the spectral region with wavelengths between 0.15 and 4.0 microns. In principle, electromagnetic radiation can be propagated with any wavelength. However, current technology limits us to detection of radiation with wavelengths in the range known as the *electromagnetic spectrum*, which is illustrated in Fig. 4.1. For convenience, different portions of the spectrum have identifying names. What we know as light is the *visible spectrum* (sometimes called the *optical spectrum*), a narrow band of wavelengths where radiant energy can produce a sensation when it impinges upon the retina of the human eye. It comprises the region from about 0.40 to 0.70 micron. Another unit of length commonly used for wavelengths is called the Ångstrom (Å), equal to 10^{-10} m. The visible spectrum is thus between 4000 and 7000 Å. Colors associated with different wavelengths range from violet and blue for short wavelengths to red for long wavelengths, as depicted in Fig. 4.1.

Since all electromagnetic waves travel through a vacuum with a speed which is independent of their wavelength (called the "speed of light" and denoted by the symbol c), a unique relationship exists between the wavelength λ and frequency v of a wave as given by

$$v = c/\lambda, \tag{4.1}$$

where $c = 2.998 \times 10^8$ m/sec. Some spectral bands are named according to their wavelengths and some according to their frequencies. The *ultraviolet* and *infrared* bands indicate their frequency regimes compared with the frequencies of visible light, whereas the radio waves are labeled according to their relative wavelengths. For a number of reasons, it will be most convenient for us to characterize radiation by the numerical value of its wavelength. Equation 4.1 affords a means for converting any wavelength into the corresponding frequency.

4.1 THE SUN

Our sun is an average star which consists mostly of hot hydrogen atoms and ions. It has a slowly rotating body with a diameter of 1.4×10^6 km, and a tenuous atmosphere that extends for several solar diameters from the surface. The sources of solar energy are thermal nuclear reactions between nuclei within its core which convert matter into energy. The temperature of about $10^{7\circ}$K within the core is high enough that the ionized nuclei have sufficient kinetic energy to overcome their mutual electric repulsion when colliding, and come together close enough to bring into play short-range nuclear forces. The resulting nuclear interactions convert mass into energy according to Albert Einstein's formula $E = mc^2$, where E is the energy released by annihilating a quantity of material of mass m. By converting mass into energy at a rate of about 4×10^6 metric Tonnes per second, the sun produces energy at a rate of 3.90×10^{26} watts (one Tonne is 10^3 kilograms or about 1.1 U.S.A. short ton). Most of this energy is emitted from the interior in the form of high-energy radiation (gamma rays), which are quickly absorbed by overlaying layers of solar matter. By convection and radiation, energy is conveyed toward the cooler exterior.

Electromagnetic radiation reaching earth from the sun emanates from a shallow surface layer about 400 km thick called the *photosphere*. The photosphere is opaque to radiation from below. The temperature at the base of the photosphere is about 8000°K and, at the top, 4700°K. Observations of this layer have revealed a granular texture with bright patches some 700 to 1500 km in diameter separated by darker lines of a typical width of 300 km. Evidence suggests that the bright granules denote material which is rising from the deeper regions by convection,

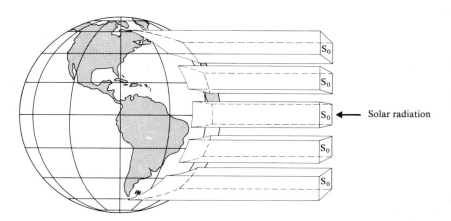

Fig. 4.2   Schematic of the inflow of solar energy at the time of the equinox. S_0 is the rate at which solar radiant energy is incident on a unit area of a surface perpendicular to the light beam. This energy is spread over a greater surface area of the earth at high latitudes, so the intensity of solar radiation on the ground is less. (J. A. Day and G. L. Sternes, *Climate and Weather*, Reading, Mass: Addison-Wesley, 1970.)

and the dark boundaries represent the subsiding material. The hot gases within this layer, a small fraction of which are ionized, continually lose energy by radiating it into space.

Solar constant

The total output of radiation from the sun is remarkably independent of time. The rate at which energy is incident on the upper extremity of the earth's atmosphere is 1.34×10^3 watts/m^2 (1.92 cal/cm^2-min), a quantity which is established with an accuracy of 2 or 3%. This quantity has come to be known as the *solar constant* and is denoted by the symbol S_0. As illustrated in Fig. 4.2, the solar constant represents the energy incident on a unit area of surface oriented perpendicular to the direction of the incident solar radiation. Since the cross-sectional area of the earth which intercepts the solar energy is πR_e^2, where R_e is the earth's radius, the rate at which the earth receives energy is $\pi R_e^2 S_0$. Dividing this quantity by the total surface area $4\pi R_e^2$ gives us the average rate at which energy is incident on each square meter of the earth's surface, $S_0/4$ or 335 W/m^2. A variation of $\pm 3\%$ about this mean intensity arises from the slight ellipticity of the earth's orbit about the sun; the Southern Hemisphere receives more energy during its summer period than does the Northern Hemisphere during its summer. But this is a small effect compared with the seasonal variation of solar radiation from winter to summer in each hemisphere.

The spectrum of radiation emitted by the sun is illustrated in Fig. 4.3. Measurements show that there is a maximum intensity of this emission spectrum in the center of the visible band, at about 0.50 micron as shown. No significant changes with time have been detected in the incident energy spectrum between wavelengths of 0.35 and 2.40 micron. But during periods of solar activity, indicated by the appearance of sunspots or flares in the photosphere, there are large increases in the far ultraviolet and X-ray portion of the spectrum and also in the very long wavelength radio band. These variations, which are a fraction of a percent of the total incident energy flow, appear to have little to do with weather but are responsible for ionization in the upper atmosphere, the aurora phenomena, and disturbance of radio transmission.

In Fig. 4.3 two smooth curves are shown which have approximately the same shape as the solar emission spectrum. They represent the emission spectrum of what is known as a *black body*. Such a body has a surface which absorbs all incident electromagnetic energy (thus it has the interesting characteristic that it cannot be seen in reflected light). The discussion in Appendix D shows that the surface of a black body emits radiation with a unique spectrum which depends only upon the temperature of the body. The spectrum is quite independent both of the body's composition and of whether it is a solid, liquid, or gas. The black body spectrum corresponds to the maximum amount of radiant energy which any body in equilibrium can emit at each wavelength. We see that the comparison in Fig. 4.3 indicates that the shape of the solar spectrum corresponds quite

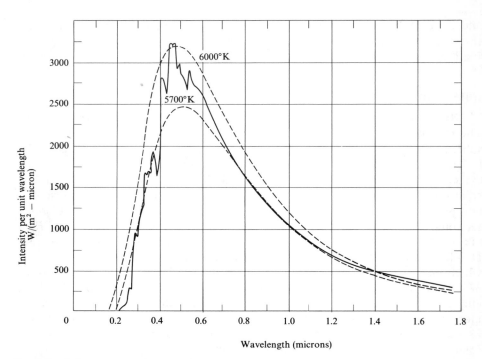

Fig. 4.3 Solar spectrum (solid line) and black body radiation at 5700°K and 6000°K. (After R. G. Fleagle and J. A. Businger, *An Introduction to Atmospheric Physics*, New York: Academic Press, 1963.)

accurately to the emission spectrum of a black body having a temperature of about 5800°K. Thus the sun is said to be a black body radiator. We shall find that to a good approximation the earth and its atmosphere are also black body radiators, and this makes the concept of black bodies an important one when dealing with the energy balance of the earth.

The reason why the earth and other objects radiate energy is that the thermal agitation of their molecular and atomic constituents causes their electrons to accelerate as they respond to the mutual buffeting. It is these accelerating charges that radiate electromagnetic waves. The spectrum of this so-called thermal radiation depends upon both the temperature of the body and the chemical and physical characteristics of its surface. Only in the special case when the surface is perfectly absorbing is the spectrum uniquely defined by the temperature alone.

Since the rate at which energy is emitted at each wavelength from the surface of a black body depends only upon the temperature of the body, so does the *total* rate at which energy is emitted. This rate is found to increase very strongly with increasing temperature of the body; from each unit area of its surface, energy is radiated at a rate proportional to the fourth power of the temperature:

$$E_{bb} = \sigma \, T^4. \tag{4.2}$$

This relation was discovered empirically by Josef Stefan in 1879 and was derived subsequently from general thermodynamic principles by Ludwig Boltzmann. But it was not until the turn of the century that the fundamental reason for this formula became known, and then it was the success of the quantum theory advanced by Max Planck[1] that explained both the emission spectrum and Eq. (4.2). The proportionality constant in this equation has the value $\sigma = 5.670 \times 10^{-8}$ W/m^2-deg^4; and we emphasize that the temperature should be expressed as the absolute temperature (in Kelvin degrees). The sun, with an effective surface temperature of about 5800°K as indicated by the shape of its black body emission spectrum, would radiate about 2×10^5 times more energy per square meter of surface than the earth with a mean surface temperature of only 288°K. The earth's radiant output would be approximately 370 watts emitted per square meter of surface.

The observed solar constant S_0, together with Eq. (4.2), the observed radius of the sun $R_s = 6.95 \times 10^5$ km, and the mean distance between the sun and the earth $D_e = 1.49 \times 10^8$ km, permits us to obtain another estimate for the temperature of the photosphere. We use the fact that the rate at which energy is radiated from the sun's surface must be equal to the solar constant multiplied by the area of a sphere of diameter D_e; for essentially no energy is lost as radiation travels outward in all directions from the sun for a distance equal to D_e. Thus we have:

$$4\pi R_s^2 \sigma T^4 = 4\pi D_e^2 S_0. \tag{4.3}$$

The temperature T which is the solution to this equation is found to be 5800°K, in excellent agreement with the temperature of the black body spectrum which fits the shape of the observed solar spectrum.

Another significant feature of black body radiation is the location of the peak in its spectrum. The maximum rate of energy emission occurs at a wavelength λ_m which is inversely related to the temperature, as given by a relation known as the Wien displacement law:

$$\lambda_m = \frac{L}{T}. \tag{4.4}$$

The constant L is equal to 2.897×10^{-3} m-deg (or 2897 when λ_m is expressed in microns and T is in Kelvin degrees). Thus higher temperatures shift the peak toward shorter wavelengths. For the sun, the observed $\lambda_m = 0.50$ micron implies that the corresponding black body temperature is 5800°K, in good agreement with previous determinations. However if the earth were to radiate as a black body of temperature 288°K, the peak in the emission spectrum would be shifted into the longer wavelength spectral region of the infrared band. That is why we cannot see terrestrial black body radiation. The earth would have its peak radiant intensity at a wavelength of about 10 microns.

4.2 RADIATION BALANCE

We are now in a position to estimate the mean surface temperature T_* of the earth from a simple balance of incoming and outgoing radiant energy. If we assume that the earth is a black body so that it absorbs all of the incident solar energy, the rate at which energy is absorbed is $\pi R_e^2 S_0$, where the radius of the earth is $R_e = 6378$ km. (The earth cannot be a perfect black body, for then we could not see the land, oceans, or vegetation by reflected light. However, the fraction of light which these objects reflect is so small that to a good approximation the surface of the earth can indeed be considered a black body.) To maintain an energy balance, the rate at which energy is absorbed must be equated with the mean rate at which it is radiated:

$$\pi R_e^2 S_0 = 4\pi R_e^2 \sigma T_*^4. \tag{4.5}$$

This predicts a mean surface temperature given by the condition:

$$T_*^4 = \frac{S_0}{4\sigma}. \tag{4.6}$$

Taking the fourth root of both sides yields $T_* = 278°$K, a value remarkably close to the observed value of about $288°$K. But unfortunately the calculation is wrong; it neglects the important effects of the atmosphere.

A closer approximation to the actual situation can be made by anticipating the discussion of the following sections. Anyone who has flown in an airplane at high altitudes is aware that clouds reflect a substantial portion of the incident light and therefore reduce the amount of radiant energy which reaches the surface of the earth. It is found that a fraction $a = 0.34$ of the incident solar energy is reflected toward outer space; then the right-hand side of Eq. (4.6) must correspondingly be adjusted by a multiplicative factor of $(1 - a)$. With this, the new solution for the mean temperature is $T_* = 245°$K, well below the freezing point of water ($273°$K). Clearly the result is unrealistically low if it represents the mean surface temperature of the earth.

Yet measurements by high-altitude satellites of the earth's spacebound radiation do indicate that the equivalent black body temperature is about $255°$K,[2] in fair agreement with the radiation energy balance result of $T_* = 245°$K. We must conclude that most of this radiation detected by the satellites had originated in the upper atmosphere where the temperature is lower than it is on the surface of the earth; the actual surface temperature is not predicted by our simple concepts of radiation balance with $a = 0.34$. Thus the atmosphere must play a role in compensating for the fraction of solar energy which is reflected and, by so doing, makes this planet a hospitable place for us.

Absorption spectrum

The atmosphere has a pronounced effect on the energy balance because its constituents selectively absorb radiation within certain wavelength bands. For

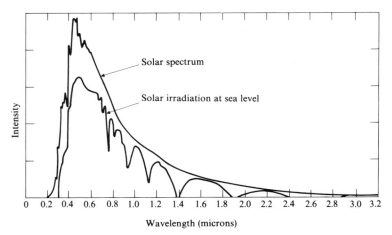

Wavelength (microns)

Fig. 4.4 Solar spectrum outside the atmosphere (upper curve) and the spectrum measured at sea level (lower curve). (*Handbook of Geophysics*, Rev. ed., U. S. Air Force, New York: Macmillan, 1960.)

example, the solar energy spectrum is altered by absorption as the incident radiation penetrates the earth's atmosphere. At ground level, the received spectrum exhibits several absorption bands as illustrated by the lower curve in Fig. 4.4. From the ratio of the received energy at each wavelength to the incident energy in the upper atmosphere, the percent of absorption due to the atmosphere can be calculated. The resulting *absorption spectrum* is shown in Fig. 4.5(b), for the ultraviolet, visible, and infrared portions of the electromagnetic spectrum. To determine at which altitudes the absorption occurs, we can compare the incident solar radiation at the extremities of the atmosphere with that observed at an altitude of 11 km; the resulting absorption spectrum for atmospheric constituents above 11 km is illustrated by Fig. 4.5(c). As a wavelength reference, we also give in Fig. 4.5(a) the black body emission spectra corresponding to the temperature of the sun and one corresponding to the radiating temperature of the earth-atmosphere system.

Through comparison of the absorption spectra in Figs. 4.5(b) and (c) with measurements in the laboratory on known gases, it is possible to identify the constituents of the atmosphere which cause the absorption at each wavelength. The absorption characteristics of the individual gases contributing to the spectrum in Fig. 4.5(b) are shown in Fig. 4.5(d).

Of particular interest is the fact that atmospheric absorption is not strong near the peak of the solar emission spectrum at 0.50 micron and thus permits transmission of this intense radiation to the surface of the earth. Approximately 40% of the solar radiant energy is concentrated in the region from 0.40 micron to 0.70 micron.[3] It is perhaps no coincidence but an example of human adaptation that the visible portion of the spectrum so important to our everyday activities coincides with the spectral region of maximum transmitted solar radiation.

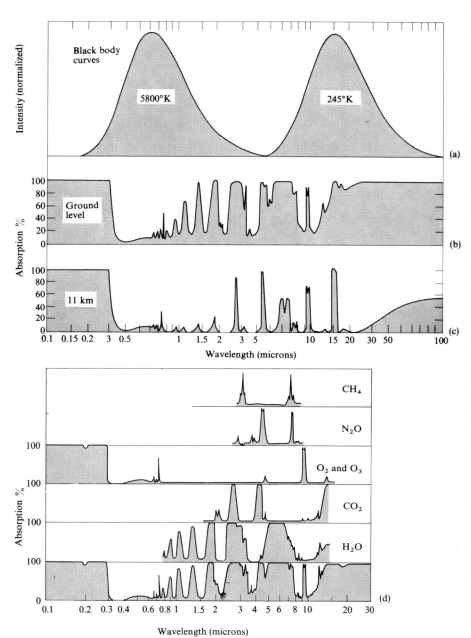

Fig. 4.5 (a) Black body emission spectra for the sun and earth, normalized to represent the same rate of energy emission. (b) Atmospheric gaseous absorption spectrum for a solar beam reaching ground level. (c) Atmospheric gaseous absorption spectrum for a beam reaching to an altitude of 11 km. (d) Absorption spectra for important atmospheric gases, with the total absorption described by the lowest curve. (From R. M. Goody, *Atmospheric Radiation*, Oxford: Oxford University Press, 1964, and R. G. Fleagle and J. S. Businger, *Introduction to Atmospheric Physics*, New York: Academic Press, 1963.)

Many isolated bands of absorption appear at wavelengths which are longer than those of the visible spectrum. As indicated in the figure, these bands are attributed principally to water, water vapor, and carbon dioxide in the lower atmosphere. It is, as we shall soon see, these absorption features which most importantly influence the energy balance of the earth. However, absorption of wavelengths shorter than those of the visible spectrum appears to be caused by different components of the atmosphere, ones which are located above 11 km. The almost continuous absorption just below 0.3 micron is due to the presence of ozone, and absorption below about 0.24 micron, to the presence of both ozone and oxygen. This absorption is not directly important to the energy balance of the earth, for less than 1% of the solar radiant energy is concentrated in this region. However, we shall see that this absorption has several significant indirect effects which benefit man.

Ultraviolet absorption

Ozone is not a normal constituent of the lower atmosphere except in trace amounts. The strong absorption in the ultraviolet, as well as near a wavelength of about 9 microns, indicates that it must occur in substantial quantities in the upper atmosphere. In the following section, we shall discuss how it is produced by photochemical reactions involving molecular and atomic oxygen. The resulting ozone protects biological activities on earth by absorbing harmful ultraviolet radiation. As anyone who has fallen asleep under an ultraviolet sun lamp realizes, absorption of short wavelength radiation can lead to physiological change. In fact the carbon-hydrogen bond in organic molecules can be disrupted by radiation with wavelengths shorter than 0.2 micron. Some processes may even cause dissociation when organic molecules are subjected to radiation at wavelengths as long as 0.3 micron. The water molecule, too, can be dissociated by ultraviolet radiation. Thus ozone in the upper atmosphere is a protective blanket.

4.3 OZONE IN THE UPPER ATMOSPHERE

Ozone is formed in the upper atmosphere at an altitude of about 30 km principally by the photodissociation of molecular oxygen. Absorption of short wavelength electromagnetic radiation excites the electrons of the oxygen molecule into orbits of high energy in which they no longer can bind together the two atoms, and the molecule then dissociates. The reaction proceeds in the presence of ultraviolet radiation whose wavelength is less than 0.24 micron. If E represents the energy of the absorbed radiation, the first step in the production of ozone is:

$$O_2 + E \rightarrow O + O. \tag{4.7}$$

Each of the resulting atomic oxygens quickly combines with molecular oxygen to form ozone:

$$O + O_2 + M \rightarrow O_3 + M, \tag{4.8}$$

where M is any other molecule (such as O_2 or N_2) which interacts momentarily with the colliding O and O_2 and takes up some of the excess energy released in the reaction. It might be supposed that the presence of M is superfluous and that the reaction could proceed as a two-body collision; however, conservation of momentum and energy conditions for such a two-body interaction are so restrictive on the initial velocities that for the reaction to proceed at an appreciable rate, the extra degrees of freedom introduced by a third body are essential. Without the third body to absorb some of the energy, O and O_2, upon coming together in a collision, would not react.

The rate of a chemical reaction between two components, which depends upon a two-body collision, is equal to the rate at which two of the appropriate molecules collide in a gas, multiplied by the probability that the molecules react during the collision. Since the collision rate is proportional to the product of the concentrations of the two components, so is the reaction rate. Similarly, the reaction rate of Eq. (4.8) is proportional to the product of the concentrations of the three components on the left-hand side of the equation. As a consequence, the rate of such three-body reactions decreases rapidly with decrease in the atmospheric density, and thus at very high altitudes this reaction proceeds only very slowly.

Ozone is destroyed both by radiation and by collision with free atomic oxygen. The two important processes which tend to reduce the concentration of ozone are

$$O_3 + E \rightarrow O + O_2 \tag{4.9}$$

and

$$O_3 + O \rightarrow 2O_2. \tag{4.10}$$

The disruptive radiation for the first equation must have a wavelength less than 1.1 micron. Ozone therefore is slightly unstable in visible sunlight, decomposing into molecular oxygen and atomic oxygen. It is especially a strong absorber for wavelengths less than 0.30 micron. This ultraviolet absorption together with the photodissociation of O_2 according to Eq. (4.7) and the excitation of electrons in molecular oxygen account for the complete absorption of solar radiation below 0.30 micron which we have previously depicted in the absorption spectra of Fig. 4.5.

At altitudes above 80 km, the rate of formation of ozone is reduced not only because of the lower atmospheric density, as for Eq. (4.8), but because of the increasing importance of a competing reaction:

$$O + O + M \rightarrow O_2 + M, \tag{4.11}$$

where M again represents any other molecules involved in the three-body collision.[4] At these altitudes, the photodissociation of molecular oxygen caused by the intense unfiltered ultraviolet radiation is so important that the consequent high concentration of atomic oxygen results in a perceptible reduction in the molecular

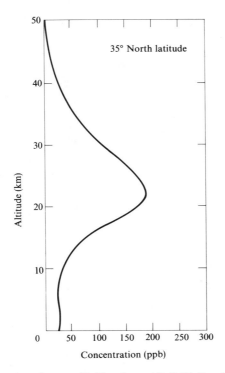

Fig. 4.6 Vertical distribution of ozone. (S. Manabe and R. T. Wetherald, *Journal of Atmospheric Sciences*, **24**, 241, 1967; courtesy of the American Meteorological Society.)

weight of air. This high concentration of atomic oxygen makes Eq. (4.11) relatively more important than Eq. (4.8). At altitudes above 100 km, three-body reactions are relatively unimportant and the corresponding ozone concentration insignificant.

Below 30 km, the rate of formation of O_3 should decrease as well, since absorption of ultraviolet radiation at higher altitudes leads to a lower rate for dissociation of molecular oxygen according to Eq. (4.7). Therefore the concentration of ozone should reach a maximum value at about 50 km. However, the observed concentration profile, illustrated in Fig. 4.6, has a maximum at a lower altitude than is predicted by calculations based upon the above chemical reactions, including the effects of radiative heat transfer. The observed maximum concentration of about 0.2 ppm is observed at an altitude of about 26 km at the equator and 18 km at the poles.

The reason for the discrepancy is that convection processes convey ozone from the regions where it is formed toward lower altitudes. Advection also causes a lateral spreading from low latitudes, where the solar radiation is most intense and ozone production is highest, toward the poles. At lower altitudes, a high concentration of ozone can be sustained despite the absence of ultraviolet radiation for its production; this causes the maximum in the concentration profile. The

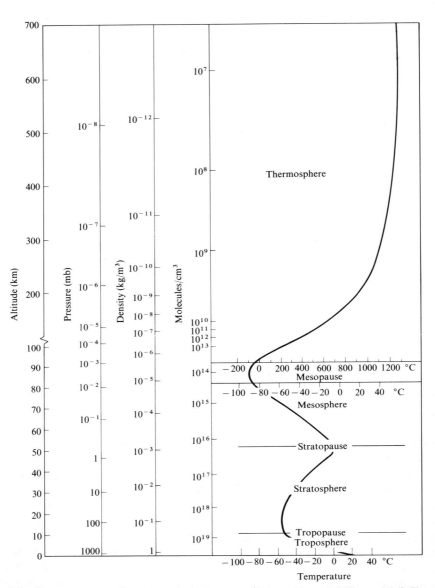

Fig. 4.7   Temperature profile and atmospheric strata. (Adapted from A. Miller and J. C. Thompson, *Elements of Meteorology*, Charles E. Merrill Publishing Co., Columbus, Ohio, 1970.)

reason is that atomic oxygen released by the reaction in Eq. (4.9) quickly combines with the high ambient concentration of molecular oxygen according to Eq. (4.8) to form ozone again. Reaction (4.10) competing against the formation of ozone is not favored because the concentration of O_2 greatly exceeds the concentration of O in the lower atmosphere. Eventually, as ozone is convected downward, it makes contact with aerosols, vegetation, or the ground, where it is eliminated.

Retainment of ozone at altitudes below 20–30 km does not represent a photo-chemical equilibrium, because the initial production occurred at higher altitudes. To illustrate this, let us consider the "half-restoration period" for ozone forma-tion, or the time required for the combination of reactions and convective processes to restore the concentration of ozone to its observed value at a given altitude. At about 50 km, and for all latitudes and seasons, this is about 1 hour. At 35 km, the period is lengthened to about 5 days at 60 deg latitude during summer, and 100 days during winter. The ultraviolet radiation responsible for ozone pro-duction is more effectively absorbed at higher altitudes in the latter season owing to the greater incident angle of sunlight. At 20 km, the period is in the order of years. Thus, at lower altitudes, the concentration of ozone is relatively indepen-dent of temperal variations of the incident solar radiation. In fact, the ozone content at low altitudes, about 20 ppb, is a reasonably well-conserved property of air, except in regions of photochemical air pollution.

4.4 ATMOSPHERIC STRATA

The photochemical reactions involving ozone and other gases in the upper atmos-phere release energy which locally increases the temperature. Figure 4.7 is an average composite of the observed temperature variation with altitude. The figure represents the average profile on a long-term basis; the temperature profile which would be observed on a given day may differ considerably from this composite. The observed temperature generally reaches a relative maximum near the upper regions of high ozone concentration where the reactions take place. At higher altitudes, the temperature profile exhibits a relative minimum and then a steady increase with height. These variations in temperature with altitude suggest that the atmosphere could be considered as a sequence of strata, with boundaries defined by the positions of the relative minima and maxima of temperature. As indicated in the left portion of the figure, higher strata are also characterized by a lower concentration of molecules and a lower pressure, in accordance with the hydrostatic condition.

The lowest stratum which extends upward from the surface of the earth is known as the *troposphere*. Weather is formed within this zone. Radiation from the sun penetrates the atmosphere to warm the oceans and ground, and the surface of the globe in turn transfers energy to the low levels of the troposphere. Convection and radiation transmit this energy upward, so there is an average decrease in atmospheric temperature with increase in altitude. On the average, the temperature profile shows a decrease in temperature of about 6.5°C per km. The upper boundary of the troposphere, above which the temperature is often observed to be independent of altitude for 10 or 20 km, is called the *tropopause*. Its altitude varies with latitude and longitude, on the average sloping downward from a height of 12 km at the equator to about 8 km at the poles. The tropopause also shows diurnal and seasonal variations, with the most pronounced variations in temperature and height occurring above the temperate zone. Often the boun-dary is ill defined, particularly in the polar regions during the winter night.

The temperature of the tropopause is often found to be about 220°K. As a result, this cold region serves as an important barrier preventing the escape of water vapor from the troposphere. Without this low temperature region, water molecules could be conveyed upward by convection until they reached an altitude where ultraviolet radiation would cause photodissociation. The hydrogen component would escape from the earth and its atmosphere in the same way that the initial concentrations of hydrogen and helium were depleted. However, the low temperature of the upper troposphere causes water vapor to condense or freeze before it reaches an altitude to which ultraviolet radiation penetrates. The heavier mass of the condensed droplets and ice particles is sufficient to prevent further upward motion. Only a small fraction of water in this region, equivalent to the water vapor associated with the saturated vapor pressure at 220°K, will continue the upward convection. Thus the low temperature of the tropopause preserves water on the earth.[5]

Above the tropopause is the *stratosphere*. This often shows two regions of distinct temperature variation. The lowest may have a temperature essentially independent of altitude, a continuation of the condition at the tropopause. However, the upper region of the stratosphere has a temperature which increases with height as a consequence of the ozone formation at about 50 km. Very little water vapor (perhaps only 5 ppm) is found in the stratosphere owing to the barrier imposed by the temperature of the tropopause. At about 50 km the temperature reaches its maximum value of approximately 270°K. This marks the upper boundary of the stratosphere which is called the *stratopause*.

Just above 50 km is a region known as the *mesosphere* in which temperature again decreases with increasing altitude. Atmospheric circulation in this rarefied mixture of gases is relatively ineffective; the transfer of energy is mostly by radiative processes. There is only a slow downward migration of ozone. The density of air in the upper mesosphere is insufficient to support effective photochemical reactions, so the temperature decreases steadily with altitude, reaching a value of about 170°K at its upper boundary, the *mesopause*.

The region overlying the mesopause, extending upward from about 100 km, is again a portion of the atmosphere in which solar energy is converted into sensible heat. This is the *thermosphere*, where there are fewer than 10^{19} molecules per cubic meter as compared with about 2.5×10^{25} at sea level. The intense ultraviolet radiation penetrating to the lower thermosphere causes photodissociation of O_2 and photoionization of N_2 and atomic oxygen. The temperature reaches a value which may be as high as 1500°K at altitudes on the order of 1000 km. Energy is also absorbed from high-energy particles emitted from disturbances within the solar photosphere and its overlying chromosphere. In fact, the thermosphere of the earth is often considered to blend with the corona, the outermost portion of the sun, surrounding the chromosphere.

The preceding terminology for atmospheric strata is based upon characteristic features of the temperature profile. But other criteria also could be used to define strata: The lower mesosphere and upper stratosphere where ozone is produced are known as the *ozonosphere*; the region above 60 km is the *ionosphere*, indicating the presence of ionized molecules and atoms; below 80 km where the molecular weight is uniform is the *homosphere* and above that the *heterosphere*; and the very high altitude reaches of the mesosphere from which molecules can escape the earth is the *exosphere*. These strata overlap various portions of the strata which are defined according to the temperature profile. But most of the variations of molecular weight and degree of ionization occur only at very high altitudes where they appear to have little effect on the energy balance within the lower atmosphere and play no important role in troposphere circulation. Air pollution is not influenced by processes which occur above the stratosphere. Nor has air pollution yet affected this upper region, except possibly from the exhaust of large rockets which pass through.

4.5 GREENHOUSE EFFECT

The mean surface temperature of the earth is determined by an energy balance between the incoming solar radiation and the outgoing radiation emitted by the surface and atmosphere. A complicating feature is the fact that the atmosphere reflects back into space some of the incident solar energy. It also absorbs some of the incoming energy and, more importantly, a large fraction of the outgoing terrestrial radiation. Attempts to analyze the processes in quantitative terms are difficult, because conditions on the earth are not uniform. One obvious factor is the seasonal and diurnal variation of solar energy incident upon any portion of the globe. Another is the latitudinal variation of this energy supply. Our understanding from a quantitative viewpoint is still primitive, although much progress has been achieved during the past twenty years in coming to grips with certain aspects of the overall problem.

The atmosphere and earth form a relatively stable system, despite the unevenness in heating and great variety in the features of terrain. The highest recorded air temperature is 58°C (136°F) in Aziza, Tripolitania, North Africa, which occurred in 1922; the lowest, −87°C (−125°F) was measured in 1958 in Antarctica.[6] This may appear to be a pronounced difference when measured in degrees Celsius, but it is only a variation from 331 to 186 degrees on the thermodynamic Kelvin scale. The variation is less than ±35% of the global long-term average of 288°K. And ordinary seasonal variations that occur with appreciable regularity at any one location are considerably less than ±20% of the local mean temperature. Thus, wide fluctuations in the thermodynamic surface temperature are not observed. This is due partially to the complex interplay of ocean and air currents which form a feedback system for stabilizing the energy balance and smoothing out spatial and temporal variations in temperature.

Ordinarily the effects of man would have little possibility of substantially affecting global systems of nature, for the means at our disposal could not concentrate the great quantities of energy which would be needed, short of simultaneously detonating many nuclear bombs. However, by his emission of pollutants, man has gained a leverage on nature. Some air pollutants at only relatively small concentrations can have an important effect on the transfer of energy between the earth and its atmosphere. In Appendix E we point out that this is because the molecules of these pollutants have the appropriate shape to strongly absorb electromagnetic radiation. Consequently, even minor constituents of the atmosphere (such as water, water vapor, and carbon dioxide) can influence the mean global temperature.

The essential features of the earth's energy balance can be understood with reference to Fig. 4.5 and can be stated simply: A large fraction of the incident solar radiation is in the spectral region where the atmosphere is transparent; although some is reflected, half passes through and is absorbed by the earth. The surface of the earth loses some energy by contact with the atmosphere, and convection of sensible and latent heat toward higher altitudes; but most of its energy loss is through thermal black body radiation. Since its surface temperature is only 288°K, the radiation will be in the long wavelength infrared portion of the spectrum, with a maximum intensity at about 10 microns. The atmosphere, however, almost completely absorbs this radiation; and a portion of the energy is then reradiated as thermal radiation back to the surface, with a smaller fraction being reradiated into space. By these processes of absorption and reradiation, the atmosphere provides an "insulating blanket" for maintaining heat near the surface of the earth. The warming of the earth's surface, resulting from the fact that the atmosphere is largely transparent to solar radiation but opaque to terrestrial, is called the *greenhouse effect*. It is this effect which maintains the surface temperature of the earth about 40°K higher than is dictated by simple energy balance considerations of the incoming and outgoing radiation from the earth, when assuming that 34% of the incoming energy is reflected without absorption.

Unfortunately the "greenhouse effect" is a misnomer when applied to the earth's atmosphere. It was once thought that the glass of a greenhouse acted in the same way to maintain the interior and ground surface warmer than the exterior. However, glass does not absorb sufficiently in the infrared portion of the spectrum to account for the magnitude of the warming effect. Both theory and experiment have shown that the correct explanation of the high temperature within a greenhouse is the suppression of convective cooling, not radiation cooling.[7,8] The glass roof forms a barrier to the convection and advection of heat away from the ground. It also discourages evaporative cooling of moist ground when the absence of circulation has permitted the air to become saturated with water vapor. Somehow these facts still appear to be insufficient to discourage the use of a catchy name, so we shall continue to use "greenhouse effect" in the well-established tradition.

A quantitative analysis of the greenhouse effect has been achieved only in general terms. No one has yet succeeded in predicting from first principles such details of the atmosphere as the global distribution of humidity or cloud cover. Often, models for the energy balance between the earth and atmosphere are based

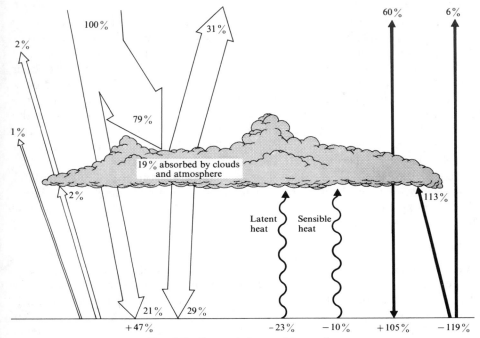

Fig. 4.8 Energy balance of the earth.

on estimated average global conditions which neglect seasonal and diurnal variations.

The results of one analysis by Houghton,[9] elaborated further by Budyko,[7] are illustrated in Fig. 4.8. The numbers in this figure indicate the average rate of energy transport by the various modes, expressed as a *percentage* of the rate at which solar radiant energy is incident downward at the tropopause. (Absorption of solar radiation at higher altitudes by ozone and oxygen leads to only a 1% difference between the solar constant and the average solar energy incident per unit area at the tropopause.) The rate at which energy is conveyed from one place to another by some modes may exceed the rate at which it is incident from the sun, so the corresponding percentage exceeds 100%.

Figure 4.8 shows that about 79% of the incident radiation is intercepted by clouds, aerosols, water vapor, and other gases, with 31% being scattered away from the earth and 29% scattered down to the surface. The remaining 19% is absorbed by constituents of the atmosphere. Thus the earth's surface receives 50% of the incident solar energy, of which only 21% is unscattered direct radiation. Most of the radiation which strikes the earth's surface is absorbed, accounting for the 47% of incident solar energy indicated in the figure.

For an energy balance, the surface must lose energy at a rate equal to this 47%. Some is removed through evaporation of water and conduction of heat to the lowest layer of the atmosphere. A total of about 33% is lost by these two processes. The black body radiation of the earth accounts for a far greater energy loss of 119%, more than twice the rate at which solar energy is absorbed. This is possible in an equilibrium condition only because most of this terrestrial infrared radiation is absorbed by constituents of the atmosphere, which because of their own temperature also emit infrared radiation, much of which is directed back toward the surface. The rate of energy return, 105%, is a far larger source of energy for the earth than is the solar radiation at shorter wavelengths, and this is the reason why the greenhouse effect enhances the surface temperature of the earth.

The overall energy balance can be verified by noting that the 34% of short wavelength solar radiation which is scattered and reflected into outer space, when added to the 66% rate of infrared energy emission from the atmosphere and earth, exactly balance the 100% rate at which solar radiation is incoming. Similarly, the incoming and outgoing energy rates at the earth's surface and in the atmosphere are in balance.

Albedo

We have noted that a certain fraction of the incident energy from the sun is reflected or scattered into space without absorption. This fraction of incident light which is reflected or scattered from an object is called the *albedo*; for the earth and its atmosphere as a unit, it has a value variously estimated to be between 0.30 and 0.39. Figure 4.8 shows a value of 0.34, which is close to the measurements

TABLE 4.1

Albedo for the spectral region from 0.3 to 2 microns*

Surface	Percent reflected
Clouds less than 150 m thick	5–60
150–300 m thick	30–70
300–600 thick	55–80
Water (normal incidence)	8
Dry clean snow	80–90
Wet snow	50–70
Yellow sand	35
White sand	34–40
Summer wheat	10–25
High standing grass	18–20
Green forest	10–20
Dry fallow field	8–12
Wet fallow field	5–7
Concrete	15–25
Asphalt road	5–10

Adapted from K. Ya. Kondrat'yev, *Radiation in the Atmosphere*, New York: Academic Press, 1969. Copyright © 1969 by Academic Press and reprinted by permission.

obtained from orbiting satellites.[2] Of this, about 0.25 is attributed to reflection from clouds and only 0.03 to reflection from the surface of the earth (thus the albedo of the surface alone is about 0.06, since only half of the incoming radiation penetrates through the atmosphere and is incident on the surface). The remainder of the albedo is believed to arise from scattering by aerosols and atmospheric gases.[7,10]

The theory of the reflection and scattering of radiation by water droplets is well understood, but the effect of haze and aerosols is not so well known, and their global distribution has been only sparsely charted. It is certain, however, that condensed water droplets are the most important constituent of the atmosphere determining the albedo. The surface of the oceans by contrast is nearly perfectly absorbing for light near normal incidence. Only when the incident light grazes the surface by 30 degrees or less does an appreciable amount suffer reflection, reaching 50% at about 16 degrees and over 99% at 10 degrees. The albedo of an ocean depends therefore on its surface roughness. Natural features on the earth are also good absorbers, with a low albedo as indicated in Table 4.1.

The absorption characteristics of complex substances such as minerals and vegetation depend upon the wavelength of the incident light. Often, natural surfaces are found to absorb a greater fraction of incident infrared radiation than of visible; this is one reason why the earth's surface emits a spectrum of thermal radiation which closely approximates a black body spectrum for a surface of the same temperature. The close parallel is illustrated by the quantities in Table 4.2, from which it can be seen that most surfaces have a peak emission intensity which differs by less than 5% from the corresponding peak intensity of black body radiation.

TABLE 4.2

Observed radiant energy from natural surfaces, expressed as a fraction of the radiant energy of a black body at the same temperature, for the spectral region from 9 to 12 microns*

Surface	Fraction
Water	0.96
Fresh snow	0.99
Dry fine sand	0.95
Wet fine sand	0.96
Dry peat	0.97
Wet peat	0.98
Thick green grass	0.99
Coniferous needles	0.97

*K. Ya. Kondrat'yev, *Radiation in the Atmosphere*, New York: Academic Press, 1969. Copyright © 1969 by Academic Press and reprinted by permission.

Direct heating

Questions have been raised as to whether man's activities are directly releasing energy in sufficient amounts to upset the natural energy balance. The release of sensible heat with the hot exhaust gases and the warm water used to cool condensers in power generating stations is one example. Direct heating of the environment has been estimated to be about 0.012 W/m^2 over the land area.[11] For comparison, we might choose a figure of 370 W/m^2 which the earth emits as thermal radiation over the entire surface, or we might choose about one-third of that as representing the difference between what the earth absorbs from all sources of incident radiation and what it emits. In either case, the conclusion is clear. Man's addition represents only a small fraction of one percent. The current levels of direct heating are negligible on a global basis. We shall note later however that this conclusion is not true for localized regions, for the concentration of heat sources within some cities may very well affect the urban climatology.

4.6 WATER DROPLETS AND VAPOR

Water in both vapor and droplet form is the principal agent for the greenhouse effect. We have pointed out that the planetary albedo is determined largely by its cloud cover. So, too, water and water vapor are the principal constituents of the atmosphere that strongly absorb the terrestrial infrared radiation, as indicated in Fig. 4.5. Carbon dioxide has only a secondary role in the greenhouse effect, partly because its concentration in the atmosphere is lower than that of water vapor so that fewer molecules are available to participate, and partly because its main infrared absorption is localized within a much narrower band of the electromagnetic spectrum.

Concern recently has been focused on the consequences of increasing the water vapor content of the stratosphere as a result of a greater number of high-altitude flights by supersonic transport aircraft. Should the greenhouse effect be enhanced materially, the rise in the earth's surface temperature may cause poleward shifts of surface isotherms (lines of constant temperature) with a consequent melting of the polar ice. If sufficiently pronounced, the accompanying rise in the levels of the oceans would cause widespread flooding. A definitive analysis of the effect resulting from the anticipated increase in stratosphere water vapor has not yet appeared, and it might very well turn out that other effects such as interference with the production of ozone are a more significant consequence than a possible perturbation of the energy balance.[12] But in any event it is worthwhile to examine one of the more recent estimates of the thermal effect, if only to expose what assumptions and difficulties are associated with such a calculation.

Humidity

One major uncertainty in estimates for changes in mean surface temperature T_* is lack of knowledge of how the distribution of water and water vapor in the atmosphere would be affected. The global mean for the relative humidity is estimated to be about 0.77 at sea level, so the content of water vapor is substantially less than would exist in a saturated atmosphere.[13] Were the relative humidity to remain constant as the surface temperature rose, then the absolute humidity or vapor content would increase with an increase in T_*. This would enhance even further the greenhouse effect through more effective absorption of the earth's thermal radiation. In addition, greater concentration of water vapor near the surface would lead to greater absorption of terrestrial radiation at low levels, accompanied by a warming of the atmosphere and discouragement of convective cooling of the surface. So the consequences of a temperature rise with no change in relative humidity result in a feedback effect through which there is a still further increase in temperature. The resulting enhancement of the equilibrium T_* can be considered a *self-amplification effect*, which magnifies the temperature change brought about by any disturbance of the energy balance, be it an increase in stratospheric water vapor, or other cause.[14]

One piece of evidence that the distribution of the relative humidity is fairly insensitive to changes in temperature is illustrated in Fig. 4.9; the summer and

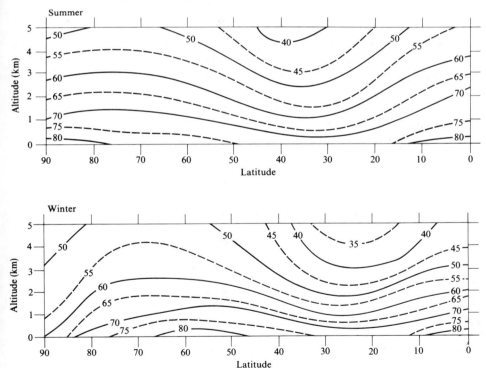

Fig. 4.9 Distribution of relative humidity with latitude and altitude in the northern hemisphere for both summer and winter. (S. Manabe and R. T. Wetherald, *Journal of the Atmospheric Sciences*, **24**, 241, 1967; courtesy of the American Meteorological Society)

winter patterns show that the zonal distribution does not vary greatly in response to seasonal changes. S. Manabe and R. T. Wetherald[13] have suggested that this may signify a mode of atmospheric stability, and that, given sufficient time, the atmosphere will establish similar patterns should it be perturbed by extraneous influences. If true, the absolute humidity in the atmosphere will be sensitive to T_*, and the self-amplification of temperature changes will be an important factor. Because of this, any calculation for predicting global temperature changes should show that T_* will be more sensitive to variations in atmospheric concentrations of CO_2, O_3, or aerosols if the relative humidity is assumed constant than if a less realistic fixed absolute humidity is assumed. The possibility that the atmosphere would however respond differently and *not* maintain the same relative humidity cannot at this time be ruled out.

Manabe and Wetherald[13] have calculated the change in T_* which results from a change in the water vapor concentration of the stratosphere, employing the following assumptions for mean characteristics of the atmosphere:

1. The relative humidity decreases with altitude according to

$$h = h_0 \frac{(P/P_0) - 0.02}{1 - 0.02}, \qquad (4.12)$$

a variation which is in accord with observations. In this expression, $h_0 = 0.77$ is the mean value of the relative humidity at the surface of the earth, P_0 is the atmospheric pressure at sea level, and P is the pressure at the altitude of interest.

2. The minimum value of h corresponds to a concentration of 5 ppm by volume of water vapor in the stratosphere and is constant at this value for altitudes above which Eq. (4.12) predicts a lower value.

3. The concentration of CO_2 is 300 ppm by volume.

TABLE 4.3

Cloud characteristics of the atmosphere[*]

Cloud		Altitude (km)	Average percent of global coverage	Albedo	Contribution to planetary albedo
High		10	0.228	0.20	0.046
Middle		4.1	0.090	0.48	0.043
Low	top	2.7	0.313	0.69	0.22
	bottom	1.7			
				Total	0.31

[*]Adapted from S. Manabe and R. T. Wetherald, *Journal of the Atmospheric Sciences*, **24**, 241, 1967; courtesy of the American Meteorological Society.

4. The vertical distribution of ozone is similar to the curve shown in Fig. 4.6.

5. Cloud characteristics are as given in Table 4.3.

6. The albedo of the earth's surface is 0.102.

These assumptions together with a simplified model for convective and radiative energy transfer provided the basis for computer calculations. The results predict that a fivefold increase in the water vapor concentration within the stratosphere, to 25 ppm by volume, would lead to a $+2°C$ increase in T_*. This assumes that the relative humidity in the troposphere is unchanged. Whether such a large increase in water vapor concentration in the atmosphere would result from operation of supersonic transports remains a matter of debate. Some calculations of horizontal convection by Manabe et al.[15] suggest that moisture may be removed by freezing out near the high equatorial tropopause. Other estimates give an increase of only 0.6 ppm in the steady-state water vapor concentration from operations of the projected fleets of aircraft.[16] Whether the assumption of a fixed relative humidity is appropriate is, of course, also debatable.

Reservations

How accurate are the results of energy balance calculations such as the one we have just described? An answer to this must await further calculations, based on different models of atmospheric behavior, which can test the sensitivity of the results to the initial assumptions. Atmospheric models have lacked several ingredients known to be important, at least in some local contexts. We will mention three in particular:

1. The processes of evaporation and condensation of water provide a means of transporting energy from one portion of the earth to another. A change in the mean temperature of the lower atmosphere shifts the locations in which these processes take place and thus alters the distribution of atmospheric temperature over portions of the globe.

2. Accompanying an alteration of the distribution of clouds will be shifts in the location of precipitation. The albedo of certain regions would be changed. This would further affect low-level temperatures.

3. The influence of the oceans also has been neglected. The thermal reservoir which they provide is known to smooth out seasonal temperature variations in each hemisphere. Their effect in stabilizing longer term temperature variations is unknown in quantitative terms.

Our understanding of the dynamics of the atmosphere and oceans systems is still in an embryonic stage. Much remains to be appreciated. No firm estimate is available for the errors which may result when latitudinal and longitudinal variations in temperature or humidity are ignored. It may be possible that a nonuniform atmosphere is for some reason more stable than a uniform one. Predictions of future temperature trends resulting from air pollution must answer these questions if they are to be convincing. With these reservations in mind, we now turn to another aspect of world-wide air pollution.

4.7 CARBON DIOXIDE

It was first suggested by J. Tyndall in 1863 and then by T. C. Chamberlain in 1899 and S. Arrhenius in 1903 that the rapidly mounting pace of industry might release sufficient carbon dioxide to cause climatic changes by affecting the surface temperature of the earth. Indeed this has become an issue of concern, for the CO_2 concentration is now observed to be increasing throughout the troposphere. The increase is attributed to the widespread dependence on combustion for the production of power and to the increasing use of automobiles for transportation. Table 4.4 illustrates that of all the pollutants emitted in sizable quantities by combustion processes, carbon dioxide is the one released in by far the greatest amount. Figure 4.10 shows an estimate of CO_2 global emissions from anthropogenic sources during the last century. This is a slight underestimate because it does not take into account the use of wood and other so-called "noncommercial fuels." Clearly evident is a steep increase in recent years. The output in 1965 was estimated by E. Robinson and R. C. Robbins[17] to be about 1.3×10^{10} Tonnes, half from coal combustion and a third from petroleum combustion. However large this output may seem, in fact, it represents less than 10% of the yearly release of CO_2 by all sources, natural and anthropogenic. A perspective is provided by the

TABLE 4.4

Constituents of the atmosphere and common pollutants. The annual emission rate on a global basis (in 10^9 Tonnes per year) is indicated for those produced by combustion, with for comparison the estimated emission rate from natural sources. The half-life indicates the estimated length of time between emission of a quantity of a given pollutant and removal from the air of half of that quantity through natural means.

Gas	Concentration(ppm)	Combustion emissions	Natural emissions	Half-life
CO_2	320.	13.	160.	2–5 years
CO	0.08	0.27	4.0	0.2 years
SO_2	0.0002	0.13	0.044	4 days
Hydrocarbons (except CH_4)	<0.001	0.1	0.2	—
NO and NO_2	0.001	0.05	0.5	5 days
H_2S	0.0002	0.003	0.1	2 days
CH_4	1.5	—	2.0	1.5 years
N_2O	0.3	—	0.6	4 years
NH_3	0.01	—	1.2	7 years
O_2	2.1×10^5	− 10.	100.	1,000 years

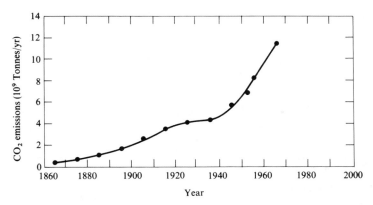

Fig. 4.10 Average CO_2 emissions to the atmosphere from fossil fuel combustion (decade average). (E. Robinson and R. C. Robbins, "Sources, Abundance, and Fate of Gaseous Atmospheric Pollutants," Stanford Research Institute, 1969, from R. Revelle *et al.* in *Restoring the Quality of Our Environment*. Report of the Environmental Pollution Panel, President's Science Advisory Committee, November, 1965.)

schematic of the global carbon cycle given in Fig. 4.11. It is difficult to estimate accurately the rate of emission and absorption at the surface of the ocean, but they are believed to be nearly equal.

The observed rise in the ambient concentration of CO_2 is illustrated in Fig. 4.12, as observed at the Mauna Loa Observatory in the Hawaiian Islands. The upper chart displays the average monthly concentration and the seasonal variation of almost 2%, as vegetation in the Northern Hemisphere responds to the

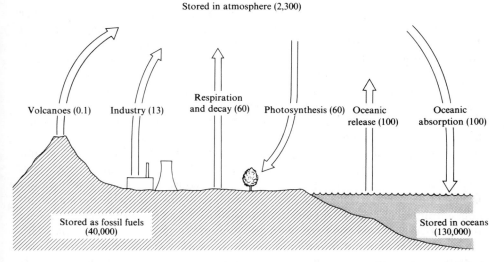

Fig. 4.11 Carbon dioxide cycle: Annual emission and absorption rate, with amounts expressed as mass of CO_2 involved, in 10^9 Tonnes.

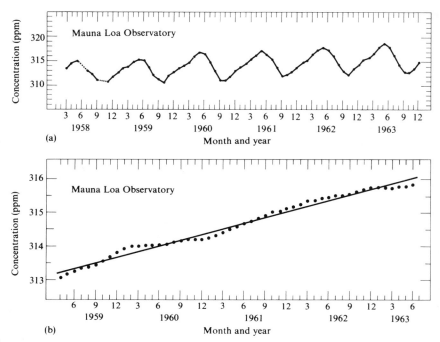

Fig. 4.12   (a) Monthly average concentration of atmospheric CO_2 at Mauna Loa Observatory in Hawaii; (b) Twelve month running mean of the concentration of atmospheric CO_2. (J. C. Pales and C. D. Keeling, *Journal of Geophysics Research*, **70**, 6053, 1965; courtesy of the American Geophysical Union.)

growing season. Such seasonal oscillations decrease in magnitude at higher altitudes, according to aircraft observations over Sweden. Averaging the Mauna Loa data over 12 months removes most of the seasonal effect and clearly demonstrates in Fig. 4.12(b) the presence of a steady rise of the mean concentration. The rate of increase amounts to 0.2% of the concentration per year or an addition of 0.7 ppm each year. This trend is confirmed by measurements in other parts of the world, including Scandinavia and Antarctica.

The amount of CO_2 accumulated in the atmosphere each year is roughly half of the estimated emissions from anthropogenic sources. Thus it appears that the natural CO_2 cycle involving land vegetation, marine phytoplankton, and absorption by the upper layers of the oceans is able to increase the rate at which CO_2 is eliminated but not to such an extent that a constant ambient concentration is maintained. It is possible to estimate on the basis of this trend what the maximum concentration of CO_2 might be if all fossil fuel reserves were combusted. The total recoverable coal and petroleum fuels amount to about 2×10^{12} Tonnes, with a yield of CO_2 of about 8×10^{12} Tonnes. Were half of this to be distributed uniformly through the atmosphere, it would ultimately increase current CO_2 levels by about a factor of 3, resulting in an ambient concentration of 900 ppm.

Carbon dioxide in the atmosphere contributes to the greenhouse effect, because it has a strong absorption band for electromagnetic radiation centered at a wavelength of about 15 microns. Absorption at 15 microns is significant, for it is close to the wavelength at which the maximum terrestrial radiation is emitted.

Manabe and Wetherald[13] have calculated the change in the mean surface temperature T_* which may be expected from an increase in CO_2, using the energy balance model outlined in the preceding section. A doubling of the CO_2 concentration to 600 ppm is estimated to produce a $+2.4°C$ increase in T_* for average cloudiness, assuming a fixed relative humidity. This is sufficient to affect the climatology of some regions of the world. However, the predicted change in T_* is about half of this value if a fixed absolute humidity is assumed. The difference reflects the importance of the "self-amplification" effect. The sensitivity of such predictions to the nature of the assumptions is illustrated by a comparison of the change in humidity necessary to counterbalance an increase in CO_2, so that the net effect would yield no change in T_*. Estimates indicate that the warming effect of a 10% increase in CO_2 concentration could be counterbalanced by a 3% decrease in the absolute humidity or a 1% increase in the average cloudiness.[14] This shows the dominant importance of water vapor and water droplets for the greenhouse effect; variations in the humidity might very well obscure the warming effect of increasing CO_2 ambient concentrations. At this time, such small variations in the average cloudiness could not be detected.

Furthermore, the role of the oceans should not be forgotten. They are capable of absorbing large quantities of CO_2, and it is estimated that there is 60 times more CO_2 stored in the oceans than in the atmosphere.[18] Equilibrium is maintained between the concentration of bicarbonates dissolved in sea water and the concentration of atmospheric carbon dioxide. The equilibrium ratio of these two concentrations is sensitive to the ambient temperature; therefore long-term temperature trends will cause absorption or release of CO_2. Since higher temperatures produce the latter, a gradual warming of the earth for *any* reason might cause a corresponding increase in the ambient CO_2 concentrations over the globe. If only the upper mixing layer of the ocean, about 100 m, contributes, it is estimated that a 1°C rise in temperature would release amounts of CO_2 sufficient to increase the ambient concentration by about 0.4%.[17] Of course this is only a gross estimate. One major difficulty in obtaining a more precise prediction is the fact that the temperature of the air and the oceans varies with position and time. However, the possibility that an increase in CO_2 concentrations in the atmosphere may be the *result* of a global temperature increase and not the *cause* of it is fair warning that we should not jump to conclusions when a correlation between CO_2 levels and mean global temperature is established. Existence of a correlation does not prove a causal relationship. But, in fact, the predicted CO_2 increase from oceanic release for a temperature increase of 1°C is much too small to explain the global trends observed in the first half of this century. To see this, let us regard the temperature data.

Figure 4.13(a) shows mean annual temperatures since 1738 for the eastern sea-

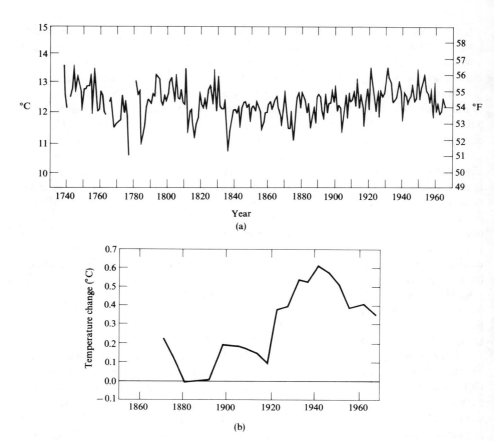

Fig. 4.13 (a) Annual mean temperatures for the eastern seaboard of the United States (H. E. Landsberg, *Science*, **170**, 1265, 1970. Copyright. 1970 by the American Association for the Advancement of Science.) (b) Change in the mean temperature of the Northern Hemisphere (from 0° to 80°N latitude) since 1880. (J. Murrary Mitchell, Jr., *Global Effects of Environmental Pollution*, S. F. Singer, Ed., New York: Springer-Verlag, 1970, and private communication).

board of the United States, using Philadelphia as the reference point. The average over such a large region is more representative of global trends than is the temperature of one city. The span of time encompasses 230 years, except for a few isolated periods for which there are insufficient data. Despite the fact that the temperature represents an average over a fairly large geographical area, the figure demonstrates that the mean temperature is subject to pronounced fluctuations. This reflects the changing aspect of the atmosphere. There are definite short periods of abnormally cold or warm temperatures. In addition, some long-term trends are indicated, almost obscured by the "noise" of the short-term fluctuations. By considering similar data from other locations about the world, we can estimate a mean global temperature. The results of such an averaging process as calculated by Mitchell[19] are depicted in Fig. 4.13(b). In view of the magnitude of the fluctuations in part (a) of the figure, it may be appreciated that the significant features in part (b) are the gross trends, not the details. Figure 4.13 reveals that there was a generally steady rise since the start of the industrial revolution in the nineteenth century. Between 1900 and approximately 1940 the mean global temperature rose at a rate of about 1°C per century. A result has been the retreat of many glaciers such as the Rhône in Switzerland.

It was commonly believed that the increase in CO_2 and the enhanced greenhouse effect was responsible for this temperature trend. However, since 1940 the mean temperature has evidently decreased. This is unexplained by the steadily increasing levels of carbon dioxide. Suggestions have been advanced that another air pollutant recently may have begun to play a more effective role in the energy balance. One class of contaminants which may have considerable leverage on the energy balance is the atmospheric aerosols, which affect the albedo. We shall take up their role in the following section.

Before we do, it should be noted that there is, in fact, no direct proof that man is affecting the mean surface temperature of the earth. The world has undergone numerous climatic changes in the past. For example, paleoclimatological studies provide evidence for several glacial periods during the past million years, as illustrated by the temperature trends in Fig. 4.14. Perhaps some of these were caused by the changing nature of atmospheric constituents, perhaps not. It may be presumptuous for man to believe that he is causing global temperature changes now. On the other hand, it would be foolhardy to discount the possibility that man could initiate an atmospheric instability which could produce much more drastic temperature variations than are predicted by contemporary calculations.

4.8 INFLUENCE OF AEROSOLS

The concentration of aerosols is on the increase in some portions of the atmosphere, as indicated by measurements over Europe, North America, and the North Atlantic; but the change does not seem to be global in scope, for it does not show in measurements over the central Pacific. In fact, the aerosol burden over some cities is actually decreasing, apparently in response to more stringent control regulations over the emissions of particulate matter.[20] It is estimated that the rate at which particulates are released into the air on a global basis is

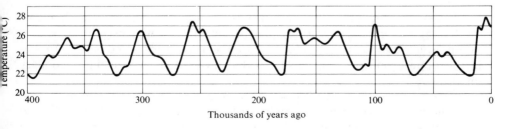

Fig. 4.14 Temperature for the surface water of the central Caribbean, indicating widespread changes and seven most recent glacial periods. (From C. Emiliani, *Science*, **154**, 851, 1966. Copyright © 1966 by the American Association for the Advancement of Science.)

about 10^7 Tonnes per day from all primary and secondary sources. Only 10% of this has an anthropogenic origin.[21] Thus, natural sources of aerosols, such as volcanic eruptions, forest and grass fires, dust storms, and sea spray, are far more important than those added by man, when gauged in terms of total amount emitted.

Aerosols can have two effects on the earth's energy balance:

1. Their scattering of solar radiation may increase the albedo, thereby decreasing the earth's surface temperature.[22]

2. Their absorption of radiation may locally increase the temperature of the atmosphere, and thus both increase the radiant energy directed toward the earth and decrease the convective transfer of energy from the earth to the warmer atmosphere. Absorption would consequently cause an increase in the surface temperature.[23]

It appears that whether scattering or absorption is the dominant effect is determined by the size, chemical composition, and distribution of the particles. Let us turn now to consider the natural and anthropogenic sources of aerosols, so that we can assess the relative importance of man's contributions.

Natural sources

Dust and sand are among the most commonly experienced atmospheric contaminants. There are many instances when strong winds or the convective upward motion of air picks up loosened soil and transports it for great distances. For example, in the arid regions of northwest India and West Pakistan, rising currents of hot air entrain fine dust from the desert floor and carry it high into the upper troposphere. The effect is pronounced in this area because the desert floor lacks pebbles and large grains of sand which in many other deserts tend to anchor the finer dust beneath. The ambient concentration of dust may reach 500 to 800 $\mu g/m^3$, far more than is ordinarily experienced in polluted urban areas.

Another example which illustrates the importance of natural sources is the accumulation of deep loess soils in China, formed over the centuries by the fallout of large particles carried by the prevailing winds from the barren deserts of central Asia. Finer particles remain airborne longer, and most appear to work out of the air over the Pacific Ocean, forming extensive deposits on the bottom.[24]

It was once thought that meteoritic dust, produced by the impact of meteors on the atmosphere and their subsequent disintegration, was a significant source of atmospheric contaminants. However, samplings by rocket and balloon probes in the upper atmosphere have shown that this is a source of only secondary importance.[25]

Forest and grass fires yield great quantities of fine particulate matter, in addition to the heavier ash that rapidly settles out of the air. It has been estimated that a grass fire burning over a hectare (1.0×10^4 m^2 or 2.5 acres) produces 5×10^{22} fine particles.[26] Forest fires in western Canada during September 1950 are thought to have released sufficient smoke to account for an unusual discoloration of the sun witnessed later in Edinburgh, Scotland. On this occasion the sun appeared to be blue, apparently a result of the preferential scattering of red light by large particles of a fairly uniform size. (How this may happen is explained in Appendix C.) The aerosols from such conflagrations must have a diverse composition. Dry distillation of wood has been observed to yield over 100 types of hydrocarbon compounds; the processes which occur in a fire with the customary extreme variations in temperature and ambient concentration of oxygen must lead to even more varied end products.

Volcanoes are also impressive contributers of aerosols, and the effects of eruption have been noticed around the world. Usually, however, their emissions are significant only on an intermittent basis. The effects of the 1883 eruption of Krakatoa in the East Indies reportedly caused spectacular atmospheric effects, such as brilliantly red sunsets, for more than two years afterwards.[27] This resulted from great quantities of fine ash being thrown up to altitudes of at least 20 km. From the duration of effects, it appears that the residence time of dust in the stratosphere is about two or three years. Measurements of radioactive fallout from high-altitude test explosions of nuclear bombs confirm this. Other major eruptions in this century include Mt. Katmai, Alaska, in 1912 and Hekla, Iceland, in 1947. An eruption may eject as much as 10^8 m^3 of fine ash into the air, although the size and composition of this debris is not known. In 1963, the eruption of Mount Agung on the island of Bali, 8°S latitude, emitted particles with such energy that substantial quantities rose to an altitude of 50 km. The increase in air contamination from this eruption is perhaps the best documented example we have. Absorption of solar radiation by the ejecta in the stratosphere caused an increase in the local temperature in the high altitude equatorial belt by 6 or 7°C as the debris drifted eastward. However, no such pronounced changes were noted at sea level. The stratospheric temperature remained abnormally high by several degrees for two or three years until the bulk of the particulate matter had settled.

A natural aerosol found over almost the entire globe is sea salt, including both NaCl and sulfates such as magnesium sulfate. Their formation begins with the burst of one of the myriad of bubbles that are found on the surface of the sea. The small droplets which are released into the air are carried to higher altitudes through the effects of atmospheric turbulence. Some of the water in the droplets evaporates as they are carried into regions of low relative humidity, the radius of the droplet decreasing by a factor of 2 or more. If the relative humidity is less than about 65%, all of the water evaporates, leaving a solid particle. Prevalent over the oceans and inland regions near seashores, sea salt aerosol may react with other atmospheric constituents such as gaseous ammonia (NH_3) to yield chlorides and sulfates; for example, $(NH_4)_2SO_4$. Over continental regions, the additional amounts of aerosols from air pollution are usually present in much greater concentrations than the natural sea salt aerosol.

We should mention several other natural aerosols, although they are generally more important from the standpoint of local effects than their global influence. Pollens and spores are an example, the former causing discomfort to numerous individuals with specific allergies. Pollens may also have a debilitating effect on asthmatics. Organic molecules emitted by vegetation such as coniferous trees are another example. The air over some forests may have such high concentrations that the molecules in the presence of sunlight photochemically react to generate clouds of aerosols with a particle diameter generally less than 0.2 micron. Such small particles scatter blue light effectively and produce a well-known haze, as is evident over the forests in southern United States, the jungles of Central America, and the Amazon basin.

Anthropogenic sources

Many ways by which man releases aerosols are well known. Rock-crushing, trash incineration, and open burning of field refuse are examples. Also important are the emission of sulfur oxides, particularly SO_2 from smelters, power plants, and heating furnaces. Both SO_2 and hydrogen sulfide (H_2S) from decaying vegetation are eventually oxidized into sulfates, perhaps with the participation of gaseous NH_3 and water vapor. Another anthropogenic source is photochemical smog which produces a fine aerosol as a secondary pollutant. Because of the extensive nature of the sources, large geographical areas may be affected as the aerosol drifts with the wind.

The residence time of aerosols in the troposphere is three to five days, if emitted at low levels, much shorter than the 30 days residence in the upper troposphere or the two or three years in the stratosphere. Thus the lower troposphere is responsive to changing emissions and cleansing by natural processes of scavenging; however, the residence time in this region is sufficiently long that aerosol concentrations can build up to markedly pronounced levels as primary pollutants are added to incoming air that has already been polluted by upwind communities. An illustration of this point is provided by the hypothetical example in Fig. 4.15. We imagine that as air moves eastward across the United States it gathers a burden of pollutants when traversing each large city; subsequently, natural scavenging processes reduce this burden. But if time is not sufficient for total cleansing, the city downwind will not receive clean air and, on the average, the particulate concentration increases. This is perhaps most apparent near the industrialized east coast. In Chapter 11 we shall examine the characteristics and dynamics of aerosols in more detail; and for the moment we shall limit our attention to the effect they may have on the energy balance.

Turbidity

A commonly used measure of the amount of particulate matter in the air is a parameter known as the *turbidity*. Measurements of the turbidity for example enabled ground-based observers to follow the movement of the ejecta from the Mt. Agung eruption as it circulated about the world. Turbidity, as the name

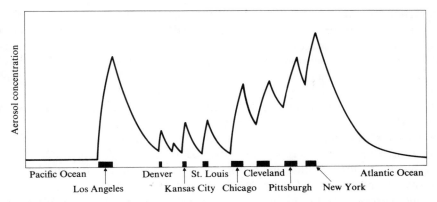

Fig. 4.15 Build-up of aerosol concentration as an air mass is imagined to make a journey across the United States passing over the indicated cities. (After M. Neiburger, *Bulletin of the American Meteorological Society*, **50**, 957, 1969.)

suggests, is related to the amount of radiant energy removed from a beam of light through scattering or absorption as it traverses the atmosphere. In a scientific sense, it has several possible definitions. One common and practical definition is based upon the intensity of solar radiation observed when an appropriate measuring device is pointed at the sun; this device receives only the radiation which is incident from the direction of the center of the sun, called the *direct radiation*. As the ambient concentration of aerosols in the beam path increases, the amount of light scattered away from the forward direction increases, and thus the received intensity decreases. A comparison of the received intensity with the intensity measured on a clear day is a measure of the turbidity and therefore is an indication of the atmospheric burden of aerosols.

The effect aerosols have on the measured direct radiation can be derived by a technique similar to the one we used in Section 2.3. We consider the change in light intensity ΔI of solar radiation as the solar beam traverses a short pathlength Δz of the atmosphere in the vertical direction. A fraction α_a of the intensity will be Rayleigh scattered or absorbed by dry atmospheric gases; a fraction α_w, by water vapor; and a fraction α_p, by aerosols. (The measurement is assumed to be made on a cloudless day.) We can neglect the small amount of radiation incident from the side which is scattered toward the instrument, because the direct radiation is so much more intense. The change in the intensity of the direct beam upon traversing a distance Δz is thus

$$\Delta I(z) = - \left[\alpha_a + \alpha_w + \alpha_p \right] I(z) \, \Delta z. \qquad (4.13)$$

If α_a, α_w, and α_p are independent of altitude z, comparison with Eqs. (2.8) and (2.9) indicates that the solution is a decreasing exponential. The solution of Eq. (4.13) thus has the form

$$I(0) = I_0 e^{-(\alpha_a + \alpha_w + \alpha_p)mL}, \qquad (4.14)$$

where $I(0)$ is the intensity of radiation received at ground level ($z = 0$). The parameter I_0 is the intensity of the incoming radiation in the outer atmosphere, and (mL) is the effective path length through the atmosphere along which the light travels before reaching the ground. The exponential decrease of intensity with path length is commonly known as *Beer's law*, although strictly speaking this name should be applied only to the attenuation of monochromatic light. If we let L be the effective vertical path length through the atmosphere, then it is a straightforward application of geometry to generalize Eq. (4.14) to account for almost any solar elevation above the horizon by defining $m = \operatorname{cosec}(\phi)$, where ϕ is the solar elevation above the horizon, expressed in angular degrees. This is a good approximation so long as ϕ is restricted to the angular interval $10° \ll \phi \ll 90°$; that is, so long as the sun is not too near the horizon. With m defined in this way, the products $(\alpha_a L)$, etc. are dimensionless numbers.

The expression in Eq. (4.14) is deceptively simple as written. This is because it was derived under the assumption that the concentration of aerosols and other scattering and absorbing constituents of the atmosphere are independent of altitude. In general, this is not correct, for a greater concentration of aerosols is found near the ground. Thus in most practical cases, a measurement which yields the parameters $\alpha_a L$, $\alpha_w L$, and $\alpha_p L$ gives us only an empirical determination of the total scattering and absorption; it cannot be directly interpreted as yielding an extinction coefficient α_p, etc. For this reason, a common definition of turbidity is simply the empirical parameter $\alpha_p L$. The value of the atmospheric turbidity can be evaluated from measurements of the direct radiation if $\alpha_a L$ and $\alpha_w L$ are known from other measurements or are calculated from theory.

But another definition of turbidity is also common. It is given in terms of the relation

$$I(0) = I_0 \, 10^{-(\tau_r + \tau_z + B)m} \tag{4.15}$$

as described by F. Volz.[28] This equation is equivalent to Eq. (4.14) but the parameters in the exponent represent a different grouping of the mechanisms which cause an attenuation of the direct radiation. The fact that Eq. (4.15) is written with a decade base for the exponents rather than the natural base of Eq. (4.14) is merely a computational convenience; it does not imply that any new physical effects are envisioned. The parameter τ_r represents the effect of Rayleigh scattering by pure air, and τ_z, the effect of absorption by ozone in the stratosphere and mesosphere. The parameter B is called the turbidity, and represents scattering and absorption from both aerosols and water droplets. Both $I(0)$ and I_0 should be adjusted to represent the respective quantities for the mean distance between the earth and sun. The value of τ_r is calculated to be 0.0634, and τ_z is 0.004.[28] Over most of the United States the turbidity is found to be 0.2 in the summer and about 0.1 in winter. Thus the turbidity factor B can be by far the most important one in the exponent of Eq. (4.15). The minimum turbidity over the United States on fairly clear days is considerably lower, about 0.02 to 0.03,[29] corresponding to an aerosol burden of about 10 $\mu g/m^3$. By comparison, the cleanest air from remote oceanic region has been found to have a turbidity of only

0.003, in which case most of the observed light attenuation is due to the gaseous components of the air.[30]

The parameters of Eq. (4.15) can be related to those of Eq. (4.14) by taking the logarithm (to the base 10) of both equations and setting the right-hand sides equal to each other. Assuming that the effect of ozone can safely be neglected, we find the relationship

$$B = (\alpha_w + \alpha_p)L \log_{10}e$$

Measurements of turbidity indicate that tropical air masses from maritime regions are generally more turbid than polar air masses from continental regions. This suggests that high humidity contributes to high turbidity, probably because of aerosol formation through the nucleation of water droplets on sea salt and other hygroscopic aerosols.[29]

A distinction should be made at this point concerning turbidity and a parameter which is more relevant to the energy balance of the earth's surface, the *total radiation*. Whereas turbidity refers to *direct radiation* along the line of sight from the sun, the total radiation is the amount which is received on a square area of horizontal surface without regard to the direction of incidence. In turbid atmospheres, a greater proportion of the radiant energy incident upon a horizontal surface such as the ground will be diffuse radiation; that is, scattered radiation incident from a direction other than the line of sight from the sun. The distinction is important, for a given reduction in the direct intensity as indicated

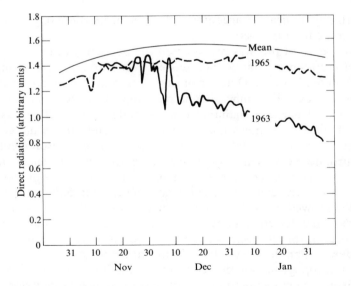

Fig. 4.16 Direct solar radiation on Amundsen-Scott station, Antarctica. The mean curve represents data for the period 1957 through 1962, and the rapid decrease beginning in November, 1963, is attributed to increased turbidity from the Mt. Agung debris. By 1965 the turbidity had nearly returned to the long-term mean. (H. J. Viebrock and E. C. Flowers, *Tellus*, **XX**, 400, 1968.)

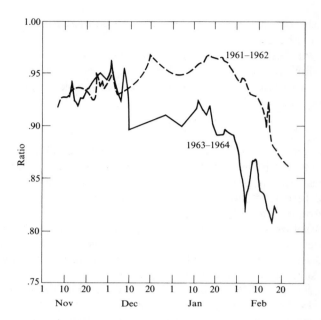

Fig. 4.17 Ratio of the observed total solar radiation on a horizontal surface to the theoretical values for Rayleigh scattering from atmospheric gases, as measured at Amundsen-Scott station, Antarctica. (H. J. Viebrock and E. C. Flowers, *Tellus*, **XX**, 400, 1968.)

by the turbidity is usually accompanied by a proportionately much smaller reduction in the total intensity.

An illustration of this was provided on the occasions when debris from the Mt. Agung eruption of March 1963 passed over turbidity monitoring stations across the United States in October of that year, and also by a circuitous route reached Antarctica in December. Numerous supplementary experiments showed that the cloud of particles maintained an altitude of about 20 km as it circled the globe. Figure 4.16 shows the effect of that cloud on the direct radiation received at a monitoring station in Antarctica.[31] The measurements showed that the turbidity remained anomalously high for several years. During late November and early December, the intensity of direct radiation decreased by about 30%; however, separate measurements of the total radiation, as illustrated in Fig. 4.17, showed a much smaller reduction of only 5%. It is of course the total radiation that determines how much solar energy reaches the earth's surface and therefore, in this respect, is the more relevant parameter for gauging the effects of aerosols on the energy balance of the earth.

Since anthropogenic sources contribute only 10% of the global aerosols, fluctuations in the emissions from natural sources—such as occasional volcanic eruptions or extensive forest fires—can far outweigh in contributions and effects what man's emissions apparently have done. Nevertheless, anthropogenic aerosols are not a negligible fraction of the total and are steadily increasing.

As long ago as 1784, Benjamin Franklin suggested that atmospheric dust may affect the temperature of the earth, and he may not have been the first.[32] Little, however, could be done at that time to estimate the importance of the effect. Unfortunately the current lack of information about the type and distribution of aerosols still does not permit a quantitative assessment of their influence on the energy balance of the earth and its atmosphere. An increase in the albedo due to increased light scattering is one possibility. Estimates for the temperature drop which might result from a violent volcanic eruption, range from 0.2°C to over 1°C, for various portions of the globe.[33] Indeed, the effects are believed by some to have been even more pronounced: Humphreys suggests that volcanism has been so significant as to change the earth's climate.[32]

But particulate matter may also absorb some of the incident solar radiation and thereby contribute to heating of the atmosphere.[23] How much the primary and secondary pollutants absorb is unknown, again because of our current ignorance of the optical properties of the aerosols themselves and of their global distribution. At this time, one can only speculate about the possible effects. It is not yet clear whether global cooling or heating would result from substantial increases in the atmospheric concentration of fine particulate matter.

4.9 LATITUDINAL ENERGY BALANCE

From geometrical considerations, it is clear that the tropical portions of the globe receive more incident solar radiation per unit surface area than polar regions (Fig. 4.2). However, a higher mean surface temperature also prevails in the equatorial latitudes, so the thermal radiation from the surface is also greater [according to the Stefan-Boltzmann law in Eq. (4.2)]. The question arises as to whether there is a radiation energy balance between incoming solar radiation and outgoing terrestrial radiation at the equatorial troposphere, or whether there may be a net absorption or loss of radiant energy. Either possibility could be consistent with the energy balance illustrated in Fig. (4.8), for the numbers in this figure are for global averages and do not apply to every locality.

An answer to this question is provided by the results of calculations by H. G. Houghton and others, as well as by recent radiation measurements by instrumented satellites.[7,9] It is found that the higher temperature in the tropics is accompanied by a sufficiently large enhancement of the absolute humidity that the greenhouse effect more than compensates for the greater thermal radiation emitted from the warmer surface of equatorial latitudes. This is an example on a local scale of the self-amplification effect which we discussed in Section 4.6. Thus there is a net absorption of radiant energy by tropical regions of the earth and the overlying atmosphere.

On the other hand, the cold polar regions are found to have the opposite imbalance: The low moisture content in the atmosphere permits a greater fraction of the earth's thermal radiation to escape into outer space, thus producing a net outflow of energy. This is schematically illustrated in Fig. 4.18. A more

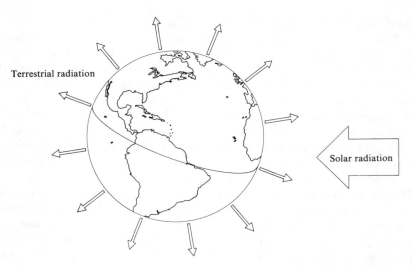

Fig. 4.18 Latitudinal imbalance of incoming and outgoing radiation. Tropical regions of the earth and atmosphere absorb a greater amount of energy than they radiate, whereas the polar regions have the opposite imbalance.

quantitative comparison is provided by recent measurements by satellites of the earth's albedo, and intensity of emitted infrared radiation. From these, the latitudinal imbalance between incoming and outgoing radiant energy can be

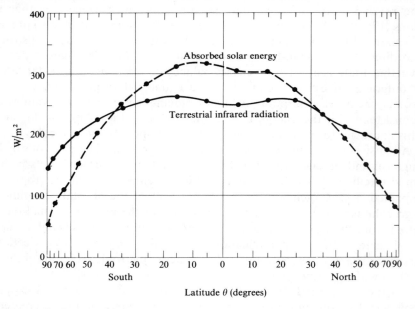

Fig. 4.19 Latitudinal imbalance of radiant energy absorbed and radiated by the earth, as obtained from measurements by satellite. (T. H. Vonder Haar and V. E. Suomi, *Journal of the Atmospheric Sciences*, **28**, 305, 1971; courtesy of the American Meteorological Society.)

deduced. The two curves in Fig. 4.19 illustrate the results of a five-year average of satellite data. The horizontal scale in this figure is arranged to convey an impression of the relative amount of surface area of the earth which lies between a given latitude θ and a nearby latitude $\theta + \Delta\theta$. This is done by fixing the scale so that the distance along the axis between the index marks for θ and $\theta + \Delta\theta$ is proportional to $\cos\theta$.

To achieve an overall energy balance, there must be a transport of energy from tropical to polar regions by means other than radiation. The mechanism is, in fact, largely provided by atmospheric motion, and the accompanying convection and advection of energy in both sensible and latent form, from equatorial to polar regions. The rate of energy flow attains a maximum of about 4.6×10^{15} watts (10^{20} calories per day) crossing at the 40° latitude. Ocean currents also contribute to the northward flow of energy, but their contribution in the mid-latitudes accounts for only about 20% of the total, and above 60° latitude less than 10%.[33] Thus large-scale energy conveyance is a driving influence for global circulation of the atmosphere; this is fortunate, for one of the resulting benefits is the widespread dispersal of pollutants.

We have seen in the previous sections how important are the mechanisms of thermal radiation and atmospheric absorption for the energy balance of the earth. The minor constituents—water vapor, water, and carbon dioxide—play the dominant role, whereas the major constituents—nitrogen and oxygen—have essentially no effect. The reason is to be found in terms of the shape of the molecules and the fact that absorption of radiant energy by a molecule is a quantum phenomenon. The interested reader will find the explanation for this in Appendix E.

SUMMARY

The sun is the main source of energy which causes atmospheric circulation. The spectrum of incident solar radiation is centered in the visual band of the electromagnetic spectrum and is transmitted with little absorptive loss through the atmosphere. Ozone and oxygen in the upper atmosphere, however, absorb the incident ultraviolet radiation, thus providing a protective shield against these harmful rays. Only about 34% of the solar energy incident on the troposphere is reflected into space by the atmosphere and surface of the earth; about 47% is absorbed by the earth, and the balance is absorbed by the atmosphere. The earth's surface frees some energy through the evaporation of water and conduction of heat to the adjacent air; but by far more important is the energy it radiates in the infrared portion of the spectrum. The fact that water, water vapor, and carbon dioxide absorb virtually all of the earth's radiation gives rise to the "greenhouse effect" by which thermal radiation from the warmed atmosphere transmits a large portion of the energy back to the earth. The net result is that the equilibrium surface temperature is higher than it would be if all the earth's thermal radiation were lost to outer space. The earth's energy balance and its mean surface temperature are sensitive to the albedo of the planet (largely determined by the amount of cloud cover) and the water vapor content of the atmosphere; the balance is also affected by the carbon dioxide concentration and aerosol burden, but in a less sensitive way. Both of the latter factors may be influenced by global aspects of air pollution.

NOTES

1. M. Planck, *Annalen der Physik*, **4**, 553 (1901); English translation appears in D. ter Haar, *Old Quantum Theory*, Oxford: Pergamon, 1967, paperback.

2. E. Raschke and W. R. Bandeen, "The Radiation Balance at the Planet Earth from Radiation Measurements at the Satellite Nimbus II," *J. Appl. Meteorol.*, **9**, 215 (1970).

3. F. S. Johnson, "The Solar Constant," *J. Meteorol.*, **11**, 431 (1954).

4. R. A. Craig, *The Upper Atmosphere; Meteorology and Physics*, New York: Academic Press, 1965.

5. H. C. Urey, "The Atmosphere of the Planets," *Handbuch der Physik*, Vol. 52, Berlin: Springer-Verlag, 1959, p. 363.

6. R. W. Longley, *Elements of Meteorology*, New York: Wiley, 1970.

7. K. Ya. Kondrat'yev, *Radiative Heat Exchange in the Atmosphere*, Oxford: Pergamon Press, 1965.

8. R. G. Fleagle and J. A. Businger, *An Introduction to Atmospheric Physics*, New York: Academic Press, 1963.

9. H. G. Houghton, "On the Annual Heat Balance of the Northern Hemisphere," *J. Meteorol.*, **11**, 1 (1954).

10. K. Ya. Kondrat'yev, *Radiation in the Atmosphere*, New York: Academic Press, 1969.

11. G. J. MacDonald, "Energy and the Environment," Forum on Energy, Economic Growth and the Environment, 1971.

12. H. Johnston, "Reduction of Stratospheric Ozone by Nitrogen Oxide Catalysts from Supersonic Transport Exhaust," *Science*, **173**, 517 (1971).

13. S. Manabe and R. T. Wetherald, "Thermal Equilibrium of the Atmosphere with a Given Distribution of Relative Humidity," *J. Atmos. Sci.* **24**, 241 (1967).

14. F. Möller, "On the Influence of Changes in the CO_2 Concentration in Air on the Radiation Balance of the Earth's Surface and on the Climate," *J. Geophys. Res.*, **68**, 3877 (1963).

15. S. Manabe, J. Smagorinsky, and R. F. Strickler, "Simulated Climotology of General Circulation with a Hydrologic Cycle," *Mon. Wea. Rev.*, **93**, 769 (1965).

16. H. Harrison, "Stratospheric Ozone with Added Water Vapor: Influence of High-Altitude Aircraft," *Science*, **170**, 734 (1970).

17. E. Robinson and R. C. Robbins, "Sources, Abundance, and Fate of Gaseous Atmospheric Pollutants Supplement," Stanford Res. Inst. Rep., June, 1969.

18. R. Revelle and H. E. Suess, "Carbon Dioxide Exchange Between Atmosphere and Ocean and the Question of an Increase of Atmospheric CO_2 During the Past Decades," *Tellus*, **9**, 18 (1957).

19. J. M. Mitchell, Jr., "A Preliminary Evaluation of Atmospheric Pollution as a Cause of the Global Temperature Fluctuation of the Past Century," in *Global Effects of Environmental Pollution*, S. F. Singer, Ed., New York: Springer-Verlag, 1970.

20. R. Spirtas and H. J. Levin, "Characteristics of Particulate Patterns 1957–1966," U.S. Dept. of Health, Education and Welfare, NAPCA Publication AP-51, March 1970.

21. G. M. Hidy and J. R. Brock, "An Assessment of the Global Sources of Tropospheric Aerosols," 2nd Clean Air Congress, IUAPPA, Washington, D.C., Dec. 1970.

22. R. A. McCormick and J. H. Ludwig, "Climate Modification by Atmosperic Aerosols," *Science*, **156**, 1358 (1967).

23. R. J. Charlson and M. J. Pilat, "Climate: The Influence of Aerosols," *J. Appl. Meteorol.*, **8**, 1001 (1969); also S. I. Rasool and S. H. Schneider, "Atmospheric Carbon Dioxide and Aerosols: Effects of Large Increases on Global Climate," *Science*, **173**, 138 (1971); and P. F. Gast, *Science*, **173**, 982 (1971).

24. R. W. Rex and E. D. Goldberg, "Quartz Contents of Pelagic Sediments of the Pacific Ocean," *Tellus*, **10**, 153 (1958).

25. N. Bhandari, J. R. Arnold, and D. Parkin, "Cosmic Dust in the Stratosphere," *J. Geophys. Res.*, **73**, 1837 (1968).

26. H. Neuberger, *Mech. Eng.*, **70**, (3) 221 (1948).

27. Royal Society of London: "Eruption of Krakatoa and Subsequent Phenomena." London, 1888.

28. F. Volz, "Photometer mit Selen-Photoelement zur spektralen Messung der Sonnenstrahlung und zur Bestimmung der Wellenlängen-abhängigkeit der Dunsttrübung," *Arch. Meteor. Geophys. Bioklim.*, **B10**, 100 (1959).

29. E. C. Flowers, R. A. McCormick, and K. R. Kurfis, "Atmospheric Turbidity over the United States, 1961–1966," *J. Appl. Meteorol.*, **8**, 955 (1969).

30. W. M. Porch, R. J. Charlson, and L. R. Radke, "Atmospheric Aerosol: Does a Background Level Exist?" *Science*, **170**, 315 (1970).

31. H. J. Viebrock and E. C. Flowers, "Comments on the Recent Decrease in Solar Radiation at the South Pole," *Tellus*, **20**, 400 (1968).

32. W. J. Humphreys, *Physics of the Air*, New York: McGraw-Hill, 1940.

33. J. M. Mitchell, Jr., "Recent Secular Changes of Global Temperature," *Annals N.Y. Acad. Sci*, **95**, 235 (1961).

34. E. Palmen and C. W. Newton, *Atmospheric Circulation Systems*, New York: Academic Press, 1969.

FOR FURTHER READING

H. ZIRIN   *The Solar Atmosphere*, Waltham, Mass.: Blaisdell, 1966.

N. ROBINSON   *Solar Radiation*, Amsterdam: Elsevier, 1966.

R. M. GOODY   *Atmospheric Radiation*, London: Oxford Univ. Press, 1964.

K. YA. KONDRAT'YEV   *Radiative Heat Exchange in the Atmosphere*, Oxford: Pergamon Press, 1965; *Radiation in the Atmosphere*, New York: Academic Press, 1969.

W. D. SELLERS   *Physical Climatology*, Chicago: Univ. of Chicago Press, 1965.

J. M. MITCHELL, Ed   *"Causes of Climatic Change," Meteorological Monographs*, **8**, No. 30, Boston: American Meteorological Society, 1968.

S. F. SINGER, Ed   *Global Effects of Environmental Pollution*, New York: Springer-Verlag, 1970.

"INADVERTENT   Report of the Study of Man's Impact on Climate (Cam-
CLIMATE MODIFICATION"   bridge, Mass.: M.I.T. Press, 1972).

R. U. MUNN and   "Global   Air   Pollution—Meteorological   Aspects;   A
B. BOLIN   Survey," *Atmos. Environ.* **5**, 363 (1971).

QUESTIONS

1. Why do all objects radiate energy?

2. Why does a surface which absorbs all incident radiation have a unique spectrum for the energy which it radiates? Upon what parameters of the material forming the the surface does this spectrum depend?

3. Why are nighttime temperatures on earth generally lower when the humidity is low than when it is high? Why do desert regions usually experience wide differences between daytime and nighttime temperatures?

4. If you were to construct a scale model of the earth starting with a globe 1 m in diameter, how far would each of the following extend above or below the surface?
   a) The Trieste depth
   b) Mt. Everest
   c) The level below which 99% of the atmosphere is found
   d) The tropopause
   e) The stratopause

5. The radiation balance of the earth for an albedo of 0.34 indicates what the mean radiation temperature of the earth should be. Which layer of the atmosphere has this temperature? Is the altitude of this radiating layer uniquely determined?

6. What reactions are responsible for the temperature profile in the stratosphere? What correspondence is there between this profile and the ozone concentration profile?

7. By what factor does the rate of emission of thermal radiation by the earth's surface exceed its rate of absorption of solar radiation?

8. The oceans have absorbed great quantities of carbon dioxide since their origin. How may the CO_2 content affect temperature changes of the surface of the earth?

9. What is meant by the "self-amplification" factor in reference to the effect that humidity variations have on the mean surface temperature of the earth?

10. What atmospheric constituent plays the most important role in the greenhouse effect, and in what ways does it influence the surface temperature?

11. Suppose that you wished to measure the black body temperature of the earth's surface from a high-altitude satellite. Considering the absorption spectrum of the atmosphere, which wavelength band for your measurements would give the best results?

PROBLEMS

1. What fraction of the total solar radiant energy does the earth intercept?

2. If the mean distance between the sun and moon is 1.5×10^8 km, what should be the average temperature of this body as indicated by an energy balance of radiant energy? Assume that the albedo of the moon is 0.07 and the diameter 3476 km. At what wavelength is the peak of the moon's thermal radiation located?

3. The thermal radiation emitted into space from the earth cannot truly have a black body spectrum, because some is radiated by the surface of the planet and some by the atmosphere, and the temperatures of the two differ. In which portion of the electromagnetic spectrum might you expect to find the greatest deviation from a 255°K black body spectrum were you to observe the earth from a high-altitude satellite?

4. If a field has an albedo of 0.05 and is assumed to be a gray body, what is the proportion of reflected energy to thermal radiated energy per unit area of surface if the temperature of the ground is 290°K? (See Appendix D.)

5. Suppose that the earth were overlaid at an altitude of 20 km by an atmosphere of negligible thickness that passes solar radiation but totally absorbs the thermal radiation of the earth. What is the mean temperature of the surface of the earth if it completely absorbs all incident radiation?

6. Suppose for a simple model that we imagine the fraction of terrestrial thermal radiation escaping to outer space is given by a formula of the form

$$\text{fraction} = \exp(-\beta n),$$

where n is the concentration of water vapor in the atmosphere and β is a constant. If the fraction is 0.05 for a mean surface temperature of 285°K, what is it for a mean temperature of 290°K, assuming (a) a fixed relative humidity and (b) a fixed absolute humidity? Assume that the proportional change in water content is uniform throughout the atmosphere and is equal to the change at the surface of the earth.

7. By how much is the intensity of solar radiation reaching the ground reduced if the turbidity of the atmosphere is increased from $B = 0.025$ to $B = 0.25$? Calculate the fractional reduction when (a) the sun is directly overhead and (b) the sun is 30° from the zenith.

8. The turbidity of the atmosphere over cities is often found to be related to reduced visibility in the lower atmosphere. If the turbidity B of Eq. (4.15) is found to increase by a factor of 2 during a day when pollutants accumulate uniformly throughout the atmosphere, by how much would you expect the prevailing visibility to be changed?

9. Estimate the rate at which a human body radiates energy, assuming that its absorption coefficient is unity. Would you expect that the actual rate would be higher or lower than this estimate?

10. About 6×10^{15} liters of water are evaporated annually from the earth's surface into the atmosphere. How much energy is absorbed each second for this purpose? About what percentage of the total solar energy absorbed by the earth does this represent?

11. Show that about 40% of the CO_2 emitted from anthropogenic sources remains in the atmosphere, if the CO_2 concentration of ambient air is observed to increase at an annual rate of 0.2%.

CHAPTER 5

MACROSCALE MOTION

A glance at the daily weather map of a newspaper is clear evidence that the circulation of air over the globe forms complex patterns. The molecules in any volume of air have performed many functions and traveled long distances during the past year. At one time they may have been transporting dust from the Sahara desert eastward across the Atlantic Ocean, or participating in a Pacific typhoon, or carrying away exhaust from the combustion chamber of a coal-fired power plant, or propelling a sail boat on Lake Geneva.

Although air circulation on a local scale is complex, certain average patterns are evident. The wind at any place can be considered as a composite of several individual scales of motion, each of which is caused somewhat differently. The smallest scale which comprises flow patterns whose characteristic dimensions range from a few millimeters to a few kilometers is known as the *microscale*. Motion is primarily influenced by terrain roughness and variation in temperature. Irregular small-scale motion is often called turbulence. An intermediate scale, the *mesoscale*, comprises motion whose characteristic dimensions are from 1 to 100 km. Terrain and temperature variations are also important, but flow patterns are more regular and temporal variations slower than is the case for microscale effects. Circulation may form closed flow patterns or convective cells, such as occur in a sea breeze. The largest scale of air motion, with dimensions of hundreds and thousands of kilometers, comprises what is known as *macroscale* motion. Patterns of this motion may persist for days or even months and are influenced by large features of the earth's surface (oceans, continents, and high mountain ranges) and large-scale variations in temperature (such as those caused by the latitudinal dependence of the radiation energy balance).

Macroscale motion is concerned with the movement of large air masses. These are relatively homogeneous volumes of air which have been conditioned over a period of days by processes of radiation, convection, evaporation, and condensation so that they acquire characteristics of a certain region of the world. Unfortunately, meteorological processes are far too complicated for a complete mathematical description of the motion of air masses. In addition, there is a serious lack of quantitative information concerning such features as the variation of surface temperature over the globe to enable detailed day-to-day analysis, even if the calculations were possible. Nevertheless, the major features of macroscale motion can be understood on the basis of fundamental thermal and mechanical effects. Application of a few simple principles is particularly successful for descriptions of average circulation patterns. Since such macroscale motion has profound influence on problems of local ventilation and episodes of air pollution, we shall now consider its more relevant aspects.

5.1 GENERAL CIRCULATION

The average patterns of air movement over the entire globe is called the "general circulation." It is determined by averaging wind observations over long periods, perhaps decades or longer. Seasonal variations of the general circulation may be obtained by averaging over a number of years the behavior during a given season. This average flow must provide a mechanism for transporting energy from the tropics toward polar regions to maintain the overall balance of the earth. At no instant is the actual circulation as simple as is portrayed by the average trends; however, many important characteristics of climate and regional weather are predicted by features of the general circulation.

The simplest pattern of global circulation which conveys sensible energy from equatorial to polar regions would be a simple convective cell in each hemisphere, as first proposed by Sir George Hadley in 1735. Figure 5.1 is a partial cross section through the earth showing the two Hadley cells, each of which has simple flow along the lines of longitude. We can most simply consider the dynamics

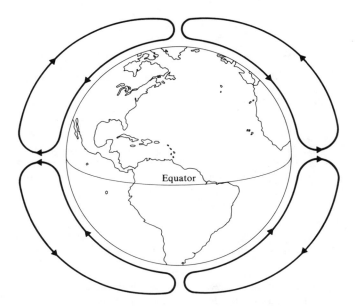

Fig. 5.1 Hadley's suggestion for the general circulation.

of this circulation by first focusing our attention on the tropics. Here the intense sunlight warms the ground and ocean surfaces, and the adjacent air with which they are in contact. For the moment we shall rely on the reader's knowledge that a volume of air which is warmer than the surrounding air at the same level will rise as a result of its buoyancy, the cooler air flowing underneath to take the place of the warmer. This process occurs at the equator of the earth where the warm air rises and is replaced at low levels by air which flows in from temperate regions. According to Hadley's hypothesis, once the warm air had reached a high elevation, it would necessarily flow toward the poles and convey sensible heat from the equatorial regions. When the air had lost this heat at the poles through radiative cooling and convection to the ground, it would subside and commence flow toward the equator to complete the convective cell.

The flow envisioned by Hadley involves purely longitudinal motion. Unfortunately, however simple and appealing such an atmospheric model may be, it

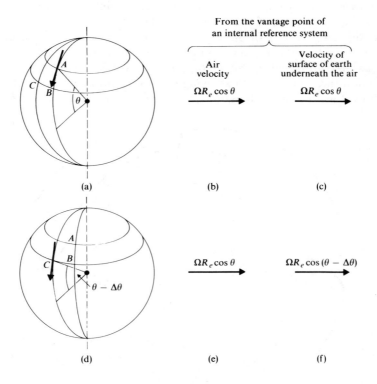

Fig. 5.2 (a) Air moving initially southward at a latitude θ with a speed v_S with respect to the earth; (b) the eastward velocity of the air with respect to a fixed inertial system; (c) the velocity of a point on the surface of the earth directly under the air. Parts (d), (e), (f) are the same conditions viewed a short time later after the air has moved to a new latitude $\theta - \Delta\theta$.

cannot correspond to the actual circulation of the atmosphere. There are several reasons for this; a key one is a consequence of the rotation of the earth—*the Coriolis force.*

5.2 CORIOLIS FORCE

The earth's rotation has a profound effect on macroscale patterns of air currents. This is because our reference system, based on the features on the surface of the globe, does not comprise an inertial reference system; because of the earth's rotation, each of these features is accelerating inward toward the axis of rotation. One consequence of using an accelerating reference system, as we noted in Section 3.3, is the appearance of inertial forces which appear to influence the way air moves. These forces account for the tendency of any object to continue its present motion in a straight line without acceleration. Such straight line motion appears quite different when viewed from an accelerating reference system: The object itself appears to be accelerating, but in the opposite direction, and thus it appears to be under the influence of a force. We call this apparent force an "inertial" force. The centrifugal force is an inertial force which depends only upon the position of the object on the globe. In contrast with this, the Coriolis force is an inertial force that affects only objects which are *moving* over the globe.

In 1835, Gaspard Gustave de Coriolis developed the equations of motion for fluid motion on a rotating earth and uncovered the inertial force that bears his name. Some twenty years later, William Ferrel applied the concept to predict consequences of Hadley's hypothesis for the general circulation. To see how the circulation is affected, let us consider a simple case illustrated in Fig. 5.2. We suppose that wind at point A, at a latitude θ, is directed southward toward point B with a velocity v_S with respect to the earth. It is also useful to regard the situation with respect to an inertial reference system, one which is fixed with respect to the stars. With negligible error, we can neglect the effect of the earth's motion about the sun and assume that the earth does not move with respect to the inertial system; the earth only rotates with the angular velocity Ω. Thus both the air at point A and the surface of the earth under the air are moving eastward with the velocity $v_E = R_e\Omega \cos\theta$ with respect to the inertial system, as illustrated in parts (b) and (c) of Fig. 5.2.

During a short time interval Δt, the air originally at A will have moved southward to a lower latitude $\theta - \Delta\theta$, where $v_S\Delta t = R_e\Delta\theta$ is the distance toward the equator that the wind has traveled. Because of inertia, the wind maintains its same eastward velocity v_E in the inertial system. However, the earth's surface at latitude $\theta - \Delta\theta$ is rotating faster in the inertial system, at a speed $v_E' = R_e\Omega \cos(\theta - \Delta\theta)$. *The wind moving toward the equator therefore lags behind the earth as the air moves toward lower latitudes.* Thus an earth-bound observer would conclude that during the interval Δt the wind had acquired a *westward* component of velocity $v_E' - v_E$.

This is not the complete picture, however, because the southward component of the original velocity v_S after the interval Δt is no longer directed southward when

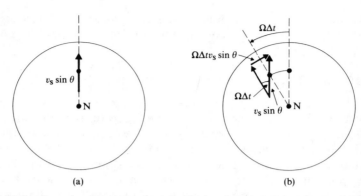

Fig. 5.3 (a) Southward component of wind velocity viewed directly over the North Pole. The apparent length of the velocity vector is $v_s \sin \theta$ in such a projection. (b) Because of inertia, the wind continues in a straight line as the earth turns by an angle $\Omega \Delta t$ beneath, and thus the new westward component of velocity seen from the rotated earth is $v_s \Omega \Delta t \sin \theta$.

seen from the earth. It, too, contributes a component to the westward motion as viewed from the earth. Consider Fig. 5.3 which shows the situation when viewed from a vantage point on the inertial system overhead of the North Pole. As the earth turns by an angle $\Omega \Delta t$, v_S continues to point in its original direction because of inertia; and consequently an earth-bound observer would conclude from this effect alone that the wind had acquired a westward velocity $v_s \Omega \Delta t \sin \theta$.

The total westward component of velocity which the wind has relative to an earth-bound observer after the interval Δt is thus $v'_E - v_E + v_s \Omega \Delta t \sin \theta$. The westward *acceleration* of the wind, as viewed from the earth, is therefore $(v'_E - v_E)/\Delta t + v_s \Omega \sin \theta$. This acceleration, which we denote a_C, can be expressed in terms of Ω, v_S, and θ by the following algebraic development:

$$a_C = (v'_E - v_E)/\Delta t + v_s \Omega \sin \theta$$

$$= \frac{R_e \Omega}{\Delta t}[\cos(\theta - \Delta \theta) - \cos \theta] + v_s \Omega \sin \theta$$

$$= \frac{v_s \Omega}{\Delta \theta}[\cos \theta \cos \Delta \theta + \sin \theta \sin \Delta \theta - \cos \theta] + v_s \Omega \sin \theta.$$

We are interested in small intervals of time Δt, corresponding to small angles $\Delta \theta$, so we can make the approximations $\cos \Delta \theta = 1$ and $\sin \Delta \theta = \Delta \theta$. Then we find:

$$a_C = 2v_s \Omega \sin \theta. \tag{5.1}$$

This is the Coriolis acceleration; for an air mass moving toward the equator the acceleration is directed toward the west. When multiplied by the mass of the object which it affects, Eq. (5.1) gives the Coriolis force. From our derivation, we see that it arises simply because the air lags behind the earth's rotation, and

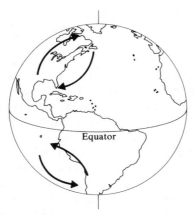

Fig. 5.4 Deviation of winds in the Northern and Southern Hemispheres is a result of the Coriolis force.

the equatorward component of velocity appears to veer westward, when the wind is initially directed toward the equator. Similarly, a wind directed toward the pole would soon find itself advancing ahead of the rotating globe and would therefore acquire an eastward component of velocity relative to the earthbound reference system.

In Appendix F we show that Eq. (5.1) is also valid for *any* wind direction (not just motion directed toward the equator), in which case v_S is the magnitude of the total wind velocity, or the wind speed, which we denote by the symbol v:

$$a_C = 2v\Omega \sin \theta. \tag{5.2}$$

The Coriolis acceleration is always directed perpendicular to the wind velocity in the sense that an observer with his back to the wind will see it veer toward the right (in the Northern Hemisphere) or to the left (in the Southern Hemisphere). This is illustrated in Fig. 5.4. As Eq. (5.2) indicates, there is no Coriolis acceleration at the equator ($\theta = 0$).

In the preceding discussion, we neglected the consequences of friction between the moving air and the earth's surface. Any frictional forces must be added to the Coriolis force to predict the dynamic response of air masses. Friction has the effect of pulling the air along with the earth as it rotates. It most strongly influences the lowest layer of air, but has little effect on the air at high altitudes.

The Coriolis force is a weak force, so its effects are only observed in large-scale wind patterns, where over a long time the small acceleration can produce an appreciable change in wind velocity. As an example, a wind of 10 km/hr at a latitude of 40° will experience a transverse acceleration of 2.6×10^{-4} m/sec^2, corresponding to a circular orbit with a radius of 30 km. Thus it would require 2.4 hours of movement before the wind velocity would be rotated by 90 degrees. Although this illustration is useful to demonstrate the scale of the effect which the Coriolis force could produce, such circular wind trajectories are not observed because other forces due to nonuniform pressures also influence the motion of air.

Several effects

It should be remarked that some ocean currents and not only the wind are observed to respond to the Coriolis force. The clockwise circulation of the Gulf Stream and the Japanese current in the Northern Hemisphere are examples. The predominance of cold water off the coasts of California, Peru, and northern Chile is another consequence. This coastal phenomenon arises because the low-level wind tends to be deflected toward the equator by coastal mountains and other effects so that it blows parallel to the shore. Surface water responds to the driving wind by accelerating toward the equator, but as soon as it acquires an appreciable speed the Coriolis force deflects it westward, sweeping the warm surface water away from the coasts. Deeper, cold water then wells up to replace the warm water. An accompanying feature is that the deeper water is comparatively well fertilized compared with the surface water, affording an advantage to fishermen and birds.

As a result of the Coriolis acceleration, a model of general circulation which has only meridional flow is incorrect. In Hadley's cell, the low-level winds in the Northern Hemisphere would veer toward the west at all latitudes and the high-level winds, toward the east. But this should have an observable effect on the earth, because friction between the low-level atmospheric winds and the earth's surface, and the resulting net westward force exerted by the air on the earth's surface, would slow the earth's rotation. However, since the earth's angular velocity is known to be essentially constant, Hadley's model cannot be correct. Furthermore, the observed low-level winds are not westward at all latitudes, but are eastward in mid-latitudes, suggesting that the general circulation is more complicated than the single-zone model.

5.3 THREE-ZONE MODEL

In the mid-nineteenth century, Ferrel pointed out that three circulation cells in each hemisphere would explain many global climatological features. Since 1930 a model has been developed, notably by T. Bergeron and C. G. Rossby, that does indeed account for the observed features of general circulation. The model incorporates a three-zone structure as illustrated in Fig. 5.5. To either side of the equator is a Hadley cell, in which tropical air rises as a result of its absorption of heat and gain in buoyancy. However as this air commences its journey toward the poles at high altitudes, it is cooled by the emission of thermal radiation toward outer space. A decrease in temperature of about 1° to 2°C per day causes a loss of buoyancy and the poleward journey is terminated by a subsidence of the air in the subtropics at about 30° North and South latitude.[1] As we shall see, this subsidence has important consequences not only on the patterns of general cir-

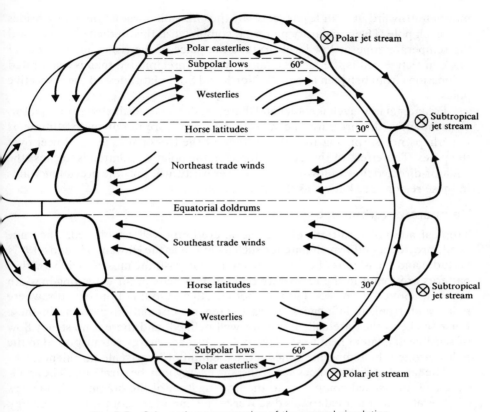

Fig. 5.5　Schematic representation of the general circulation.

culation, but also on the potential for air pollution in the coastal communities beneath. Once the subsiding air reaches low levels, it takes either an equator-directed path to complete the Hadley cell or a poleward route.

The Coriolis force acting on the flow toward the equator would therefore cause northeast or east winds, in agreement with the direction of the well-known trade winds of the tropics. (Note that in meteorology, winds are labeled according to the direction from which they come.) The region of convergence of the tropical Hadley cells causes the equatorial doldrums, a belt in which the winds are extremely gentle or nonexistent for long periods of time. There is very little circulation across the doldrums. The low-level air near the equator has a high relative humidity, so when it rises and cools, clouds are formed with a resulting heavy precipitation. There is in fact more precipitation than evaporation in these equatorial latitudes.

According to the three-zone model of Fig. 5.5, a fairly simple circulation also occurs in both polar regions. Air which has been warmed in the mid-latitudes moves northward at high levels, cools by the emission of radiation, and subsides near each pole. The circulatory pattern is completed as this air then moves toward the temperate zone at low levels to take the place of the rising air. A result of this convective cell is the fact that more water is precipitated than is evaporated at locations from between about 40° North and South latitude and the respective poles.

Polar air at low levels is cooler and therefore denser than air above the equator. Consequently, the pressure and density decrease more rapidly with altitude. It is perhaps not surprising that the altitude of the tropopause is also lower over the poles. The effect of the Coriolis force on the polar circulation is to cause the equator-directed flow of low-level air to veer westward; thus the prevailing winds in these regions are known as the polar easterlies.

Temperate turbulence

Tropical and polar patterns of circulation are fairly well established. The trade winds provide a uniform climate for such locations as the Hawaiian Islands (19° N latitude) and the islands of the Caribbean. By contrast, the most variable weather patterns are found in the temperate regions, two belts of air extending in width between about 40° and 55° latitude, both north and south of the equator, where polar and tropical influences interact. General circulation patterns in these Ferrel cells, as they are called, are not well defined, and there is no steady flow of sensible and latent heat toward the poles. Instead, energy is transported to the poles through the temperate zone by the medium of large-scale turbulence.

The fact that disturbances can transport energy may be surprising. The application of high-speed computers to the investigation of the dominant influences of general circulation led to this discovery. The importance of the kinetic energy associated with these disturbances and its northward transfer was shown by calculations performed notably by N. Phillips, H. L. Kuo, and J. G. Charney. The features of large-scale turbulence are responsible for the highly variable weather in the temperate zone, as warm air makes its way to the poles, and cold air, to the equator.

Some important conditions for the creation of turbulence have been demonstrated by simple laboratory experiments. D. Fultz and coworkers in the 1950's conducted these with a rotating, flat-bottomed pan of water which was heated at the rim and cooled at the center.[2] The arrangement thus approximated the supply and loss of radiant energy in one of the hemispheres of the earth. The water responded to the horizontal temperature gradient by setting up flow patterns to convect heat from the rim to the center. A relationship was found between the patterns of flow and a dimensionless parameter $\bar{u}/\Omega L$, where $\bar{u}$ is a characteristic speed of the water convection current, Ω is the rotational angular velocity of the pan, and L is a characteristic length, such as the diameter of the pan. For low angular velocities, fairly steady flow patterns approximate those of a Hadley cell, as illustrated in Fig. 5.6(a). However, for high angular velocities, the steady flow patterns break down into irregular convection, including nonmeridional

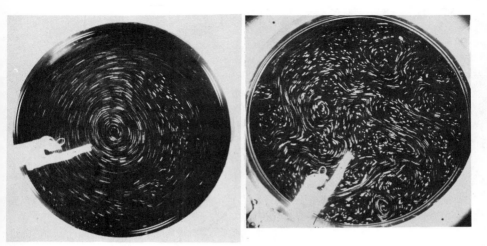

Fig. 5.6   (a) Steady flow patterns in a slowly rotating pan of water, heated at the rim and cooled in the center. (b) Turbulence develops when the pan is rapidly rotated. (D. Fultz, Hydrodynamics Laboratory, Department of Geophysical Sciences, University of Chicago.)

cells of closed circulation and discontinuities in density. The resulting turbulence is illustrated in Fig. 5.6(b). The significant aspect of these results is that a similar variation between steady flow and turbulence is found in the earth's atmosphere. In equatorial regions, where the vertical component of the earth's angular velocity is negligible, we have the Hadley cells with their steady trade winds; and at higher latitudes, where the vertical component is larger, the circulation is more irregular.

Despite the irregular weather patterns normally experienced in the mid-latitudes, there is an average trend as revealed by the three-zone model. The mean movement of low-level air to the poles, shown in Fig. 5.4, is affected by the Coriolis force; and the net result is a prevalence of west or southwest winds in the Northern Hemisphere; of northwest in the Southern Hemisphere. These contrast with the easterlies of the tropics and polar regions. But the irregular features of macroscale patterns over land often dominate and obscure the average trend.

Convergence and divergence

Several features of the pattern of general circulation merit further comment: The latitudinal boundaries between different zonal convection cells have distinct characteristics. We have already mentioned what causes the doldrums at the equatorial boundary of the two Hadley cells. The poleward boundary of these cells, at about 30° latitude, is also distinguished by calm air, where the subsiding air diverges at low levels. These regions often becalmed Spanish sailing vessels plying their trade with the New World; and sailors were occasionally forced to throw overboard some of their horses to lighten the load, hence the name "Horse latitudes." The highest rates of evaporation occur in this region over the oceans,

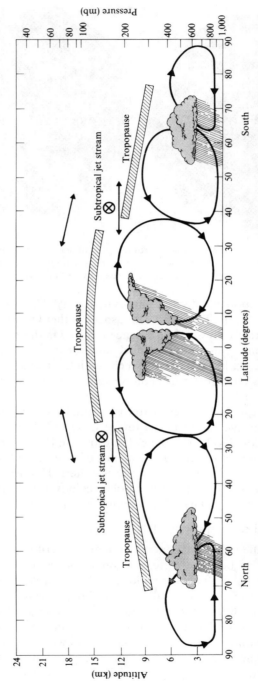

Fig. 5.7 Schematic of the circulatory system in the troposphere during summer in the Northern Hemisphere, showing the discontinuity in the troposphere at the subtropical jet stream.

for the subsiding air is dry, having previously lost its moisture during its ascension in the temperate and equatorial zones. Net evaporation occurs from about 10° to 40° N and S.[1] Thus the ocean gives up energy in an amount equivalent to the latent heat. This energy is conveyed both north and south to zones where net precipitation is observed, with an accompanying conversion of latent to sensible heat upon condensation.

In contrast with this diffuse boundary between the low-level Hadley cell circulation and the temperate zone circulation, there is a marked boundary at the low-level confluence of the polar zone and temperate zone, where cold polar air meets the warmer air from mid-latitudes. This boundary, across which pressure and temperature vary markedly, is known as the *polar front* and displays seasonal advances and retreats. It is not necessarily continuous about the polar region because it is influenced by the local meteorology of continental and maritime regions. The polar front exerts major influences on the turbulent nature of circulation within the temperate zone; but it would not be appropriate for us to consider these in detail.

Another well-defined convergence is observed at latitudes of about 30° N and S where tropical and temperate air meet in the upper atmosphere. Throughout the troposphere there is a rapid decrease in temperature if one proceeds toward the poles at any altitude; one of the results is a high-speed wind known as the *subtropical jet stream*. The tropopause suffers a discontinuity in the region of the jet stream as illustrated in Fig. 5.7, and it is through this discontinuity that much of the circulation appears to occur between the stratosphere and troposphere. Thus much of the aerosol matter injected into the stratosphere during violent volcanic eruptions must work its way horizontally through this break in the troposphere before it can be brought to earth through the convective motion within the troposphere. The location of the subtropical jet stream varies considerably from day to day.

Another boundary in the upper atmosphere occurs closer to the poles where the polar and temperate air masses diverge, but the boundary is less distinct. The *polar jet stream* occurs in winter in the Arctic stratosphere near this divergence and wanders considerably more in latitude, especially in response to the effects of continental land masses or changes in the position of the polar front. It is common that wind speeds of the jet streams are measured at 150 kilometers per hour, with the subtropical jet usually being the stronger.

The three-zone model for general circulation thus accounts for the observed direction of prevailing winds at low levels and provides a mechanism for the required transport of energy between the tropics and polar regions. This is accomplished largely by a flow of sensible heat and only a minor role is played by the flow of latent heat. The observed net precipitation near the equator and near the polar front is explained by rising and cooling air, whereas a strong subsidence with generally clear skies occurs in the subtropics. We shall see later that this subsiding condition contributes to severe air pollution episodes in the urban areas under its influence.

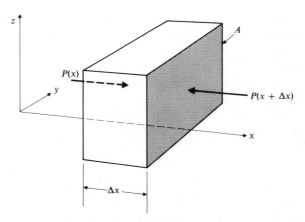

Fig. 5.8 Net force along the x-direction on the volume of air between x and $x + \Delta x$ is the difference between $AP(x)$ and $AP(x + \Delta x)$, where A is the area of the face of the volume perpendicular to the x-axis.

5.4 GEOSTROPHIC BALANCE

In the preceding section, we considered in general terms how air moves over the face of the globe in response to variations in surface temperature, radiational cooling, and the Coriolis force. Now we shall examine in more detail what causes the winds and how their direction is influenced by pressure variations, the Coriolis force, and frictional interaction with the ground. Variations in the atmospheric pressure at sea level are caused by an uneven distribution of the number of molecules in the overlying mass of air, a result ultimately of uneven surface temperatures. These pressure differences cause horizontal forces on air masses according to Newton's second law and thus produce the winds.

A representative situation is indicated in Fig. 5.8, where a volume of air (which for convenience we imagine as having a rectangular shape) is located on the horizontal x-axis of our reference system between x and $x + \Delta x$. If the pressure from adjacent molecules striking against the left face of our volume of air $P(x)$ exceeds the pressure on the right face $P(x + \Delta x)$, the air in the volume will experience a net force in the positive x-direction and will therefore be accelerated. If ρ is the density of air and A is the area of the face of the cube perpendicular to the x-direction, Newton's second law relates the acceleration $\Delta v/\Delta t$ of the volume of air to the pressure force F_p according to the equation:

$$\rho A \Delta x \left(\frac{\Delta v}{\Delta t} \right) = F_p$$

$$= A \left[P(x) - P(x + \Delta x) \right] \qquad (5.3)$$

$$\rho \left(\frac{\Delta v}{\Delta t} \right) = - \frac{\Delta P}{\Delta x},$$

where $\Delta P = P(x + \Delta x) - P(x)$. The quantity $\Delta P/\Delta x$ of course depends upon the orientation of the x-axis with respect to the pattern of pressure variations in the horizontal plane. If the direction of the x-axis is chosen so as to maximize the force in the positive x-direction, $-\Delta P/\Delta x$ takes on its maximum value and is called by meteorologists the *pressure gradient*. With this definition, we can imagine the pressure gradient on a weather map as being represented by an arrow which points in the direction of the force exerted by the pressure gradient on a volume of air. Thus air is accelerated from a region of high pressure toward a region of low pressure, if a nonuniform distribution of pressure is the prime influence.

On weather maps, it is convenient to designate the pressure distribution at a given altitude by contours of uniform pressure called *isobars*. The pressure changes most rapidly with position along a line perpendicular to the isobars, so the pressure-gradient force is also perpendicular to the isobars. If the pressure differences between all adjacent isobars are chosen to have the same value, the pressure gradient is inversely proportional to the distance between isobars; thus the closer the spacing, the greater the pressure-gradient force. In cases where pressure gradients are the most important influence determining air motion, air flow in trajectories would be expected to intersect isobars at nearly right angles in responding to the direction of the maximum pressure-gradient force. Such behavior is observed for some local situations such as the initial stages of formation of a sea breeze; however, it is not generally observed for the average macroscale motion of air masses. Such masses are usually found to move nearly *parallel* to the isobars!

The reason is that the Coriolis force F_C cannot usually be neglected when considering macroscale motion. An air mass which is accelerated from rest by the force from the pressure gradient will be deflected toward the right (in the Northern Hemisphere) as it acquires a velocity, as illustrated in Fig. 5.9. The air mass will

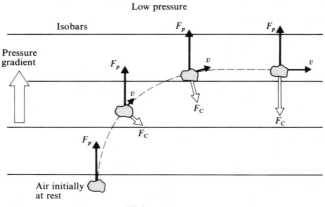

Fig. 5.9 Air mass responding to a pressure-gradient force F_p is imagined to accelerate initially from rest. Once it gains a velocity v, the Coriolis force F_C deflects it until a force balance is reached in geostrophic flow.

cease to be deflected after a sufficiently long time has elapsed that the two forces balance, and this is possible when F_p and F_C are oppositely directed and the air mass is moving with the proper speed parallel to the isobars. Then stable equilibrium is achieved. If the wind speed were too high so that the Coriolis force deflected the velocity slightly in a direction opposite to that of the pressure gradient, the force F_p in Eq. (5.3) would slow the wind until it were directed once again parallel to the isobars. By equating the magnitude of the pressure-gradient force F_p on a unit area of a volume of air and the Coriolis force [Eq. (5.2) multiplied by the density of air ρ], we obtain a condition for the wind speed at equilibrium:

$$v_g = \frac{\Delta P/\Delta x}{2\rho\Omega\,\sin\,\theta}. \tag{5.4}$$

A wind that satisfies this condition is said to be in *geostrophic balance*. The geostrophic wind speed is therefore proportional to the pressure gradient. Note that this is quite a different statement from that of Eq. (5.3), which says that *in the absence of* F_C, the *acceleration* is proportional to the pressure gradient. We shall later investigate several instances which illustrate the importance of the concept of geostrophic balance. One example in the upper atmosphere are the jet streams, which flow in approximate geostrophic balance. The high speed of these winds confirms the existence of large pressure gradients in their locality.

Equation (5.4) is not valid near the equator between about 10° N and S where the Coriolis effect is negligible. There, only the pressure gradient is influential in determining horizontal motion.

The prevalence of geostrophic winds in macroscale motion led C.H.D. Buys Ballot to formulate the following rule in 1857: If in the Northern Hemisphere you stand with your back to the wind, the high pressure is on your right and the low pressure on your left; in the Southern Hemisphere, the high pressure is on your left and the low pressure on your right.

Friction

For an accurate description of the motion of air at low levels, friction with the earth cannot be neglected. Air immediately next to the ground is hindered in its motion by surface irregularities. It gives some of its momentum to the earth through several complex mechanisms which are phenomenologically described by a frictional force. On the average, momentum in the direction of the prevailing wind is transferred to the earth; because the momentum of the air in this direction is reduced, the direction of the frictional force must be considered as opposing the wind velocity. This retarding effect is transferred upward into the lower reaches of the atmosphere by the viscous nature of air. In Chapter 7 we shall point out that it is actually through the effects of turbulence that the frictional effect of the earth is transmitted upward. Turbulent patterns of small-scale air motion, or eddies,

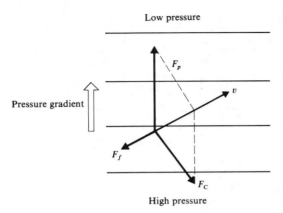

Fig. 5.10   Force balance between the pressure gradient force F_p, the Coriolis force F_c, and the frictional force F_f (which must be directed opposite to the wind velocity v).

arise from two causes: (1) thermal effects from uneven solar heating of the ground and subsequent air convection, and (2) the mechanical effects when air is diverted around obstructions such as hills and buildings. Such thermal and mechanical turbulence can affect winds up to altitudes as high as 500 to 1000 m. The effect varies greatly and depends upon such variables as the roughness of the surface, mean wind speed, and the rate of change of the horizontal wind speed with altitude.

    In steady-state flow, the net force on a mass of air must be zero; therefore the Coriolis force F_C added to the pressure gradient force F_p must be equal and oppo-site to the frictional force F_f, as illustrated in Fig 5.10. Since F_p must be parallel to the pressure gradient $-\Delta P/\Delta x$, it will always be perpendicular to the isobars. Therefore the force balance can be accomplished only if the wind velocity points slightly toward the region of low pressure. Hence the effect of ground friction causes what we might intuitively expect: Air moves from regions of high pressure toward those of lower pressure.

    The angle between the wind direction and isobars increases as the frictional force increases. Close to the ground, where the frictional force is largest, the wind direction may be at a sizable angle with respect to the isobars. Over grasslands and ocean, the angle commonly is in the order of 10° to 20°. The surface wind, by convention the wind measured at a height of 10 m, is about 0.9 of the geostrophic wind speed at high altitudes. Over cities, the rough nature of the terrain reduces the speed at low levels to about 0.5 of the speed at high altitudes, and the angle between the low-level wind and the isobars may be 30° to 50°.[3]

    Thus, local effects may be highly influential in determining mean wind direc-tions at low levels. At higher altitudes, the frictional effect is less, since air has a small viscosity. As a result, the wind direction will differ from the low-level pattern, being more nearly in geostrophic balance. Wind direction commonly spirals with altitude, an effect known as the Ekman spiral after the Swedish oceanographer

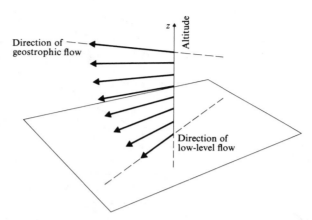

Fig. 5.11   Ekman spiral showing the variation of wind velocity with altitude caused by the deviation from geostrophic flow at low altitudes. Wind speed increases with altitude.

V. W. Ekman, who at the beginning of this century discussed an analogous effect for wind-driven ocean currents. A schematic example of the Ekman spiral is given in Fig. 5.11. Both the direction and speed of the wind vary with altitude. When either or both types of variation are observed, the wind is said to exhibit a *shear*.

5.5 VERTICAL MOTION

The preceding discussion has focused on how air responds to horizontal forces. Now we shall investigate its vertical motion, which we shall find depends upon the temperature profile of the atmosphere. We shall first take up the familiar effect of *buoyancy*, the tendency of an object to rise in a surrounding fluid medium if the object is less dense on the average than the medium. A boat floats on the surface of water as a result of the same buoyant effect that causes a volume of warm air to rise when surrounded by cooler air.

It is useful to deal with what we shall call a *parcel* of air. This is a fixed number of molecules occupying a contiguous volume, and the temperature is essentially uniform throughout. A parcel consists of uncountable numbers of molecules which are continually colliding with other molecules of their surroundings; furthermore, it is not a solid but can readily be deformed or compressed. Such characteristics notwithstanding, a parcel does constitute a fairly well-defined object. It is an object in the sense that only a very few molecules in the boundary of a large parcel may mix with the surroundings, leaving the bulk relatively undisturbed. Thus in many cases of vertical air motion, the exchange of heat between the interior of a parcel and the surroundings is negligible.

In Fig. 5.12 we illustrate an air parcel which for the sake of argument is warmer than its surroundings. As it warms, perhaps by contact with the earth, it expands in all directions according to the ideal gas law $P = \rho R_a T$ [Eq. (3.9)], so as to reduce its density ρ and therefore reduce its pressure to that of the surroundings.

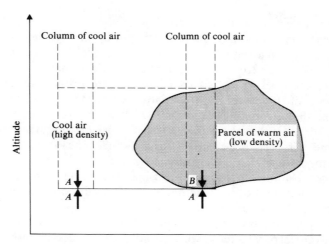

Fig. 5.12   Upward buoyant force on the underside of a parcel of warm air is caused by the higher pressure at low levels of the surrounding cool air. This pressure is transmitted horizontally through the cool air, so that it also affects the underside of the parcel.

As an incidental, extraneous effect, this expansion slightly compresses the surrounding air. The expanded, low-density parcel is then not in an equilibrium with respect to vertical motion. The reason is that at the level of the bottom of the air parcel (A) the pressure of the surrounding cooler, and therefore denser air is greater in the upward direction than the pressure exerted downward at the same level by the air at the lower boundary of the parcel (B). This must be the case, since the pressure at the bottom of a column of air is equal to the weight of the overlying air per unit horizontal area. The warmer, less dense air within the parcel contributes less weight than an equal height of the cooler air. The net upward pressure on the molecules of the parcel causes them to rise and therefore gives the parcel buoyancy.

Although it is the lower density of a warm parcel relative to its cooler surroundings that produces the buoyancy, the inverse relationship between ρ and T in the perfect gas law permits us to simplify our discussion by considering only the relative temperature of a parcel and its surroundings. We can thus say that "warmer air rises." The buoyancy of warmed air is a familiar phenomenon, employed every day in home heating systems where the warm air is introduced at a low level in each room. The rising air in such situations provides a fairly uniform heating. A more vigorous phenomenon is familiar to anyone who has flown in an airplane at low altitudes over hot desert or semi-arid regions. Rising warm currents of air known as "thermals" may produce violent motion of any aircraft that encounters them. Of course, as warm air rises, the surrounding cooler air must fill in below; and it thus appears that cooler air descends.

In Section 3.6 we applied the hydrostatic condition to predict the pressure and density profiles of the atmosphere. In equilibrium the density of course must decrease with increasing altitude, as is implied by the hydrostatic condition Eq. (3.10). Hence a rising parcel of air will continually find itself proceeding into regions of lower pressure and lower density, provided that the ambient temperature does not increase greatly. As a result of the lower pressure in the surroundings, the molecules of the parcel push outward, expanding the size of the parcel and compressing the surrounding air to some small extent. An important consequence of this expansion is that the parcel performs work on the surrounding air, transferring some of its sensible heat to the surroundings. The loss of this energy implies that the temperature within the parcel decrease.

The decrease in temperature of a rising, expanding air parcel is an essential feature of vertical air motion; whether a parcel remains warmer than its surroundings and continues to be buoyed upward as it rises depends, therefore, upon the *ambient* temperature profile. Consequently the variation of temperature with altitude is a crucial parameter by which conditions for good or poor vertical mixing of air pollutants may be predicted. We turn now to an examination of the thermodynamics of the vertical motion of air parcels to establish quantitatively the criteria governing vertical mixing.

Adiabatic process

To calculate the temperature decrease of a rising parcel, we note first that conduction or convection of heat across the boundaries of a large air parcel is so slow that in most cases of vertical motion it is completely negligible. The dependence of its temperature T on a change in altitude can be nicely approximated by assuming that there is no exchange whatsoever of heat between the parcel and its surroundings. A process or change under such conditions is said to be *adiabatic*. The thermodynamic consequences of adiabatic motion can be predicted from the application of two basic concepts: (1) the ideal gas law, and (2) the conservation of energy. Using these, we shall now derive an expression for the rate at which the temperature T of a parcel varies with altitudes as it moves; the final result will be given by Eq. (5.12).

In the most general case, conservation of energy requires that any heat flow, denoted by ΔQ, across the enclosing boundary into an air parcel is equal to the sum of ΔU, the increase of the internal energy of the gas, and ΔW, the energy which is lost by the parcel to the surroundings in the form of work done to change the configuration of its boundaries. In symbols, this is expressed as the first law of thermodynamics.

$$\Delta Q = \Delta U + \Delta W. \tag{5.5}$$

The work done when a parcel changes its size is equal to the force exerted by the confined air multiplied by the distance through which it acts. Thus, for a small change in volume ΔV, the work performed by the air in the parcel is

$$\Delta W = P \Delta V. \tag{5.6}$$

For typical atmospheric pressures and temperatures, a change in the internal energy ΔU is just a change in the mean kinetic energy of molecular motion and therefore is proportional only to the temperature change ΔT. If energy is supplied to a parcel whose volume is by some means held constant ($\Delta V = 0$), then,

$$(\Delta Q)_{V = \text{constant}} = \Delta U = C_V \Delta T. \tag{5.7}$$

The proportionality constant C_V is called the *specific heat at constant volume* for the parcel. (It also can be written as the product of the mass of air in the parcel $M = \rho V$ and the specific heat at constant volume per kilogram of air, commonly denoted by c_V.) From our preceding remarks, we conclude that even if the volume of the parcel is allowed to change, the following remains true:

$$\Delta U = C_V \Delta T. \tag{5.8}$$

We are interested in what happens to an air parcel precisely when its volume is allowed to change upon ascending or descending in the atmosphere. To find the change in T dictated by Eq. (5.5) it is more convenient to express $P\Delta V$ in a form in which pressure and temperature are the independent variables which appear in the equation, rather than the volume. This can be accomplished by application of the ideal gas law $PV = MR_aT$, to develop a relation between ΔP, ΔV, and ΔT (where again M is the total mass of air in the parcel, and R_a is the gas constant per kilogram of air). Most generally we have

$$V\Delta P + P\Delta V = MR_a\Delta T;$$

so that the work done can now be expressed as

$$\Delta W = MR_a\Delta T - V\Delta P. \tag{5.9}$$

Now we are in a position to find the dependence of T upon altitude for an adiabatic rise or descent of our air parcel. Equation (5.5) for the adiabatic condition $\Delta Q = 0$ yields

$$0 = \Delta U + \Delta W.$$

Using Eqs. (5.8) and (5.9), we can rewrite this as

$$0 = C_V\Delta T + (MR_a\Delta T - V\Delta P). \tag{5.10}$$

The volume V can be eliminated from this expression by substituting the ideal gas law; this gives

$$0 = (C_V + MR_a)\,\Delta T - MR_aT\frac{\Delta P}{P}.$$

Since we seek to learn what happens to a rising parcel, we wish to eliminate pressure from this equation in favor of altitude z as the independent variable. When the pressure profile is given by the hydrostatic condition, Eq. (3.11), we find for the change in temperature per change Δz in altitude:

$$\frac{\Delta T}{\Delta z} = -\frac{Mg}{C_V + MR_a} \tag{5.11}$$

$$= -\frac{g}{c_V + R_a}$$

When this is evaluated for dry air, the change in temperature with change in altitude is found to be

$$\frac{\Delta T}{\Delta z} = -9.86°C/km. \tag{5.12}$$

Therefore the temperature of the parcel decreases by about 10°C for every kilometer of rise.

Our derivation shows that the basic reason for a temperature decrease of this magnitude is the high compressibility of air; that is, a given decrease in the pressure experienced by a rising parcel of air results in a fairly large increase in its volume. Water, on the other hand, is only slightly compressible. A rising parcel of water in a lake would suffer a temperature decrease of less than 0.2°C/km.[4]

Lapse rate

The above derivation indicates that a rising parcel of dry air will cool at a rate of about 10°C for each kilometer of ascent, while a descending parcel warms at a rate

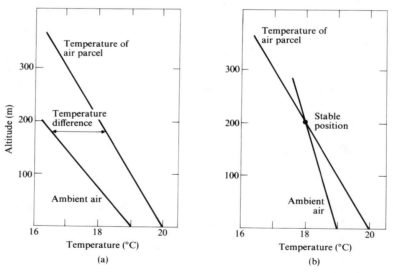

Fig. 5.13 Temperature profiles. (a) A rising parcel of air, initially at 20°C when the surrounding air at ground level is at 18°C, will continue to be buoyed upwards if the ambient lapse rate exceeds the dry adiabatic lapse rate because the temperature difference increases with altitude. (b) The same parcel will cease its upward motion if at some point its temperature is identical to the ambient temperature and the ambient lapse rate is less than the adiabatic rate.

of 10°C for each kilometer of descent. At first thought, one might suppose that the vertical motion of air parcels would establish a similar variation in the ambient temperature, and in some cases this indeed occurs. However, other factors such as the prevailing wind and the intensity of sunlight can have a more important effect on the temperature profile, so the observed variation of *ambient* temperature with altitude at any instant of time generally is quite different. In Fig. 4.7, for example, we show a temperature profile in which temperature decreases with altitude in some layers of the atmosphere but increases with altitude in others. The rate at which the ambient temperature is found to *decrease* with altitude is called the *temperature lapse rate* (or more simply, the *lapse rate*). If meteorological conditions are such that the atmospheric lapse rate is the same as that for an adiabatically rising parcel of dry air, about 10°C/km, it is said to have a *dry adiabatic lapse rate*. Usually, however, the atmospheric lapse rate does not satisfy this condition, since it is determined by factors which are more important than the adiabatic convection of air parcels.

5.6 LAPSE RATES AND STABILITY

The value of the lapse rate in the lower portion of the troposphere has a profound influence on the vertical motion of air. Good mixing in the vertical direction minimizes the immediate ground-level effects of air pollutants, since the contaminants may be quickly diluted through their dispersal into higher regions. On the other hand, if air does not vigorously mix upwards, pollutants which are released at low levels tend to remain there.

Vigorous mixing will occur if a rising parcel of warmer air continues to find itself buoyant even though it cools as it rises. Suppose that a warm parcel commences to rise in a dry atmosphere which for some reason has a lapse rate exceeding the adiabatic rate (the temperature decreases more rapidly than 10°C/km) as illustrated in Fig. 5.13(a). The air parcel cools at the adiabatic rate of 10°C/km as it rises. Then the temperature *difference* between the rising parcel and its surroundings *increases* with altitude, despite the fact that the temperature of both decreases. Thus buoyancy increases with altitude, and upward motion is further encouraged. A rising parcel continues to find itself buoyant if the ambient lapse rate exceeds the adiabatic rate.

Similarly, air in a parcel which is cooler than the surroundings would commence to descend, and by compression be warmed at the adiabatic rate. However, the temperature of the surrounding air increases even faster with decrease in altitude, so the downward motion receives even more impetus. Since vertical motion in both directions is encouraged, the atmosphere is said to be *unstable*. Small fluctuations in the temperature of an air parcel will immediately cause it to rise or fall, and the motion will be accelerated by the ever-increasing temperature difference with the surroundings. The lapse rate for such instability is called *superadiabatic*. Thus when the lapse rate exceeds 10°C/km, there is generally good vertical mixing. A superadiabatic rate can be established by intense solar heating of the ground when there is little or no advection through the atmosphere to carry away the warming air adjacent to the ground.

If the ambient lapse rate is less than the adiabatic rate, the air is *stable* and discourages vertical currents. This can be seen by arguments similar to those we have just given: A rising warm-air parcel cools more rapidly with altitude than the temperature of its surroundings. A point is reached where the temperature of the parcel and of its surroundings are equal, and the parcel ceases to rise, as illustrated in Fig. 5.13(b). Any further upward motion would be opposed by negative buoyancy since the parcel would be slightly cooler than the surroundings, and so the parcel eventually comes to an equilibrium where its temperature equals the ambient temperature. An analogous argument shows that subsidence of an air parcel is also discouraged. Any fluctuations of the temperature of an air parcel will cause it to rise or fall, but only for a very short distance, until it reaches a point of equilibrium. This is why the air is said to be stable. Some of the most severe episodes of air pollution occur in stable air, or in air above which for some reason there is a stable layer of air.

Extreme cases of atmospheric stability are characterized by a negative lapse rate, where the temperature increases with altitude. Since this is the inverse of the common temperature profile, the region where the lapse rate is negative is given a special and well-known name: a *temperature inversion*. Atmospheric strata with temperature inversions are called *inversion layers*. Because vertical air motion within such layers is practically nil, they often have a well-defined base and top. Depending upon how they are formed, inversion layers may have their base touching the ground, or they may be elevated. We have already met one example of an elevated inversion: The upper portion of the stratosphere has an inversion layer whose base over some regions of the globe is at 20 km or higher. Because of the

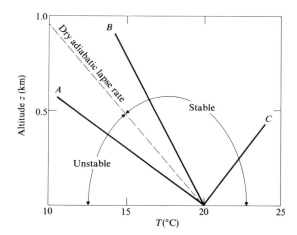

Fig. 5.14 Temperature profiles which are examples of (*A*) unstable, (*B*) stable, and (*C*) very stable inversion lapse rates in a dry atmosphere.

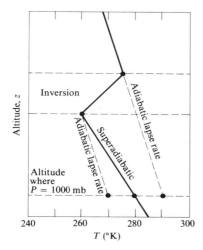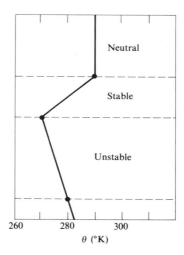

Fig. 5.15 Profiles for the absolute temperature and potential temperature for a hypothetical situation with an elevated inversion layer. The dashed lines in the left figure show how the temperature of two parcels of air would change if the parcel were lowered adiabatically to an altitude where the pressure was 1000 mb.

stability of the stratosphere there is little vertical convection and the residence time for aerosols is long, amounting to several years.

An intermediate case of atmospheric stability occurs when the ambient lapse rate equals the adiabatic rate. Since this neither encourages nor discourages vertical motion, the atmosphere is then in a *neutral* condition. The temperature profiles for various cases of stability are illustrated in Fig. 5.14.

Meteorologists frequently describe the state of air by what is known as its *potential temperature*. This is the temperature which a parcel of air would have if brought adiabatically from its original pressure in the atmosphere to a standard pressure of 1000 mb. Thus for a dry atmosphere which has the dry adiabatic lapse rate, all portions of the atmosphere have the same potential temperature. The formula which relates potential temperature Θ to the thermodynamic temperature T (Kelvin scale) can be approximated by

$$\Theta = T + \Gamma z, \tag{5.13}$$

where Γ is the dry adiabatic lapse rate of about $\Gamma = +10°C/km$ and z is the altitude. A useful feature of this relationship is the fact that for dry air the rate at which potential temperature changes with altitude is

$$\frac{\Delta\Theta}{\Delta z} = \frac{\Delta T}{\Delta z} + \Gamma, \tag{5.14}$$

and thus the sign of $\Delta\Theta/\Delta z$ indicates whether the atmosphere is unstable. If Θ increases with height ($\Delta\Theta/\Delta z > 0$), the atmosphere is stable; if it decreases ($\Delta\Theta/\Delta z < 0$), the atmosphere is unstable. For neutral stability, Θ does not vary with altitude. An example which compares profiles for T and Θ is shown in Fig. 5.15. It includes an elevated inversion layer.

The elevated inversion is an important example, because it illustrates a temperature profile commonly found over many communities during daylight hours. The superadiabatic condition is established in the lowest layer of air by solar heating of the ground and by the subsequent transfer of sensible heat to the adjacent air. The lapse rate causes this air to circulate vigorously in the vertical direction and convect energy upward. It is in this region that pollutants are generally emitted, and therefore the low-lying unstable layer is called the *mixing layer*. Although pollutants may be rapidly diluted in the mixing layer, they cannot penetrate upward into the stable inversion layer. Thus the amount of air available for diluting pollutants is limited since the inversion layer acts as a "blanket" to hold them within the mixing layer. It is important to realize that the entire troposphere, from the ground level up to the tropopause, is *not* available for diluting pollutants in such circumstances. Only the mixing layer, which in many cases may be only a few hundred meters thick, can serve this function.

Moist air

So far the discussion in this chapter has been limited to a description of the motion of dry air. If water vapor is also a component gas, a rising parcel of air may cool to an extent where the pressure of the water vapor equals the saturation vapor pressure of water. The air is then at the dew point and, if there are sufficient aerosols present to encourage nucleation, some of the water vapor will condense and precipitation may occur. Release of an amount of energy equivalent to the latent heat of

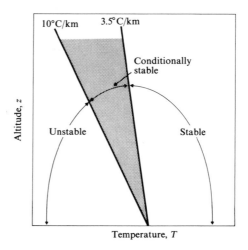

Fig. 5.16 Whether moist air is stable depends upon both the ambient lapse rate and its relative humidity. Warm saturated air can be neutrally stable with a lapse rate as low as 3.5°C/km; cooler saturated air is neutrally stable at higher lapse rates.

the condensing vapor enhances the kinetic energy of the air molecules in the parcel as it rises, with a result that the temperature decrease upon increase in altitude is less than for dry air.

If L is the latent heat per kilogram of water, then the release of this heat represents a contribution of the left-hand side of Eq. (5.10) that is equal to $ML\Delta r$, where r is the ratio of the mass of water vapor to the mass of dry air in a given volume of air and Δr is the amount of vapor which is condensing as the temperature changes by ΔT. Carrying this new term through the balance of the derivation yields an expression for the *wet adiabatic* temperature change of the parcel:

$$\frac{\Delta T}{\Delta z} = - \frac{g}{c_v + R_a} - \frac{L}{c_v + R_a}\left(\frac{\Delta r}{\Delta z}\right). \tag{5.15}$$

Since $\Delta r/\Delta z$ for a rising parcel is negative, the effect of the last term in the equation, as we have mentioned, is to make a less marked cooling of rising air. Because the saturated vapor pressure of water increases very rapidly with increase in temperature (Fig. 3.2), the ability for the atmosphere to absorb water does also, and the quantity $\Delta r/\Delta z$ will depend strongly upon the temperature. Therefore, unlike the dry adiabatic lapse rate, the wet adiabatic lapse rate for saturated air is not a constant independent of altitude. In warm tropical air, the wet adiabatic rate is approximately one-third of the dry adiabatic rate, whereas in cold polar regions where there is little water vapor there is very little difference. The average lapse rate within the troposphere is about 6.5°C/km.

Strictly speaking, since condensation may lead to precipitation of water from a rising moist parcel of air, the process is not adiabatic. Nevertheless, the precipitating water has a negligible effect on the temperature. Whether moist air is in a stable or unstable condition will depend on the lapse rate and relative humidity. Figure 5.14 must be appropriately amplified to include a region of "conditional stability," as we show in Fig. 5.16. The lowest lapse rate for this region is somewhat arbitrarily denoted by 3.5°C/km.

Global view

With the preceding in mind, we can now see how significant it is that the atmosphere is largely transparent to solar radiation and that most of this radiation is absorbed by the surface of the earth. Consequently the earth is the source of most of the energy absorbed by the atmosphere. Through direct contact with the earth's surface or through absorption of terrestrial radiation, the atmosphere is warmed from below and therefore generally has a conditionally stable lapse rate in the lower troposphere which encourages vertical convection. In fact, the name of this stratum is derived from the Greek *tropos*, meaning "turn." Because of the high transparency of the atmosphere, the world does not suffer from perpetual pollution at low levels.

This feature is in marked contrast with the behavior of ocean circulation, where solar radiation is absorbed in the upper layer; the ocean does not develop appreciable instability within the depths. Only the top 300 m to 700 m or so has

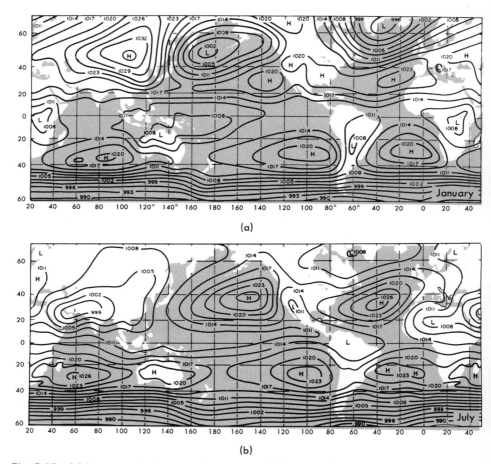

Fig. 5.17 (a) Mean sea level pressure in January. (b) Mean sea level pressure in July. Pressure is expressed in millibars. (From G. T. Trewartha, *An Introduction to Climate*, McGraw-Hill, New York, 1968, by permission)

appreciable vertical circulation, while by and large, the lower layers maintain a stable configuration. This is evidenced by measurements of lead concentrations at various depths which show that a markedly large concentration occurs only in the upper layer, where it has accumulated as a result of sedimentation and washout from the polluted atmosphere. It has been estimated that 1500 years must elapse before there is a complete turnover of ocean water.

5.7 CYCLONES AND ANTICYCLONES

The dynamic response of the atmosphere as it warms and cools over the differing environments presented by oceanic and continental regions of the globe will not be uniform. The patterns of air motion which were described by the long-term averages in our consideration of the general circulation do not represent the true state of the atmospheric circulation on any given day. Nor is the general circulation representative even of a monthly average. The irregular patterns formed by the land masses, with the vertical obstructions presented by high mountain ranges, and the variable and responsive surface temperatures are responsible for disrupting smooth circulation over the globe.

This is perhaps most clearly evident from the patterns of atmospheric pressure at sea level as shown in Fig. 5.17 for January and July monthly averages. The pressure at sea level is indicated by isobars labeled in millibars. Although these contours encircle regions of relatively high or relatively low pressure, it should be appreciated that the maximum variation in the barometric pressure is actually a small fraction of the global mean. For example, there is only a 3% difference between the highest and lowest pressures indicated in Fig. 5.17. The importance of continental effects is underscored by the irregular patterns of the isobars in the Northern Hemisphere as compared with the strong zonal patterns found in the Southern Hemisphere where the area of land mass is minimal.

The apparently chaotic patterns of high- and low-pressure regions does, in fact, have a correspondence with the general circulation. Perhaps the most obvious feature relating to the general circulation is the belt of low pressure at the equator where the tropical air of the Hadley cells rises. The relatively low density of this warm air means that fewer air molecules exist in the column of air overlying the equatorial surface as compared with the adjacent subtropical surface and therefore, according to the hydrostatic equation, Eq. (3.10), the sea level pressure is relatively low. Low-lying air which responds to this pressure gradient between tropics and subtropics moves toward the equator as part of the Hadley cell circulation.

Perhaps less obviously related to the three-zone model for general circulation are the two belts of high-pressure areas in the subtropics, at about 30° N and S latitude. Figure 5.5 shows that these are located where the poleward flow of upper air in the Hadley cell converges with the equatorward flow from the temperate zone. Both upper air masses have been cooled by expansion during their previous rise; and during their journey toward the subtropics they have been further cooled by radiating energy into space. By the time they converge, they have lost buoyancy and have begun to sink back down toward earth. The substantial concentration of cool, dense air in the upper atmosphere therefore produces a high pressure at sea level. Consequently, there is a divergence of air at low levels from this belt of high pressure, with air which is affected by ground friction moving both poleward and equatorward toward lower pressure.

It is apparent from Fig. 5.17 that land masses exert a countervailing influence (especially during the summer season) which serves to break up the zonal pattern of the general circulation. For example, the belts of subtropical high pressure are separated into individual regions which are centered over the oceans. Continental

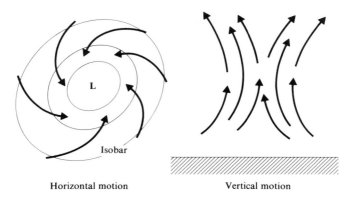

Horizontal motion   Vertical motion

Fig. 5.18 Low-level counterclockwise spiral of winds which converge in a cyclone in the Northern Hemisphere. The vertical motion of the air is depicted at the right.

low-pressure regions are especially evident over the southwestern United States, central Asia, north and South Africa, and Australia during their respective summer seasons. In these cases, the arid deserts of these regions are warmed sufficiently to produce strong thermals which rise to very high altitudes and impede the subsidence of air in the middle and upper troposphere. The tall column of warm, rising continental air is less dense than the surrounding air over the oceans, so a local low-pressure region develops. The seasonal response of continental air will be explored further in Section 6.1 when we discuss monsoons. For the moment, it will suffice to emphasize that the continents play a major role in breaking up zonal patterns into continental and oceanic regimes.

Cyclonic motion

Another influence which serves to break up zonal patterns is the Coriolis force: Air which converges at low levels toward regions of low pressure must also execute a circular motion in response to the deflection of the wind velocity by this inertial force. Our discussion of geostrophic balance between the pressure force and Coriolis force indicates that wind would parallel the isobars if it were not for the frictional force of the earth's surface. The effect of friction is to tip the balance so that wind at low levels is directed slightly toward the region of lower pressure, thereby accounting for an inward spiraling motion toward the center of a low-pressure region. The direction of the motion is counterclockwise in the Northern Hemisphere (and clockwise in the Southern Hemisphere), as given by Buys Ballot's rule. The vortex-like motion of the wind gives rise to the name *cyclone* for these low-pressure regions. An example of the low-level flow around a cyclone is shown in Fig. 5.18 The correctness of the direction of the wind's rotation around the low-pressure area can be verified by comparing this figure with Fig. 5.10.

On the other hand, the low-level diverging flow from a high-pressure region

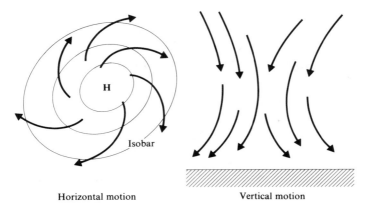

Horizontal motion   Vertical motion

Fig. 5.19   Clockwise diverging spiral of winds from an anticyclone in the Northern Hemisphere. The vertically subsiding motion of the air is shown at the right.

will spiral outward in a clockwise direction in the Northern Hemisphere (and counterclockwise in the Southern). Such high-pressure regions with their spiraling winds are called *anticyclones*. An example is depicted in Fig. 5.19. Some types of anticyclones have a nearly permanent existence, but others do not. In the temperate zone, wind patterns are more chaotic than is suggested by the long-term average illustrated in Fig. 5.17. Cyclones and anticyclones are born in one part of the world and migrate elsewhere before meeting their demise. This behavior exerts a major influence on the weather, and we shall shortly return to examine how this occurs.

The regions of large-scale spiraling air motion of cyclones and anticyclones should not be confused with the hurricanes over the Atlantic Ocean or the similar but even more destructive typhoons of the Pacific and the cyclones of the Indian Ocean. The influence of hurricanes is usually more localized, although the center of the hurricane is also a low-pressure region. Typical wind speeds in a hurricane may reach hundreds of kilometers per hour, whereas the normal winds in a cyclone are only 30–50 km/hr. The high winds of a hurricane extend from the center for a distance of perhaps a hundred kilometers, whereas the dimensions of a cyclone are commonly from hundreds to a thousand kilometers. The distinction between the two scales of motion appears to be that the main source of energy for a hurricane is the latent heat of water released by the rapid condensation resulting from saturated air being drawn up and cooled near the hurricane's center. Such storms are most frequently found in warm tropical regions, and rapidly die when they move over land areas. Cyclones are a phenomena of the temperate zones and easily move over both land and water. Their energy is derived from the sensible heat of warm rising air and the potential energy of sinking cold air. Both cyclones and hurricanes are transient phenomena which may last for a week or more.

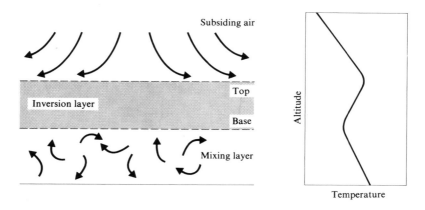

Fig. 5.20   Subsidence inversion formed by compressional heating of air above an unstable layer.

Semipermanent anticyclones

Members of the anticyclone group having a semipermanent existence are the subtropical anticyclones. These are the high-pressure regions centered over the major oceans, as illustrated in Fig. 5.17. The subtropical anticyclones are almost always evident, but shift their positions slightly toward the poles during the summer months and toward the equator during the winter, in response to the seasonal differences in the mean level and distribution of solar radiation. They exert a profound effect on the local climatology and incidence of air pollution.

One reason is that the cold subsiding air aloft, descending at a rate of about 1 kilometer a day, is warmed by compression as it descends and achieves a temperature higher than that of the air underneath. Thus the subsiding air establishes a temperature inversion with respect to the low-lying air. Figure 5.20 illustrates this condition and temperature profile for a so-called elevated *subsidence inversion*. In mid-latitudes, the inversion layer approaches closer to the ground with increasing distance eastward from the center of high pressure. For example, the base of the Pacific anticyclone commonly lies at an altitude of several thousand meters over the Hawaiian Islands, but only 300–400 meters above coastal California.[5] Generally, the base of the inversion does not touch the ground during the day, partly because solar heating of the earth's surface and subsequent transfer of some of this heat to the lower atmosphere leads to an unstable mixing layer which permits upward convection of warm air. This warming of the underside of an inversion in essence "erodes" it away. At night, with a cessation of solar heating and upward convection, the inversion base may descend close to the ground, severely restricting the volume of the underlying mixing layer. West-coast cities which lie under the regime of semipermanent anticyclones and therefore suffer from poor vertical ventilation include Los Angeles; Santiago, Chile; Casablanca, Morocco; and Perth, Australia.

The inversion over the eastern portion of each semipermanent anticyclone is intensified as the equator-directed motion of air brings cool, fairly dry, low-level air from temperate regions. Especially in coastal areas, the low-level air has been further cooled through contact with cold water, and the incoming cool air mitigates to some extent solar heating and upward convection in the mixing layer which would otherwise tend to weaken the inversion. Clear skies are predominant in the subsiding air aloft, so sunlight can penetrate to the earth's surface with little absorption and scattering. A consequence of these features is the occurrence of major deserts on the west coasts of the continents. These include the deserts of southern California and the southwestern United States, the Sahara of North Africa, the coastal plains of Peru and northern Chile, the Kalahari desert of South

Fig. 5.21 Montage of satellite photographs of cloud cover over the Southern Hemisphere. (Courtesy of Environmental Sciences Service Administration.)

Africa, and the great western desert of Australia. Coastal regions, however, may be subject to low-level clouds from onshore breezes bringing humid maritime air.

By contrast, the western regions of the semipermanent anticyclones have a much weaker inversion, because the low-level air is from the tropics. The warmth of the low-level air reduces the difference between its temperature and that of the subsiding air aloft. Furthermore, this air is moist and readily forms clouds and precipitation as it is cooled on its northward journey. The eastern edges of the continents in the subtropics and mid-latitudes are correspondingly cloudy and humid.

Migratory cyclones and anticyclones

The semipermanent anticyclones are few in number. More numerous are the migratory cyclones and anticyclones of the temperate zone. Inhabitants of the mid-latitudes are well aware of the procession of highs and lows that make their way in a generally eastward direction. An example of the turbulent patterns of air for the Southern Hemisphere is shown in Fig. 5.21. These mobile pressure regions are formed by the give and take between arctic and tropical influences. In the Northern Hemisphere many cyclones are formed where cold Siberian air flows out over the relatively warm Kuroshio current north of Japan. No two anticyclones or cyclones are identical; they vary in extent, shape, and detailed wind patterns. Often they are elongated. If the ground-level pressure distribution were regarded as a three-dimensional topographical map, with pressure plotted in the vertical direction, elongated anticyclones could resemble high-pressure "ridges" whereas cyclones could resemble low-pressure "troughs." This terminology is frequently used in describing meteorological conditions. Most cyclones and anticyclones have a lifetime of only a few weeks and drift with the westerlies at an average speed of about 800 km/day. Through the large-scale turbulence as-

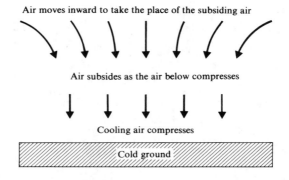

Fig. 5.22 Formation of a cold anticyclone. The process begins at ground level and proceeds upward. A high pressure develops at ground level as the mass of overlying air increases, subsequently resulting in an outward or diverging horizontal flow at low levels.

sociated with this motion, energy is transported from the subtropics through the temperate zone to the polar regions as the warm air is moved toward the poles.

It is common knowledge that cloudy skies and precipitation often accompany low-pressure regions. This feature is a result of the rising air of the cyclone which, upon expanding and cooling, becomes saturated. On the other hand, the high-pressure anticyclones consist of subsiding air in the upper regions and are therefore generally accompanied by clear skies and fair weather. This is not an inviolable rule since advection resulting from peculiar local conditions may bring clouds into an anticyclone where they can remain because radiational cooling of the cloud top overcomes compressional heating. Clouds are especially common in the mixing layer.

Migratory anticyclones are also responsible for the stagnant conditions in which pollutants in community air may accumulate to abnormally high concentrations. One reason for this is the presence of an elevated inversion layer which is established aloft by the subsiding air and is precisely analogous to the elevated inversion in the semipermanent anticyclones. Another reason is that winds which accompany anticyclones are usually observed to be gentle, at least more so than in most cyclones. There are two types of migratory anticyclones: For simplicity, they may be classified as either *warm or cold* anticyclones. The origins of warm anticyclones are not completely understood. Many are spawned by the semipermanent highs, such as the Azores high or the Pacific high. Once separated from the parent, they drift eastward, generally remaining in subtropical latitudes. If the trajectory passes over cold land in winter, a warm anticyclone may be converted into a cold one.

Most cold anticyclones are forged in subpolar regions, such as northwestern Canada and northern Siberia, where the lower-level air within these pressure regions is cooled by contact with the ground. The oblique angle of incidence of the solar radiation in winter and the high albedo of snow combine to minimize energy absorption and heating of the surface. The coldness is intensified by the nocturnal and daylight energy loss through radiation, which is particularly effective because the low humidity weakens the greenhouse effect. As illustrated in Fig. 5.22, an anticyclone is formed as the low-level air cools and compresses and thereby allows air aloft to subside. High-level air filling in the region vacated by the subsiding air thus increases the ground-level pressure, and causes the low-level air to diverge toward surrounding lower pressures. Under these conditions, the extreme coldness of surface will intensify the strength of the inversion.

These high-pressure air masses may be drawn toward lower latitudes on the tail of a cyclone. The Siberian highs generally migrate across northern China and out over Japan. Highs formed over western Europe move on an eastward path over central Asia. In response to the general patterns of circulation over North America, the Canadian highs commonly move southeastward over the Great Lakes and out over the Atlantic. If the territory over which the anticyclone moves is cold, the inversion will remain strong.

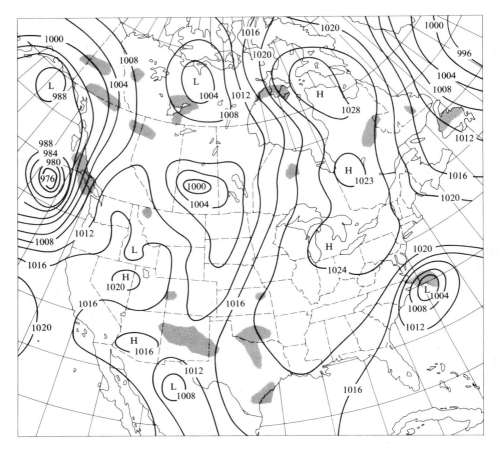

Fig. 5.23 Weather map for December 1, 1962, 1300 Eastern Standard Time. Precipitation areas are stippled.

Stagnation

The most serious episodes of air pollution in the eastern United States and Europe have occurred when an anticyclone temporarily ceases its eastward drift and stagnates for a few days. The classic episodes in Donora, Pennsylvania, (October, 1948) and London (December, 1952), and a milder one in New York (December, 1962) occurred under these circumstances. A serious build-up of pollutants resulted from the combination of an overhead inversion and light winds. Figure 5.23 illustrates the situation on December 1, 1962, in the middle of a seven-day episode which affected a very large region of eastern United States. During this time the high-pressure region was nearly stationary and the winds exceptionally light as is illustrated by the wide spacing of the isobars.

Geographical regions prone to stagnating anticyclone have a serious potential for air pollution. This includes the Great Basin between the Rocky Mountains

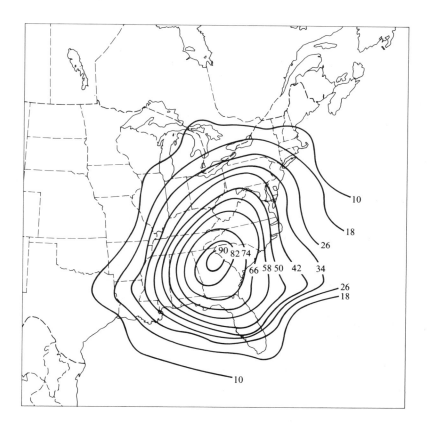

Fig. 5.24 Frequency of stagnation of anticyclones east of the Rocky Mountains. The contours are isopleths which indicate the number of times an anticyclone stagnated for four or more successive days within the enclosed region during the period 1936–1965. (J. Korshover, "Climatology of Stagnating Anticyclones East of the Rocky Mountains, 1936–1965," National Center for Air Pollution Control, 1967)

and Sierras in western United States and the central basin of California. Figure 5.24 shows the frequency of stagnation over the eastern part of the country. The isopleths (contours of equal frequency of occurrence) indicate the number of occasions that a stagnating high affected the indicated location for a period of four days or more during the interval from 1936 to 1960. Anticyclones stagnate with greatest frequency over the southern Appalachian region, in western North Carolina and eastern Tennessee. This is no doubt a contributing factor to the frequent atmospheric accumulation of hydrocarbons given off by the vegetation, especially trees. The resulting haze is responsible for the name of the Great Smoky Mountains in eastern Tennessee and the Blue Ridge Mountains of the Virginias and North Carolina. The isopleths indicate that there is an average of three occasions for stagnation per year at most.

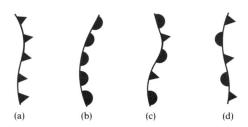

Fig. 5.25 Symbols which denote fronts. (a) Cold front advancing toward the right; (b) warm front advancing toward the right; (c) occluded front in which one type has overtaken the other; and (d) a stationary front.

The details of the subsidence inversion within an anticyclone vary considerably, but the base is often found at an altitude of 500 m, with the inversion extending upward another 500–1000 m. During nighttime conditions, the inversion will descend and may even touch the ground. The rise in temperature across an inversion layer can be on the order of 10°C, a value similar to the inversion strength of some regions of the eastern side of the semipermanent anticyclones.

5.8 INVERSIONS

Inversion layers play an important role in influencing the dispersion of air pollutants by restricting vertical mixing. The elevated subsidence inversion is one common example. We shall consider in this section additional mechanisms responsible for causing *frontal, advective*, and *radiational* inversions. In Chapter 6 several other mechanisms will be discussed, but they depend upon specific features of terrain, which is not necessarily the case with the three we consider here.

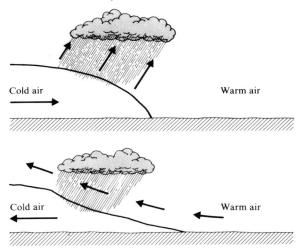

Fig. 5.26 Advancing cold front (upper) and warm front (lower). The vertical scale is exaggerated for clarity.

Frontal inversions

Frontal inversions arise in conjunction with the interface between two air masses of quite different temperature, humidity, and perhaps pressure. This interface is known as a front. Within the front, these parameters change rapidly with location and altitude. In Section 5.3 we remarked that such a boundary often divided the polar air from the more temperate air of the mid-latitudes, and was consequently known as the polar front. In the region of rapid changes of pressure, the wind direction may also change markedly.

If the colder air mass is advancing, the front is known as a cold front; if the warmer air is advancing, it is a warm front. The location where the front touches ground is called the surface front and is often depicted on weather maps by a heavy, solid line. The direction of advance of a cold front is indicated by the sharp points of triangles placed along the line, and the direction of advance of a warm front by the curved surfaces of semicircles, as illustrated in Fig. 5.25.

The structures of the two types of fronts are shown in Fig. 5.26, from which it is evident that the shapes of the fronts differ qualitatively; frictional drag tends to extend and sharpen the wedge of cool retreating air of a warm front, but blunts the leading surface of a cold front. In both cases, the warmer air, being less dense, overrides the cooler air. The slope of the interface is exaggerated in this figure, for clarity. Ordinarily the front rises with a slope from about 1 in 50 to as little as 1 in 250, and extends from the surface front for a width of about 100 km. The rise of warm air over an advancing cold front and the subsequent expansive cooling of this air leads to predominant cloud cover and precipitation which *follows* the position of the surface front. On the other hand, precipitation is commonly found *in advance* of a warm front, as should be apparent from the structure illustrated in Fig. 5.26.

Although the motion of a cold or warm front is predominantly horizontal, an inversion exists for both. The strength of the inversion depends upon the temperature difference between the two air masses. In fact, the inversion serves to sustain the front, for it provides an insulating layer which discourages vertical transfer of heat and moisture between the masses. This inversion will also confine air pollutants to the underlying cold air mass, and may lead to abnormally polluted air should the front stagnate. Unfortunately little is known about the importance of frontal inversions for local air pollution episodes. The meteorological effects are more transient than the more spectacular effects from stagnating anticyclones, and the resulting damage is more difficult to assess. Furthermore, the stronger winds and turbulence associated with the rapid pressure changes in a front usually encourage good horizontal ventilation. In most cases, this apparently compensates for the more limited vertical mixing.

Advective inversion

The second type of inversion which we will take up is *advective inversion*. We recall that *advection* is the horizontal transport of an atmospheric property such as sensible or latent heat, due to the motion of air. One type of advective inversion

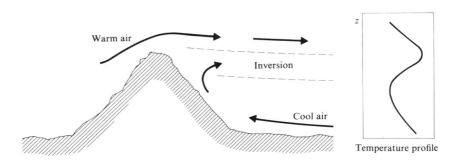

Fig. 5.27   Advective inversion formed by warm air overriding cooler air.

is formed when warm air moves over a cold surface. Convective cooling of the lowest layer of air then leads to formation of a ground-based inversion.

Another example of an advective inversion occurs when warm air is forced to move over the top of a cooler layer. An example is portrayed in Fig. 5.27. Here a mountain range forces a warm desert breeze to flow only at high levels, whereas a cool marine breeze flows at low levels in the opposite direction. There may be some mixing of the two near the mountain slopes. A consequence of the warmer air aloft is the presence of an inversion layer at the interface.

It is worth emphasizing that although an inversion is stable against air motion in the vertical direction, it does not discourage horizontal motion. Strong winds may be encountered within an inversion. In fact, as in this example, a shear may exist in which winds blow in different directions at different altitudes.

The elevated inversion east of the Rocky Mountains, affecting the city of Denver, Colorado, is frequently due to advection and wind shear. A shift in the wind from north to west with increasing altitude causes a cool polar air mass near the surface to be overlain by a warmer westerly flow from across the mountains. The inversions are strongest and most persistent in winter, especially with snow cover.

Radiational inversions

The most common form of surface-based inversion is the *radiational inversion*. It occurs wherever the surface of the earth can become cooler during the night by the thermal radiation of energy. As a result of a decrease in temperature of the ground, the lower atmosphere in contact with it loses sensible heat through conduction, convection, and, more importantly, radiation. Consequently, a temperature inversion is set up between the cool low-level air and the warmer air above. This type of inversion is most pronounced during the late night and early morning hours and is sometimes called a *nocturnal inversion*. The effects of the stable low-level air are evident early in the day when the smoke from chimneys and small fires is confined close to the ground; or if a wind is present, the smoke may form a pattern of narrow horizontal streaks as it disperses downwind.

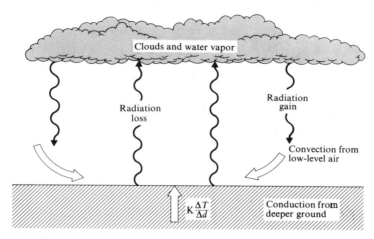

Fig. 5.28   Energy flow to the earth's surface during nocturnal radiational cooling.

Nocturnal inversions are possible because the surface of the earth can cool quickly. In Fig. 5.28 we show schematically the mechanisms by which energy flows toward and away from the surface. As the surface radiates, it also receives a flow of energy from the low-level atmosphere and from the earth below. It is reasonable to expect that the rate at which heat flows upward to the surface Q_h can be approximated by a formula that assumes a linear dependence on the temperature gradient $\Delta T/\Delta d$, where T is the temperature and d the depth below the surface:

$$Q_h = -K_h \frac{\Delta T}{\Delta d}. \qquad (5.16)$$

The constant K_h is called the thermal conductivity and depends upon the composition of the soil, its degree of compactness, and the amount of water present. It may also depend upon the temperature itself, although we shall assume for simplicity that it is independent of temperature. The negative sign is to remind us that heat flows in the direction in which T *decreases*. If the thermal conductivity is sufficiently small, very little of the underlying earth need be cooled before the surface becomes very cold. That is, the surface need radiate only enough energy to cool the upper centimeter or so, at which point the air above will commence to be cooled through its contact. This is in fact what happens. The value of K_h is about 0.2 W/m²-°C for dry soil or about 2 W/m²-°C for moist soil.[4] Therefore, if we were to assume the existence of a temperature gradient in the ground of roughly 50°C/m, which corresponds to many typical situations, the energy flow upward to the surface is about 10 W/m² for dry soil, and 100 W/m² for moist. As a comparison, we note that Eq. (4.2) for black body radiation indicates that the surface of the earth radiates energy at a rate of about 370 W/m². Under these

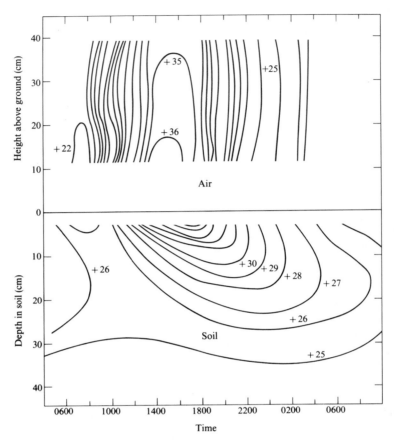

Fig. 5.29 Soil and air temperatures near ground level during a 24-hour period at O'Neill, Nebraska. Temperature is expressed in degrees Celsius. (H. H. Lettau and B. Davidson, eds., *Exploring the Earth's First Mile*, Oxford: Pergamon Press, 1957.)

conditions, the surface can easily radiate away the energy conducted upward from below and could also cool itself and the lowest layer of the atmosphere.

An example of the diurnal temperature variations within the ground and lower atmosphere is shown in Fig. 5.29. This documented case was for terrain covered by short grass. The curves are contours of uniform temperature showing the height or depth at which the indicated temperature was observed during the 24-hour period. The effects of the low thermal conductivity of the soil are evident. At depths exceeding about 0.4 m the temperature shows very little dependence on the daily variation of the surface conditions. The wind speed was fairly light, at about 5 m/sec, so the temperature of the air near the ground readily followed the surface temperature as it cooled during the afternoon from 1400 hours (2pm) until about 0600 hours (6 am) the following morning. The temperature of the air at heights of more than 50 m could not follow so readily the diurnal variation

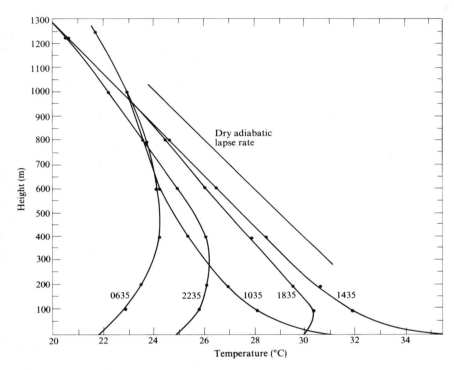

Fig. 5.30 Temperature profiles showing the erosion and formation of a radiational inversion from 0635 to 2235 on 31 August 1953 at O'Neill, Nebraska. (H. H. Lettau and B. Davidson, eds. *Exploring the Earth's First Mile*, Oxford: Pergamon Press, 1957.)

because the thermal conductivity of air is not sufficiently great. The fact that air remains warm at higher altitudes therefore is the reason for the nocturnal inversion. How this affects the temperature profile is illustrated in Fig. 5.30; clearly the rapid temperature response of the earth's surface is possible only because the thermal conductivity of the air and ground is so low. The rapid nocturnal cooling of the ground and adjacent air may lead to what are called ground-based *radiation fogs* if the air is cooled below its dew point.

The true situation in a radiational inversion is never quite as simple as our explanation would make it seem. For example, there is always absorption of the earth's radiation by the atmosphere, and the subsequent greenhouse effect, when a part of this energy is reradiated back to earth, tends to weaken the amount of surface cooling and therefore the strength of the inversion. This is particularly important during humid or overcast nights when the water vapor and water droplets close to the surface strongly absorb the earth's radiation, and the atmosphere's thermal radiation effectively counteracts the cooling mechanism. Fruit growers have long known that the greatest threat of frost comes on a cold, clear night; and in former times they tried to generate artificial clouds by the use of smudge pots in the orchards. However, the air pollution was good neither for neighbors nor for trees, so these devices have been largely replaced by orchard heaters which directly warm the air and encourage vertical ventilation or by motor-driven fans which stir the air to replace the low-lying layer by warmer air from above. It is necessary to employ such aids, because the formation of ground-based radiational inversion and atmospheric stability discourages convection which would otherwise convey heat downward from the warm air above by natural means.

The prevalence of radiational inversions is cause for many nocturnal air pollution problems. The ground-based inversion layer may extend upward for 100 m or more, depending upon advective effects, the amount of moisture in the soil, and other factors. Thus pollutants which are emitted at low levels will remain until the inversion is broken. A radiational inversion is quickly dissipated at sunrise as the ground warms, provided that a dense fog has not developed. During stagnation of an anticyclone, clear skies will encourage radiational inversions. Hence the subsidence inversion of these high-pressure regions may be reinforced by a ground-based nocturnal inversion, to the detriment of all.

SUMMARY

Global circulation of air ensures the eventual widespread dispersal of gaseous air pollutants. But features of the general circulation also contribute to local pollution episodes. The three-zone model, for example, explains the origin of the semipermanent anticyclones where air subsides from high levels over the oceans and coastal regions in the subtropics. Formation of an elevated inversion layer is a result of compressive heating of the subsiding air aloft. The nearly geostrophic flow of air in the horizontal direction, a result of a near balance between the Coriolis force and pressure-gradient force, strengthens the subsidence inversion over the eastern ends of the subtropical anticyclones. Departures from geostrophic flow, partially a result of friction with the ground, cause a net horizontal motion of air from high-pressure anticyclones toward low-pressure cyclones. The parade of anticyclones and cyclones that moves eastward through the temperate zone is responsible for the changeable weather in this zone of general circulation. Anticyclones, because of their subsidence inversion and gentle wind, are conducive to the accumulation of air pollutants at low levels, especially in winter when a low ground temperature strengthens the inversion. Serious episodes of pollution have occurred when such high-pressure regions stagnate for a few days over a locality. Other types of inversions are associated with fronts and with the advection of warmer air over a cooler layer. The most common type of inversion is formed at night by the radiative cooling of the ground and adjacent layer of air, but it is usually eroded by solar heating of the ground during the morning.

NOTES

1. E. Palmen and C. W. Newton, *Atmospheric Circulation Systems*, New York: Academic Press, 1969.

2. D. Fultz, *et al.*, "Studies of Thermal Convection in a Rotating Cylinder with some Implications for Large-Scale Atmospheric Motions," *Meteorological Monographs*, **4**, *21*, 22 (1959), American Meteorological Society, Boston.

3. H. R. Byers, *General Meteorology*, New York: McGraw-Hill, 1959.

4. R. E. Munn, *Descriptive Micrometeorology*, New York: Academic Press, 1966.

5. M. Neiburger, "The Relation of Air Mass Structure to the Field of Motion over the Eastern North Pacific Ocean in Summer," *Tellus*, **12**, 31, 1960.

FOR FURTHER READING

J. EDINGER	*Watching for the Wind*, Garden City, N.Y.: Doubleday-Anchor, 1967.
G. M. HIDY	*The Winds; The Origin and Behavior of Atmospheric Motion*, Princeton, N.J.: D. Van Nostrand, 1967.
A. MILLER and J. C. THOMPSON	*Elements of Meteorology,* Columbus, Ohio: Charles E. Merrill, 1970.
S. PETTERSSEN	*Introduction to Meteorology*, 3rd ed, New York: McGraw-Hill, 1969.
E. PALMÉN and C. W. NEWTON	*Atmospheric Circulation Systems*, New York: Academic Press, 1969.
G. T. TREWARTHA	*An Introduction to Climate*, New York: McGraw-Hill, 1968.
R. S. SCORER	*Air Pollution*, Oxford: Pergamon Press, 1968.

QUESTIONS

1. What causes the semipermanent subtropical anticyclones? In which direction would you expect them to shift as the season changes from winter to summer?

2. Trace the flow of sensible and latent heat through the various modes of atmospheric circulation from equatorial to polar regions.

3. Why do anticyclones have elevated inversion layers? What causes the "high pressure" of an anticyclone?

4. Which portion of an anticyclone is most likely to cause the greatest accumulation of pollutants near the ground, due to poor ventilation?

5. Why do migratory anticyclones and cyclones generally move eastward?

6. In the Northern Hemisphere the Coriolis force deflects winds to the right. Why then do winds blow counterclockwise around a cyclonic low-pressure region, appearing to be deflected to the left?

7. Can geostrophic wind exist at the equator? Explain your answer.

8. Consider a westward moving wind in the temperate zone of the Northern Hemisphere. Can you understand why the wind tends to be deflected northward as a result of its inertia? [Hint: Look at the inertial tendency of the wind to move in a straight line as the earth rotates by 180° about its axis.]

9. Air at low levels is found one day to have a lapse rate of the potential temperature of $-1°C/km$. Under what conditions is the air stable? Unstable?

10. Why is the elevated inversion of the semipermanent anticyclones strongest over their eastern ends? Why is it weak or nonexistent over the western ends?

11. What types of inversions are commonly found over your home town? When do they occur and how frequently? How can you tell when one is present?

12. Where over the globe is the weather least changeable? Why?

13. Why does the geostrophic wind speed increase most rapidly with height where there is the greatest horizontal temperature gradient, e.g., in the vicinity of fronts?

PROBLEMS

1. For the isobar spacing near but not at the center of the high-pressure region shown in Fig. 5.23, estimate the wind speed for geostrophic balance.

2. What is the magnitude of the force on a cubic meter of air which is in a pressure gradient of 1 mb per 100 km? What is the wind speed in geostrophic balance for this pressure gradient at a latitude of 40°N? If the pressure gradient is directed toward the south, in which direction is the wind?

3. If a low-level wind at 50°N is observed to make an angle of 60° with respect to the pressure gradient, what is the magnitude of the effective frictional force from the ground if the pressure gradient is (a) 1 mb/100 km and (b) 2 mb/100 km?

4. A parcel of air at a temperature of 30°C commences to rise as a consequence of its buoyancy. Assuming that its relative humidity is initially 80%, by how far will it have risen when the relative humidity reaches 100%? You may wish to consult Fig. 3.2. Assume that the vapor pressure of water depends only upon the ambient temperature.

5. Suppose the relative humidity of low-level air is 50% and it has a temperature of 20°C at the ground. What temperature lapse rates correspond to stable and unstable conditions? If the lapse rate at low levels is neutral and if the absolute humidity is independent of altitude, at what altitude will the atmosphere become unstable? Use the approximation indicated in the previous problem.

6. A dry exhaust plume from an oil refinery is emitted at night at a temperature 10°C warmer than its surroundings. If the low-level lapse rate is $-8°C/km$, what is the maximum height the plume may reach? What approximations did you make to obtain your answer?

7. If smoke from a burning refuse dump is 20°C warmer than the surrounding air at a height of 10 m, what is the maximum height it could attain if the low-level air is neutrally stable up to a height of 110 m, and an elevated inversion with a lapse rate of $-20°C/km$ persisted above that? Assume that the smoke and air are dry. Qualitatively, how would your answer be changed if the smoke were saturated with water vapor?

CHAPTER 6

MESO- AND MICROMETEOROLOGY

We now take up the subject of air circulation on a smaller scale. *Mesometeorology* deals with weather patterns whose extent is less than about 1000 km and, more commonly, less than 100 km. These include thunderstorms, the sea and land breezes, tornados, and mountain-valley breezes. We will consider only a few of these—the ones that bear most significantly on the phenomena of air pollution. The motion of air in patterns on an even smaller scale is the subject of *micro-meteorology*. This encompasses the many turbulent processes of air motion near the ground—evaporation, heat conduction, and diffusion of contaminants, to name but a few. Sometimes there is no sharp distinction between micro- and meso-scale phenomena, and it will serve no purpose to make a distinction in this book. Generally the mass of air involved in meso- and micro-circulation is not large; motion is on such a small scale and is so short-lived that the Coriolis force plays only a minor role. One common aspect is found in all of the phenomena included in this chapter. The dominant processes occur in the lower atmosphere where ground features and terrain are prominent influences.

6.1 MONSOON

Some features of local wind patterns are predictable from the nature of the topography. One example is the *monsoon*, a name derived from the Arabic word for season. Some monsoons are virtually continental-scale circulations and would be more properly considered macroscale phenomena. Others, however, are more local in extent. The circulation of air in a monsoon is a good example of a simple convection cell in which, by and large, the air circulates in a closed pattern. During the summer season, when land is warmer on the average than the oceans' surfaces, low-level air over the continents also tends to be warmer than that over the oceans. So the relative buoyancy of the warmer land air causes it to rise, as illustrated in Fig. 6.1. Cooler ocean air displaces the warm air at low levels, moving toward the interior to warm in turn and rise. The convective cell shown in the figure is completed as the air at high levels moves seaward and sub-sides as it takes the place of the lower-level air moving toward the land. During the winter season, the more rapid cooling of the land, contrasted with a much smaller decrease in ocean temperature, reverses the cycle.

In southeast Asia and India where the monsoon is well established, seasonal

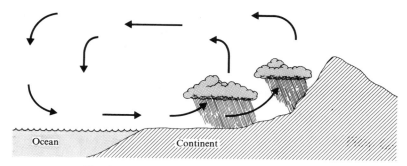

Fig. 6.1   Convection cell of a summer monsoon.

precipitation is linked with the convection. As the onshore moist air penetrates inland, it is forced to rise; the expanding air cools sufficiently for condensation, with consequent cloudiness and considerable precipitation. This is especially marked if terrain forces a rapid uplift. As a result, precipitation is especially heavy in the southern foothills of the Himalaya Range, for example Cherrapunji at about 1400 m elevation experiences an average yearly rainfall in excess of 10 m.

The winter monsoon, by contrast, is unlikely to result in precipitation over continental regions, because the air is subsiding as it moves toward the ocean. On the other hand, once the air is over the ocean, it quickly absorbs moisture and may cause heavy rainfall in the western mountains of offshore islands such as the Philippines, Japan, and Taiwan. Monsoon circulation over the continent of North America is less noticeable owing to perturbations from migrating cyclones and anticyclones in the temperate zone. However, in summer there is a tendency for air to be drawn from the Gulf of Mexico into the warm southwest, an effect which may be partly responsible for the summer thundershowers over the plateaus of New Mexico, Arizona, and southern Colorado. There is also a mild monsoon from the Pacific across southern California to the high plateau of Nevada and Arizona.

6.2 SEA AND LAND BREEZES

The sea and land breezes are convection cells akin to the monsoon, but much more localized. Only a region extending several tens of kilometers to either side of a coastline is affected. These breezes are not limited solely to seacoasts, but may take place near the shore of a large lake. They also have a diurnal rhythm instead of enduring for a few weeks or months as does the monsoon.

The *sea breeze* is formed because the surface of the ground quickly warms in response to absorption of solar radiation. The opaqueness of the soil and its very low thermal conductivity ensure that the absorbed solar energy is initially confined to the upper centimeter or so of earth, as we discussed in Section 5.8.

Consequently, the surface temperature increases rapidly. Convection and radiation then transmit a portion of this energy through the lower layer of the atmosphere, resulting in a general warming.

During this time, the surface temperature of the ocean hardly changes. Solar radiation penetrates to depths of 5 to 10 m below the surface and so is absorbed by a large volume of material. Wind-caused eddies in the upper layer can further distribute the energy to a total depth of perhaps 200 m. As a consequence, the solar energy is distributed over a much greater volume of water than of inland soil. Since the specific heat of water is also somewhat greater than the specific heat of dry ground, the temperature rise of the ocean's surface is negligible in comparison with the temperature rise of the land's surface. So the temperature of the low-level air over the ocean remains essentially constant while the temperature of the land air increases.

The heating and accompanying expansion of the land air begin a seaward movement of air extending far up into the troposphere, with resulting wind speeds on the order of 1 km/hr. However, simultaneously, the land air commences to ascend because of its buoyancy. The horizontal movement at low levels reverses as the cold and denser sea air moves inland to take the place of the rising air, as in Fig. 6.2. Close to shore this breeze may be quite steady at a speed of about 10 km/hr. In the upper atmosphere, the convection cell is closed by air returning seaward at a height of 1 km or so and with a speed of about 5 km/hr; in the tropics, the effect may be sufficiently strong to cause a cell height of 3 or 4 km. Air which then subsides over the sea may cause an elevated inversion to form. It is not uncommon for the convection cell to extend for 20 km to either side of the coast, although in some circumstances the sea breeze may extend inland for 80 km.

As day progresses, the cell widens, and air trajectories lengthen; effects of the Coriolis force become more important and the wind direction in higher regions conforms more nearly to geostrophic flow. Late in the afternoon the

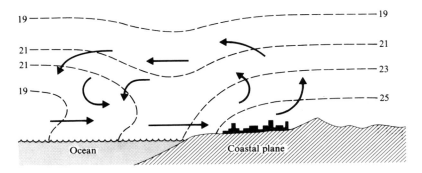

Fig. 6.2 Representative air circulation during a sea breeze. The dashed curves represent contours of uniform temperature (isotherms), and the numbers give their respective temperatures in degrees Celsius.

flow is more nearly parallel to the isobars, so the sea breeze may actually parallel the coast. In the Northern Hemisphere it would move northward along east coasts and southward along west coasts.

With nightfall, the situation may reverse. On clear nights, ground temperatures fall rapidly at the surface owing to radiational cooling. The low-level land air then cools by convection to the surface, establishing a ground-based radiation inversion, and moves seaward to displace the warmer sea air. This *land breeze* sets up a convective cell in which the flow is the reverse of that for the sea breeze in Fig. 6.2. Returning air which descends over land from high levels warms by compression and reinforces the radiational inversion. This may lead to very stable air which severely discourages vertical mixing. Owing to the stable lapse rate over land and the consequent limitation of the volume of air which can be cooled effectively by the ground surface, the land breeze is usually weaker than the sea breeze.

One of the potential adverse effects of diurnal cycles such as the land-sea breeze is that pollutants which have been removed from a particular location during the day may be blown back at night. This appears to be a real contribution to the pollution problem of some large coastal cities, particularly if an elevated inversion layer from an anticyclone caps the convection cell. Several times a year the sea breeze off Lake Michigan aggravates the pollution conditions of Chicago, Illinois.

6.3 MOUNTAIN AND VALLEY WINDS

On a sunny day, air near a heated mountain slope will warm faster through its radiative transfer and convection of sensible heat from the ground than air at the same altitude but farther away. The relative buoyancy of the warmer air near the slope causes an upslope wind, a phenomenon familiar to every mountain climber. The situation is illustrated in Fig. 6.3. The depth of the upslope wind

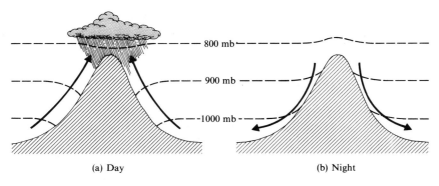

(a) Day   (b) Night

Fig. 6.3 (a) Upslope daytime winds; (b) downslope nocturnal winds. The dashed curves represent contours of uniform pressure (isobars).

varies according to the intensity of sunlight and the slope inclination, but it typically extends for several hundred meters above the terrain. It often results in clouds and showers over mountain peaks as the rising air cools by expansion.

With sunset the process reverses. Radiational cooling rapidly lowers the temperature of the slope and the adjacent air, and this less-buoyant air drains down into the valleys. The downslope flow is shallower than the upslope day breeze, because convection is discouraged by the degree of stability established in the air near the slope.[2] If the terrain has converging valleys and narrow passages, the drainage winds may attain high speeds. In some geographical areas, mountain-valley winds reinforce the strength of other types of mesoscale flow. Coastal mountains for example encourage daytime and nighttime winds which reinforce the land-sea breeze.

The cyclic nature of mountain and valley winds can encourage pollution episodes such as occur with a sea breeze. The city of Denver, Colorado, for example is situated on the South Platte River, which flows toward the northeast in the direction of generally decreasing land elevation. In the absence of weather disturbances associated with fronts, when winds are light (5–10 km/hr) for several successive days, a daily wind regime is established in which there is downslope drainage at night and in the early morning that carries pollution from the city toward the northeast. The flow however reverses during the afternoon, and the north or northeast wind carries the already polluted air back across the city. As a consequence, the onset of heavy pollution in the central city and northeast suburbs often takes place in the late afternoon or early evening.

Nighttime downslope winds encourage atmospheric stability when they converge on an enclosed valley and collect over the bottom land. Citrus growers have long recognized this as a factor increasing the chance of frost and prefer to locate orchards at higher altitudes on the surrounding slopes. The cold, dense air hugging the valley floor also aggravates conditions for serious air pollution, for it can cause a ground-based inversion or strengthen an existing radiational inversion. This is because the low-level air in the valley is surmounted by warmer air which is relatively insulated from the cold mountain slopes or valley floor. The resulting stable layer can extend upward for several hundred meters, and presents a serious hindrance to the disperal of pollutants released near the ground.

6.4 KATABATIC WINDS

Drainage winds may persist for several days if a large body of cold air over an inland plateau is set in motion in response to a low-pressure area which is down-slope. Such winds are called *katabatic* from the Greek meaning "to go down." Usually they are gentle breezes not exceeding 5 km/hr. But some of the strongest occur in Norway when drainage winds are further cooled by passage over a snow-field or glacier and are then confined to flow through a narrow canyon or fjord. In this section, we shall consider several examples of katabatic winds.

Bora

The northeast coast of the Adriatic is framed by a narrow coastal plain backed by inland plateaus. Air which has accumulated inland during prolonged periods of cold weather continually seeps outward through mountain passes toward the sea, where the warmer low-level air is less dense and therefore more buoyant. This is a good example of a katabatic wind. During periods, when a low-pressure area stands over the Adriatic, the drainage is especially pronounced and gives rise to strong, cold winds called the *bora* whose wind speeds occasionally reach 100 km/hr or more. Good ventilation of coastal regions is a result. The effect is most frequent and strongest during the later hours of the night, when reinforced by the land breeze.

Mistral

A similar wind is caused by drainage from the central highlands of France and Germany toward the Mediterranean. Ordinarily the effects of air pollution are minimal during the bora and mistral because of the vigorous mixing occasioned by the high winds. However, blowing dust can cause considerable discomfort. Also the fire danger is high, and there have been many times when the skies were blackened by smoke from widespread conflagrations.

Santa Ana

Drainage from the high inland Nevada plateau toward the Pacific Coast in southern California causes a condition known as the *Santa Ana*. The name is derived from one of the passes in the intervening mountains, where these east winds are particularly strong. Gusts up to 100 km/hr are not uncommon. Drainage is most pronounced when reinforced by geostrophic flow in response to a high-pressure region to the north and a low-pressure one to the south, as illustrated in Fig. 6.4. When the air descends from the plateau, it warms as it is compressed, and temperatures of 27°C (80°F) on the plateau are increased to 32°C (90°F) near the coast. As the temperature increases, the relative humidity of the air decreases, sometimes producing relative humidities of less than 4% over the coastal plane. Dust often accompanies the Santa Ana, and the hot dry air and high winds may be distinctly unpleasant. Fire danger is extremely high under these conditions, especially since the Santa Ana may arrive after a period of six or more dry months. On the day described in Fig. 6.4, widespread fires destroyed over 500 homes and structures in southern California. Particulate matter emitted into the air was carried for great distances, and reports more than two weeks later indicated that increased turbidity of the air was observed several thousand kilometers away.

Although excellent horizontal ventilation exists during the Santa Ana, pollutants carried by the wind may not be harmlessly dispersed. The southwestward or westward flow of the wind veers more toward the northwest as the high-pressure regions migrate eastward. Air pollutants from the metropolitan

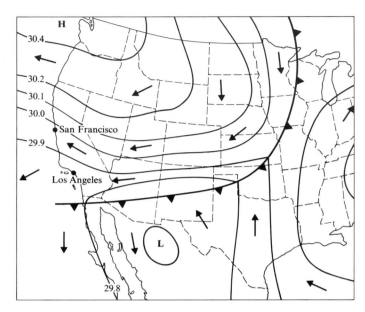

Fig. 6.4 High- and low-pressure regions and associated winds during a Santa Ana on September 25, 1970. Pressures along the isobars are stated in inches of mercury (1 inch of mercury = 33.9 mb).

Los Angeles region lie offshore or drift northwest as the winds weaken. When the Santa Ana dies and the sea breeze commences, the pollutants are sometimes blown back to the coast, affecting an area stretching a hundred kilometers or more from the city. Some of the most severe episodes of photochemical smog affect coastal cities following a Santa Ana. During the period between the demise of the Santa Ana and the development of the marine breeze, a mass of polluted air may lie offshore as a smog bank, prevented from vertical mixing because it is confined by the elevated inversion of the Pacific semipermanent anticyclone.[3]

The characteristic warming of the Santa Ana as it descends from the high inland plateau distinguishes it from a customary slow-moving, cold katabatic wind. But it is a point of similarity with our next example of mesoscale wind patterns, the foehn. Before we take this up, however, we shall mention a few other peculiar winds.

Miscellaneous cases

Many other local names have been attached to unusual winds which result from traveling low- and high-pressure regions, often reinforced by the drainage from high terrain. Most conditions are notable, because climatic characteristics of one region are transferred to quite a different one, such as in the "sirocco" which blows from north Africa over the central Mediterranean and southern Italy in front of an advancing low. Another example is the "pampero," a cold wind from the Antarctic which rushes northward over the Argentine pampas, occasionally surging northward through Bolivia and across the equator to the Amazon basin.[4]

Except when followed by a restoring condition, such as the sea breeze after a Santa Ana, most of these unusual winds assist the dispersal of pollutants from anthropogenic sources. However, the strong winds are often responsible for entrainment and transport of considerable quantities of contaminants such as sand and dust.

6.5 FOEHN AND CHINOOK

The *foehn* is a hot, dry wind which develops occasionally on the lee slope of a mountain range, particularly the northern slopes of the Alps and on the east face of the Rocky Mountains of North America. Less intense but more prevalent examples may be found on the east face of the Cascade Range in Oregon and Washington. In the American West these winds are usually called *chinooks*, for the Indian territory of the Rockies from which the local variety seemed to come. It was known to the Indians as "snow-eater" because the extreme warmth and dryness could melt as much as half a meter of snow in a day or two. In Germany these winds are called *foehns*, a term in more universal use. The extreme heat and dryness, with strong gusty winds as in the Santa Ana, may cause physiological as well as psychological discomfort.

The origin of the foehn can be seen by reference to Fig. 6.5. Winds which are uplifted by contact with the windward slopes of a mountain range will cool as the air rises and expands. For a sufficiently high relative humidity, precipitation develops over the windward slopes; and as the air ascends, it cools at the *wet adiabatic rate* of about 6°C/km. Passing over the ridge and descending on the lee side, the air warms at the *dry adiabatic rate* of 10°C/km, obtaining the benefit in thermal energy of previous liberation of latent heat from the condensing water vapor. Thus, at a given altitude, it is considerably warmer and less humid on its downward journey than on its previous ascent. The loss of moisture on

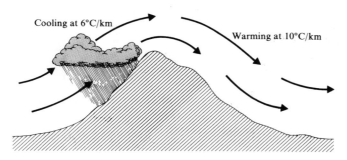

Fig. 6.5 Wind patterns during a foehn.

the windward slopes may result in appreciable climatological differences on the two sides of the mountain range if wind patterns are fairly steady throughout the year. This is so in the state of Oregon, for example. Dense forests cover the west slopes of the Cascade Range, but an arid desert region is to the east.

The wind may become especially warm if the descent on the lee side is greater than the ascent to windward. When a low-pressure region moves into the lee side, wind speeds and ventilation are enhanced. However, in less intense situations, the subsiding air on the lee may not descend to ground level. This may be the case if a shallow layer of cold air is trapped near the ground, as occurs on the east side of the Rocky Mountains on occasion after passage of a Canadian cold front. In this situation an elevated inversion may be established between the high-level subsiding air and the low-level cool air. This can effectively block upward mixing, thereby establishing conditions for the accumulation of air pollutants. The city of Denver suffers from such conditions.

6.6 CLIMATE OF A CITY

Urban regions affect the local weather. We have previously noted that mechanically induced turbulence from large structures may provide such a strong frictional effect that the low-level winds deviate by 40 degrees or more from the direction of geostrophic balance at higher altitudes. The nature of cities contributes more than a mechanical influence on the micrometeorology. The weather is also affected because a city forms a *heat island*. By this we mean that temperatures within a built-up area may be many degrees warmer than the surrounding rural environment.

The accumulation of sensible heat by urban air occurs for several reasons. An evident one is that numerous closely spaced sources emit heat. These include automobiles, factory ovens, domestic space heaters, power generating stations, and underground steam lines. Although the total heat emitted by all anthropogenic sources is negligible on a global scale when compared with the incident solar radiation, this is no longer true in urban areas. Built-up areas of Vienna reportedly release each year from $\frac{1}{6}$ to $\frac{1}{4}$ of the solar energy which is incident on that city (excluding parks). For Berlin, the fraction is estimated to be about $\frac{1}{3}$.[5] The effect of this additional energy on the ambient temperature may be accentuated in some cities which suffer from moderate or heavy air pollution. It has been suggested that pollutants such as sulfur dioxide in sulfurous smogs and ozone in photochemical smogs may intensify the greenhouse effect over urban regions, leading to even higher ambient temperatures. Calculations indicate that heating rates may reach 5°C per day.[6]

However, empirical data on the local greenhouse effect is lacking. Mitigating factors may be important; for example, a high atmospheric concentration of aerosol matter increases the urban albedo, reflecting as much as 15% of the incoming solar radiation.[7] This would decrease the amount of solar energy available for conversion into sensible heat at the ground. The absence of standing

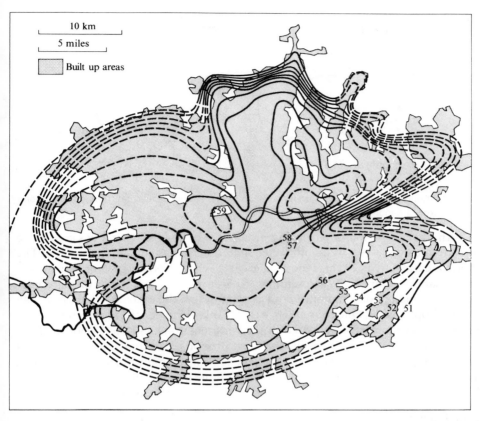

Fig. 6.6 London heat island on June 4, 1959. Temperatures along the isotherms are indicated in Fahrenheit degrees. (T. J. Chandler, *The Geographical Journal*, **128**, 279, 1962; courtesy of the Royal Geographical Society.)

water or damp ground also leads to a reduction of the relative humidity at low levels by as much as 30% in comparison with air over the surrounding rural terrain.[8] The lower concentration of water vapor would weaken the urban greenhouse effect.

Despite these mitigating factors, cities are, in fact, observed to be considerably warmer than their surroundings. Figure 6.6 shows isotherms (contours of uniform temperature) for the city of London on a summer day. The center city is some 5.5°C (10°F) warmer than the outskirts. On the day illustrated, prevailing winds were from the southwest, accounting for a slight shift of the warmest regions to the northeast of the exact center. Furthermore, this wind led to a greater separation of the isotherms in the southwest and a smaller separation in the northeast. Drainage patterns of local winds severely disturb the isotherms and produce their irregular shapes. The cooling effect of the Thames River is also evident.

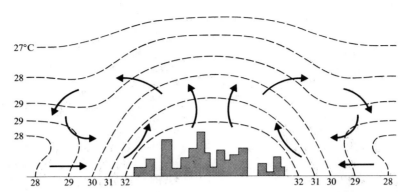

Fig. 6.7   Country breeze flowing at low levels toward an urban heat island. Temperatures along the isotherms are indicated in degrees Celsius.

Another feature of a city is the prevalence of large surface areas covered to a moderate depth with well-compacted material such as concrete or granite. The thermal conductivity of this type of aggregate is substantially greater than for soil in the suburban or rural surroundings. Therefore a greater depth of material is heated during the day and must cool at night before the surface temperature can change appreciably. One consequence is a time-lag between the rapid temperature variations of the surface material in the rural environs and the response of the surfaces in the city. As a result, the most pronounced temperature differences are observed at night when the rural surface soil cools rapidly by thermal radiation. This difference may be as much as 10°C for large cities such as San Francisco or London. The daytime temperature difference is generally less.

Local winds

An interesting feature of the urban heat island is that it encourages air circulation. As warm air rises over the city, it is replaced by cooler air blowing in from the surroundings. The incoming air is known as the *country wind*. In some cases a closed convection cell may be formed, as illustrated in Fig. 6.7. The return flow at high levels may, as it subsides, form an elevated inversion layer such as occurs in a sea breeze. The country wind should be especially noticeable in the absence of macroscale wind patterns, since the wind is fairly light.[9] However, the heat island in many cities is not sufficiently effective to ensure good ventilation. The pronounced surface roughness caused by the irregular pattern of buildings reduces wind speeds at low levels and thus discourages advection.

Expansive cooling of the ascending air over metropolitan regions appears to be responsible for causing slightly more condensation and precipitation over a city than over its environs. This may also be encouraged by a higher concentration of hygroscopic aerosols. Cities are found to have 100% more fog in winter, 5–10% more cloud cover, and perhaps 10% more precipitation than the surrounding countryside.[8,10] The role played by urban features in influencing the local climatology is incompletely understood, and the subject remains controversial.

6.7 AN AIR POLLUTION EPISODE: DONORA, 1948

Factors of meso- and micrometeorology have on occasion hindered atmospheric mixing to such an extent that episodes of severe air pollution have been experienced. An episode in Donora, Pennsylvania, in 1948 is an example. A five-day period of near-stagnation brought on such aggravated conditions that 17 deaths occurred within a 14-hour period, when only one death might have been expected under ordinary circumstances. This was the first clear instance in the United States of air pollution exacting a toll in epidemic proportions.

The episode at Donora began on Tuesday, October 26, when an east coast storm was replaced by a cold anticyclone advancing from the southwest. A cold ground intensified the elevated inversion of the anticyclone as it moved in. And then the high-pressure region stagnated over western Pennsylvania for five days until Sunday, October 31, at which time it took up its movement toward the northeast and the episode ended. During the period of stagnation, the anticyclone moved only a few hundred kilometers, and the elevated inversion layer extended over a large portion of Pennsylvania and neighboring states, as shown on the weather map in Fig. 6.8. Winds were gentle, as indicated by the wide spacing between the isobars.

Yet only the inhabitants in the immediate area of Donora experienced an abnormally high rate of morbidity and mortality. None of the other cities in western Pennsylvania, many of which were in approximately similar geographical

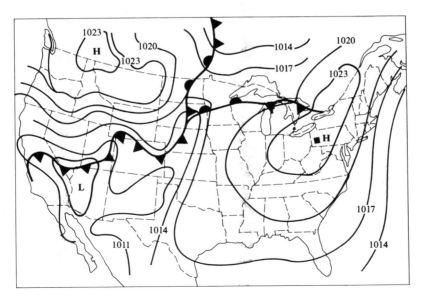

Fig. 6.8 Anticyclone stagnating over northeastern United States on October 26, 1948. The location of Donora is marked by the square. Pressures along the isobars are given in millibars.

locations, suffered similar fates. Clearly the poor ventilation was aggravated by local conditions of meteorology and pollutant emissions. Yet the 1948 episode was an unusual situation, because on at least two previous occasions Donora had been subjected to stagnating anticyclones for longer than a week, with no remarkable toll of life.

Perhaps the best way to convey a sense of what happened is to quote a portion of the description by Berton Roueché[11] based on eye-witness accounts:

> The fog closed over Donora on the morning of Tuesday, October 26. The weather was raw, cloudy and dead calm, and it stayed that way as the fog piled up all that day and the next. By Thursday, it had stiffened adhesively into a motionless clot of smoke. That afternoon it was just possible to see across the street, and except for the stacks, the mills had vanished. The air began to have a sickening smell, almost a taste. It was the bittersweet reek of sulfur dioxide. Everyone who was out that day remarked on it, but no one was much concerned. The smell of sulfur dioxide, a scratchy gas given off by burning coal and melting ore, is a normal concomitant of any durable fog in Donora. This time it merely seemed more penetrating than usual.

And the fog with its burden of pollutants did not lift for many days. The extreme persistence of conditions and atmospheric stability were largely results of the peculiar terrain and micrometeorology.

The city of Donora is located in the valley of the Monongahela River some 50 km south of Pittsburgh. About 14,000 people lived on the west bank, inside a horseshoe bend of the river. Another 1000 people lived in the village of Webster on the opposite bank, on the outside of the bend. The slopes of the valley rise sharply upward to a height of about 100 m to the east of Webster; on the west of Donora, the rise is more gentle and leads on to rolling hills. The bottom land thus forms a drainage basin for cold downslope winds at night. At the time of the episode, the ground-based inversions were strengthened by radiational cooling of the valley floor, and together they could produce a strong temperature inversion with a temperature gradient as high as $33°C/km$, as has been measured on subsequent occasions.[12]

As night fell on October 25 and the ground cooled, the high relative humidity resulting from the previous storms quickly led to a saturated condition within the lower region of the cold anticyclone. The subsequent condensation caused extensive fog over western Pennsylvania. This condition was especially pronounced in the river valleys like that at Donora, where evaporation sustained the high humidity and a high aerosol concentration in the polluted air encouraged the formation of numerous water droplets. The fog was held close to the ground by the stability of the elevated inversion layer.

Under these circumstances, the fog absorbed all of the earth's thermal radiation close to the ground and prohibited further development of a ground-based radiational inversion. However, the upper layers of the fog continued to radiate their energy into the air and thus continued to cool. Such cooling is observed also in the upper layers of elevated stratus clouds; as a result one finds there additional condensation and a higher content of water, amounting to as much as 0.5 g/m^3.[13] In elevated clouds the distribution of droplet sizes peaks at about 7 microns radius, which is similar to the size found in many fogs. As the top layer of a fog cools, the temperature profile below changes toward a more unstable condition. Since the air is saturated, an unstable lapse rate will exist if the decrease in temperature with height is greater than the wet adiabatic lapse rate of about 6°C/km. Thus at night there may be general mixing of pollutants within a fog as the cold, denser upper layers subside and are replaced with warmer air from below. But because the upper layer of the fog is cooling with respect to the overlying clear air, an elevated inversion is formed, as illustrated in Fig. 6.9(a). This stable layer acts as a blanket to confine the fog and its burden of pollutants.

Usually the fog lifts after sunrise when solar radiation penetrates to warm the ground, and the upward convection of sensible heat breaks the inversion. However this did not occur on October 26, because the fog was too dense and deep; the upper layers reflected most of the solar energy. The albedo of a stratus cloud such as fog depends upon its thickness, and some measurements indicate it may vary from 40% for a 150 m thickness to 80% for a 500 m thickness. In elevated clouds, very little radiation is absorbed, perhaps as little as 7% for thick clouds.[13] And most of this is absorbed in the upper layers. Thus if there is any appreciable warming it occurs in the upper layers of the fog, with the result that the stability of the air within the fog is increased. This is illustrated in Fig. 6.9(b). The delay in heating the ground, accompanied by stable lapse rate, discourages upward convection of energy which would otherwise serve to warm the air and evaporate fog droplets. Thus the fog is perpetuated if it is sufficiently dense and thick, as occurred in Donora. Furthermore, the daytime stability within the fog led to an accumulation of air pollutants.

Within the interior of the fog-bound valley, pollutants continued to be emitted. The economic life of the communities was centered around employment

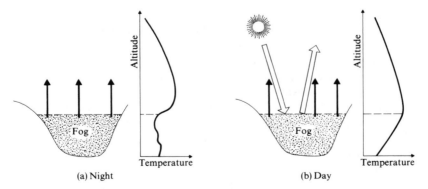

Fig. 6.9 Representative temperature profiles for a fog-filled valley (a) at night and (b) in the daytime. Broad arrows indicate solar radiation, and thin arrows thermal radiation from the upper layer of the fog.

at a large steel and wire mill, a zinc plant, and an accompanying sulfuric acid plant. Approximately 3000 people worked at these river-front establishments. The bulk of the fumes and particles from these plants was released from exhaust stacks of various heights, none exceeding 40 m. The zinc plant processed zinc sulfide ores by roasting in order to replace the sulfur by oxygen. Sulfur dioxide is a byproduct of the process and some of it was collected and further oxidized to form sulfuric acid for commercial sale. Great quantities of sulfur dioxide and particulate matter were emitted as pollutants, including substantial quantities of zinc, lead, and cadmium. Emissions from the steel mill, especially from the blast furnaces and open hearth furnaces, consisted mainly of sulfur dioxide and particulate matter. Domestic use of soft coal for space heating added to the atmospheric burden of carbon monoxide, particulate matter, and sulfur dioxide, although the daily amount of the latter was only about a quarter of the amount emitted by the zinc plant.

During the third and fourth days of the episode, as ambient concentrations of pollutants escalated, a large fraction of the population became ill. No measurements of ambient levels of pollution were taken during the episode, but a lower limit is suggested by levels monitored during a brief period of stagnation from April 20 to 21, 1949, during which however no marked increase in mortality occurred. During this latter period the mass concentration of particulate matter reached 4 mg/m^3 and the sulfur dioxide concentration 0.5 ppm.

The climax of the episode came on Saturday, October 30 when the atmospheric stability intensified. Seventeen deaths were recorded that day and two more the next. These were exclusively people over 52 years of age, many of whom had a history of cardiac or respiratory disease. During the entire episode, some 43% of the population experienced some effects from the smog. Others seemed hardly to be bothered, and a few were even unaware of the extent of the illness among the population. Of the sufferers, 17% of the population were moderately affected and 10% severely affected. The susceptibility of older people is indicated by the fact that of the group 60 years and older, a much larger percentage of about 29% suffered severely. Also, of the 50 persons who were hospitalized, more than two-thirds were over 55 years old. Most of the affected, some 90%, complained of upper respiratory symptoms such as nasal discharge, constriction of the throat, or sore throat. Some 87% reported symptoms that affected the lung.

Subsequent epidemiologic studies have shown that the incidence of illness, especially severe illness, was highest in Webster to the east and across the river from the steel mill and zinc plant. Furthermore the death rate in Webster was 6.6 per thousand, whereas for Donora it was only 0.9 per thousand, a discrepancy which can be accounted for by neither age nor state of health. The reason for this is not known with certainty. Perhaps it resulted from the prevailing northwest winds each morning, blowing at about 5 km/hr, which carried the effluent from the zinc plant over the river and directly into the town. Early mornings were a period of especially pronounced stability, a result of the nighttime drainage winds from the hillsides. Furthermore a portion of the town of Webster is on the flank of the hillside, directly exposed to pollutants emitted from smokestacks at the same level. Thus the town of Webster presumably received higher concentrations of irritating or injurious pollutants.

Alarm swept through the population on Saturday as the death rate mounted, and an emergency meeting between the city government and factory representatives was scheduled for the next day to consider means to curtail the emission of pollutants. But relief to the inhabitants of the Donora region came quickly on Sunday, October 31 by natural events when an incoming low-pressure region brought rain to clear the air.

The value of *epidemiologic* studies during an episode is underscored by the analysis of events at Donora. The incidence of morbidity is clearly directly linked with air pollution, as indicated by the type of complaints. We may therefore infer that the pronounced increase in mortality during the episode was also caused—at least indirectly—by the presence of high concentrations of air pollutants. (Unfortunately autopsies were not performed on most of the deceased.) The specific pollutant responsible for morbidity and mortality has not been identified. It is suspected that many deaths were due to heart failure, brought on by the improper functioning of the lung under the irritating conditions. Although mortality statistics on a monthly basis clearly show the effect of the episode, the statistics on a three-year basis do not. Analysis over a long term therefore would fail to reveal the incident because of the relatively small number of deaths.

Subsequently, the mortality rate of those affected by the episode has, in fact, been higher than that of the general population, partly as a result of preexisting chronic illness.[14] But the overall death rate in Donora since then has been no higher than it is for other nearby towns that did not particularly suffer during the episode.

The tragedy of Donora focused national attention on the growing problem of air pollution. It was not because of the toll, for greater mortality rates have followed more traditional epidemics. The shock came from a realization that it was apparently not abnormally high emissions of pollutants that produced the mortality and morbidity. Local sources and types of pollutants evidently had not changed substantially for decades. Rather the episode was triggered by a combination of meteorological factors which had conspired to confine the pollutants. Unless emissions were curtailed, the frequency and duration of episodes would be determined by meteorological factors that could not be controlled.

SUMMARY

The quality of the ventilation in a locality is affected by wind patterns on both large and small scales. Terrain has an important influence on meso- and micro-meteorology. Familiar examples include the mechanical effects of terrain such as the channeling of winds through valleys and the uplifting of air by mountain slopes. Thermal effects may be less obvious, but are also important. The monsoon is an example of seasonal effects, and the land-sea breeze or mountain-valley winds are illustrative of diurnal effects. To some extent, the characteristics of these winds are predictable: Some provide good ventilation; some encourage atmospheric stability. Extreme situations arise when meso- and micro-meteorological factors are reinforced by macroscale pressure patterns. The strong bora, mistral, and Santa Ana are examples. All provide good ventilation, but the strong winds may entrain dust and cause discomfort. On occasion, pollutants carried away may not be sufficiently dispersed before the wind dies, so that upon resumption of normal wind patterns the pollutants may be carried to other populated areas. The potentially serious consequences which may result when macroscale and microscale meteorology reinforce stagnant conditions is underscored by the circumstances of the 1948 pollution episode in Donora. Although a wide area of the western region of Pennsylvania was affected by a stagnating anticyclone, only the immediate environment of Donora suffered from abnormally high pollution and a marked increase in morbidity and mortality.

NOTES

1. R. P. Pearce, "Discussion on the Calculation of a Sea-breeze Circulation in Terms of the Differential Heating Across the Coast Line," *Quart. J. Roy. Meteorol. Soc.*, **82**, 239.

2. O. G. Sutton, *Micrometeorology*, New York: McGraw-Hill, 1953.

3. D. F. Leipper, "The Sharp Smog Bank and California Fog Development," *Bull. Am. Meteorol. Soc.*, **49**, 354 (1968).

4. A. Miller and J. C. Thompson, *Elements of Meteorology*, Columbus, Ohio: Charles E. Merrill, 1970.

5. P. A. Kratzer, "Das Stradtklima," *Wissenschaft Braunschweig*, **90**, 184 (1956).

6. W. T. Roach, *Quart. J. Roy. Meteorol. Soc.*, **87**, 346 (1961).

7. P. A. Sheppard, "The Effect of Pollution on Radiation in the Atmosphere," *Intern. J. Air Poll.*, **1**, 31 (1958).

8. H. E. Landsberg, in *Man's Role in Changing the Face of the Earth*, W. L. Thomas, Jr., Ed., Chicago: Univ. of Chicago Press, 1956, p. 584.

9. T. J. Chandler, Discussion of the paper by K. J. March and M. D. Foster, "The Bearing of the Urban Temperature Field upon Urban Pollution Patterns," *Atmos. Environ.*, **2**, 619 (1968).

10. H. E. Landsberg, "Man-Made Climatic Changes," *Science*, **170**, 1265 (1970).

11. B. Roueché, *Eleven Blue Men and Other Narratives of Medical Detection*, Boston: Little Brown, 1953.

12. H. H. Schrenk, H. Hieman, G. D. Clayton, W. M. Gafafer, and H. Wexler, "Air Pollution in Donora, Pa.," *Public Health Service Bull.*, 306, (1949).

13. M. Neiberger, "Reflection, Absorption, and Transmission of Insolation by Stratus Cloud," *J. Meteorol.*, **6**, 98 (1949).

14. A. Ciocco and D. J. Thompson, "A Follow-up of Donora Ten Years After: Methodology and Findings," *Am. J. Public Health*, **51**, 155 (1961).

FOR FURTHER READING:

J. EDINGER   *Watching for the Wind*; Garden City, N.Y.: Doubleday-Anchor, 1967.

R. E. MUNN   *Descriptive Micrometeorology*, New York: Academic Press, 1966.

O. G. SUTTON   *Micrometeorology*, New York: McGraw-Hill, 1953.

C. H. B. PRIESTLEY *Turbulent Transfer in the Lower Atmosphere*, Chicago: Univ. of Chicago Press, 1959.

S. PETTERSSEN   *Introduction to Meteorology*, 3rd ed, New York: McGraw-Hill, 1969.

QUESTIONS

1. Why do low-pressure regions commonly develop over continental land masses during the summer? Why does the maximum precipitation over the continents occur during this season?

2. What is the sequence of events that establishes a sea breeze? Why is a sea breeze generally stronger than the land breeze at the same location? Under what conditions might a sea breeze aggravate a situation in which pollutants accumulate over an urban region? What aspects of a land breeze encourage the accumulation of pollutants at low levels?

3. Toward what direction would the sea breeze veer along the south coast of France as it approaches a condition of geostrophic balance late in the day?

4. Why is a foehn warmer than the air which is rising on the other side of the mountain at the same level?

5. Clouds are often observed over mountain ranges in an otherwise clear sky. What are two reasons why this may occur?

6. Why do upslope daytime winds commonly include a thicker layer of air near a mountain side than the nocturnal downslope winds at the same location?

7. How might a mild foehn cause an elevated inversion over the lands immediately to the lee of a mountain?

8. Which aspects of the micrometeorology of valleys encourage the accumulation of pollutants within the valley?

9. Which factors of the micrometeorology of cities contribute to the severity of air pollution episodes and what factors enhance ventilation?

10. Which aspects of low-level fog can encourage pollution episodes? Indicate how the temperature profile can be affected by a dense fog and explain what mechanisms influence the profile.

PROBLEMS

1. Suppose that a dry katabatic wind descends adiabatically 1500 m from a high plateau to sea level. By how much is the air heated during the descent?

2. If the relative humidity of the air on the high plateau mentioned in the previous problem were 30% at a temperature of 22°C, estimate the relative humidity at sea

level assuming that the vapor pressure of water is independent of the ambient pressure, depending only upon the ambient temperature.

3. Sea breezes at 40°N latitude develop in response to horizontal pressure gradients which are commonly about 0.01 mb/km. Assuming a constant straight line acceleration of air as a result of this force alone, how far would the air have traveled before it acquired a velocity at which the magnitude of the Coriolis force (which we assume is not acting) would be equal to the pressure-gradient force? Your result indicates one reason why geostrophic flow is not commonly observed for such mesoscale circulation.

4. Estimate the geostrophic wind speed near Los Angeles from the data in Fig. 6.4. The straight line distance from Los Angeles to San Francisco is 530 km, and the latitude of Los Angeles is 32 degrees.

CHAPTER 7

EFFLUENT DISPERSAL

In former days, waste gases were vented into the air by the simplest and cheapest means. However, the development of industry and the increase in both the amount of gaseous and particulate waste from individual sources and the combined output from all sources has produced intolerable conditions in many communities when the effluent is insufficiently diluted before experienced by the population. More efficient means have been sought to clean up the effluent before it is released or to dilute the exhaust gases before they can cause harm.

It has been found that a single source can very effectively dilute its pollutants by releasing them from a sufficiently high level. This method relies on the turbulent mixing of the air to dilute the effluent, perhaps by a factor of 100 or more, before it comes near the ground. Thus was born the idea of the smokestack or exhaust stack. This device may take the form of the short fireplace chimney of domestic residences or the impressive monoliths reaching upward some 300 m high beside some modern electrical power generating plants.

The exhaust stack of course does not by itself reduce the amount of pollutants released, but only serves to encourage more dilution. A very important aspect of air pollution control is the current effort to gauge the effectiveness of exhaust stacks, both in relation to the maximum ground-level concentration of pollutants which a particular design will produce for a given source and also to the community-wide problem of pollution from multiple sources.

In this connection it is useful to consider three broad classifications for sources of air pollution, according to their geometrical shape: *point*, *line*, and *area sources*. A *point source* corresponds to the orifice of an exhaust stack, located at some geographical position and elevation. A *line source* may represent segments of a road, over which individual point sources (automobiles) are continually moving, so that the net result when averaged over a suitable interval of time can be visualized as a continuous emission of pollutants over the entire length of road. An *area source* comprises an industrialized area or community in which the emissions from numerous individual point sources and line sources mix so as to make it difficult, if not impossible, to assign responsibility for the detrimental effect of a pollutant to any one source.

The effectiveness with which the effluent is diluted in ambient air depends upon the geometry of the source; and so will the concentrations which are experienced by nearby inhabitants. In each case, the dispersal of the pollutants will be influenced by the degree of stability of the atmosphere, the horizontal wind speed, features of terrain, and other factors which we have discussed in the preceding chapters. In coming to grips with the problems of air pollution, it is essential that we understand the mechanisms by which pollutants are dispersed. Much of the current concern about individual sources and their contributions to community-wide pollution focuses upon the question of how effective these mechanisms are and with what degree of reliability their effects can be predicted.

Certain aspects of the local diffusion of pollutants are evident. Plumes which are emitted from exhaust stacks behave in a variety of ways depending, among other factors, upon the operating conditions of the source, the time of day, and local weather conditions. In open country with a moderate breeze and cloudy sky, smoke or condensed water vapor forms a fairly well-defined plume which expands as it drifts downwind. A plume will also rise if it is warmer and therefore more buoyant than the ambient air. Turbulence within the atmosphere produces an irregular dissipation of the plume as it eventually disappears from view. In light winds and under clear skies with unstable air, the behavior of a plume is more erratic as turbulence plays a greater role. This is illustrated in Fig. 7.1. At night and in the early morning, plumes may rise only slightly if constrained by a radiational inversion and drift downwind for many kilometers while retaining a well-defined form. If there is no wind, or only a gentle breeze, smoke from a ground fire will rise vertically until it is spread by stronger winds at higher altitudes

Fig. 7.1   Effluent from a cement processing plant. Turbulence causes the plume of particulates to move and spread out irregularly as it drifts downwind.

or until it meets the base of an inversion layer, at which time it will spread horizontally in the shape of a flat-topped cloud. The qualitative features of the dispersal of emissions from line and area sources are often not as evident as for point sources, yet the same principles of meteorology apply. But the rate at which the pollutants are diluted is usually not as rapid, owing to the source geometry.

We shall now consider how atmospheric conditions can affect the dispersal of a plume from a point source, a subject of the utmost practical consequences. This will be followed by a close look at processes which govern the mixing and dilution of pollutants in ambient air. With this as a background, we can proceed to develop quantitative assessments for the utility of exhaust stacks. This is a key issue facing many communities today. It is then a straightforward generalization to bring out several essential features of pollution dispersal from line and area sources, and we shall point out how they differ from the equivalent features for point sources.

7.1 TEMPERATURE PROFILES AND PLUMES

A plume from an elevated source such as a tall exhaust stack mixes vertically and horizontally with ambient air as it drifts downwind. Vertical mixing is determined largely by the degree of instability of the lower troposphere; thus it depends upon the temperature profile. Horizontal mixing can also be influenced indirectly by the lapse rate, for a movement of air in the vertical direction cannot proceed without horizontal movement somewhere. On the other hand, even if vertical circulation is suppressed by a stable lapse rate, horizontal motion is possible. Therefore the spread of a plume in the horizontal direction may be unequal to the vertical spread at a given distance downwind from the stack.

Parts (a), (b), and (c) of Fig. 7.2 illustrate three examples of a plume being dispersed when it is affected by a uniform lapse rate. In part (a), large-scale turbulent eddies cause sizable parcels of air, together with portions of the plume, to deviate from a straight downwind direction. This condition is known as *looping* and may occur when the atmosphere is highly unstable. If a plume temporarily loops downward, it can cause a momentary high concentration of pollutant at ground level near the stack. In part (b), the lapse rate is essentially neutral and the shape of the plume commonly is vertically symmetrical about what is called the *plume line*. Its conical shape suggests the name *coning*. The most confined plume is found in a stable lapse rate as shown in (c). The lapse rate suppresses vertical mixing, but not horizontal mixing entirely. The plume spreads only parallel to the ground and appears to take on a fan shape as seen from below. On occasion, plumes in a stable layer may be followed for 10 or 20 km downwind from the source. They are often observed early in the morning when the effluent is still trapped in a ground-based radiational inversion.

Two meteorological conditions of special interest are found when the lapse rate changes from stable to unstable at a height which is approximately where the

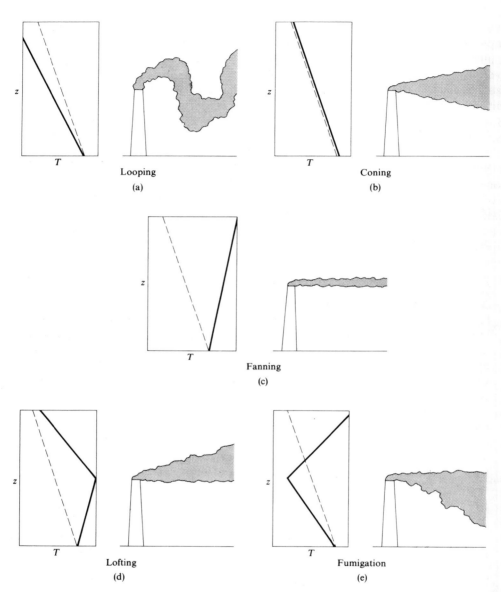

Fig. 7.2 Plume behavior influenced by the lapse rate above and below the release height. The dashed line in the temperature profiles is the adiabatic lapse rate, included for reference.

plume is released. One case is illustrated in part (d) where the lapse rate in the upper portion of the plume is unstable, and in the lower, it is stable. Mixing is vigorous in the upward direction, a circumstance known as *lofting*. This is advantageous, for pollutants can be well mixed into the upper air while barely affecting people on the ground. The second case, illustrated in part (e), poses a potentially serious air pollution situation. Here the plume is released just under an elevated inversion layer; perhaps a ground-based inversion has been eroded from beneath, as occurs in the morning as the ground warms. When the low-level unstable lapse rate reaches the plume, the effluent suddenly mixes downward toward the ground, an effect called *fumigation*. Ground-level concentrations of pollutants are high, because upward dispersal is discouraged by the inversion layer.

One case illustrating the possible widespread effects of fumigation was the cause of international litigation.[1] A smelter for processing lead and zinc had been built in 1896 in the Columbia River Valley at Trail, British Columbia, about 10 km north of the United States border. The river flows southward from this point across the border and then westward to the Pacific. The valley has flat bottom land bordered by steep sides towering upward some 600–800 m. Sulfur was removed from the ore by roasting it at sufficiently high temperatures to form SO_2, most of which was then released to the atmosphere.

By 1930 the daily emission of 600–700 tons of sulfur dioxide by the smelter had resulted in considerable damage to agriculture within the valley south of the border.[2] In 1935 an Arbitral Tribunal was charged with the responsibility of assessing damage and formulating a permanent solution. A study conducted by this commission found that ground-level concentrations were commonly highest at about 8 A.M., a condition that developed quickly and simultaneously at monitoring stations located as far as 10–55 km downstream from the source. The explanation is that the effluent from the smelter was trapped in the radiational inversion, diffusing neither upward or downward, but drifting down the valley with the drainage wind at night. After sunrise, the fairly uniform warming of the valley floor eroded the inversion from beneath, and when the layer containing the pollu-

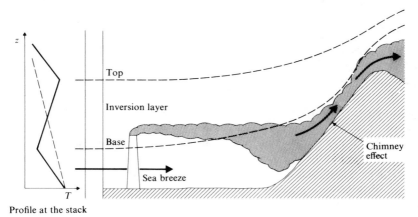

Fig. 7.3 Plume trapped within an inversion drifts inland until it emerges from the upward-sloping base of the inversion. The effluent fumigates downward and is then carried along by the upslope wind.

tants became unstable, widespread fumigation was observed along a great length of the valley.

Fumigation has accounted for some of the more spectacular effects of air pollution. These can be avoided in many cases by use of sufficiently high stacks to emit the exhaust above an inversion. A variation of the fumigation condition is observed with an upward-sloping inversion layer downwind from a stack. One possibility is illustrated in Fig. 7.3 for a coastal plane bordered by inland mountains. A sea breeze moves inland below the inversion, warming as it goes. The inversion is eroded below by the warmer inland temperatures, so that the plume at some point emerges from the underside and fumigates downward. This *could* occur in the absence of mountains. But in Fig. 7.3 we show that the plume is then carried up a mountain slope by a continuation of the sea breeze, now reinforced by a daytime upslope wind. The warm mountain slope and nearby air further help to lift the inversion. This venting of air up a mountain slope is known as the *chimney effect*. If fumigation occurs *directly* onto a hillside, high ambient concentrations can be experienced locally. Unfortunately, few quantitative facts are available to indicate the importance of mountain-slope fumigation for community air pollution. Presumably it is a phenomenon which could occur wherever industries are located in valleys or adjacent to hills.

Considerable effort has been invested in attempts to develop theories for describing the downwind dispersal of a plume, to provide a means by which the resulting ambient concentration of a pollutant at various distances from the source can be predicted before the source is put into operation. This has been only partially successful in the sense that currently accepted methods are essentially empirical and rely only in part on well-established theory. We shall examine the significance of this statement in the next few sections.

First let us note the limitations of such quantitative approaches, for almost all of them incorporate very restrictive assumptions. This survey is useful, because it will enable us to establish a perspective before we become involved in the details. Most formulas which have been advanced to describe the behavior of a plume are based on the assumption that the lapse rate is fairly uniform; that is, the temperature profile shows no abrupt changes such as we have illustrated in parts (d) and (e) of Fig. 7.2. Furthermore, the lapse rate should not vary markedly downwind from the source. The wind speed and turbulence characteristics are also assumed to be uniform, with no appreciable variation with distance from the source or with altitude. In particular, conditions of strong wind shear, such as the direction shear affecting the plume shown in Fig. 7.4, should not be included. The terrain is generally assumed to be fairly flat and featureless. In short, most plume dispersal formulas are applicable only to relatively homogeneous atmospheric conditions and terrain.

Another feature of plume behavior also deserves an introductory comment. This is the buoyancy characteristic of a warm plume. Unlike the plume illustrated in Fig. 7.1, the one in Fig. 7.4 is quite warm and continues to rise even after it has traveled almost one kilometer from the exhaust stack. There is no theory

Fig. 7.4   Directional shear of the wind causes the plume from a power plant to make almost a right-angle bend as it rises.

yet devised which has successfully described the buoyant rise of a plume with acceptable accuracy. Existing methods of estimating this are almost wholly based on empirical formulas. This is an unfortunate situation, for the accuracy with which downwind concentrations of a pollutant can be calculated will be limited by the accuracy with which the plume rise can be estimated.

In the next two sections we shall examine pertinent aspects of how gases and aerosols mix with ambient air. It is through this process of dilution that contaminants might be rendered harmless. We shall first show in Section 7.2 that molecular theory is inadequate to explain the observed mixing rates in the atmosphere and that turbulent mixing is the dominant process. However, an acceptable theory for turbulent mixing in an unstable atmosphere does not exist, so we shall explain in Sections 7.3 and 7.4 how the time-average behavior is approximated by formulas which are analogous to those applying to molecular diffusion, but which differ in having empirically determined parameters. We shall find that this approach leads to the formulation of Eq. (7.10), governing the downwind spread of a plume from a point source; solutions are provided in Eqs. (7.11) and (7.14). The implications of these important results will then be examined.

7.2 ATMOSPHERIC MIXING

One characteristic in the behavior of all matter is its tendency to establish a uniform condition. That is, when hot water is added to a pan of cold water, the result after a short period of time is a liquid of uniform temperature. The final temperature is somewhere between the two initial temperatures. The tendency toward uniformity is a statistical property of the behavior of large numbers of molecules, because a uniform condition, when possible, is the most likely condition. For our example of the mixing of cold and warm water, the mechanism is described as the transfer of a portion of the heat energy of the warm water to the cold water through the process of collisions between the water molecules, until at equilibrium (by definition) the temperature is uniform.

Analogous processes occur in other contexts, especially in the dynamics of the atmosphere. For example, in Section 5.8 we saw that during conditions of a nocturnal inversion the surface of the earth, which cools by thermal radiation, consequently absorbs energy from both the overlying warmer air and the warmer earth beneath, as Nature attempts to establish a condition of uniform temperature. An important feature in this process is that the transfer of sensible heat through the ground can be described by a parameter called the thermal conductivity of the earth. The rate at which energy is conducted is proportional to both the thermal conductivity and the temperature gradient, as expressed in Eq. (5.16). It is therefore possible to indicate the rate at which a nonuniform condition approaches an equilibrium uniform condition.

The trend toward uniformity is not exclusive to thermal processes. It also applies to the transferral of *momentum* between different regions of the atmosphere and to the *mixing* of pollutant molecules in ambient air. We shall now consider these two cases.

Viscosity

The transport of momentum commonly occurs wherever wind velocity varies with altitude, a condition we have called *wind shear*; and its effect is what we know

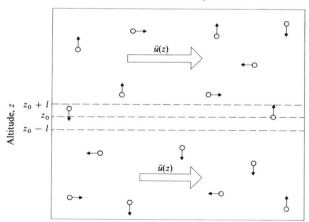

Fig. 7.5   A gas of uniform density whose molecules have an average velocity $\bar{u}$ in the x-direction (toward the right) which is greater at higher altitudes. The wind velocity (broad arrows) is greatly exaggerated in comparison with molecular velocities (thin arrows); in reality, the wind speed is a very small fraction of the average molecular speed.

as *viscosity*. The simplest case of shear is when merely the wind speed depends upon altitude, the direction being uniform. In this instance, air molecules within the stronger wind have a larger average momentum in the direction of the wind than do molecules within the adjacent slower-moving air. Figure 7.5 illustrates an example in which the average wind speed $\bar{u}$, taken to be along the x-direction, increases with altitude. The wind speed is equal to the average velocity of the air molecules. We assume that the concentration and average temperature of the air are essentially uniform, so that the only important variation is the average velocity. Air molecules are, of course, moving in all directions. Under these conditions, the rate at which molecules pass upward through any horizontal level, say $z = z_0$, is equal to the rate at which they pass downward. However, because the downward-moving molecules have a greater velocity in the x-direction than the upward-moving ones, there will be a net flow downward of x-directed momentum. The atmosphere thus conveys momentum to regions where the momentum is less, and attempts to establish a uniform distribution of momentum in the x-direction, independent of altitude.

In fact, the atmosphere never achieves a uniform wind speed, because the lowest layer of air similarly transfers momentum to the surface of the earth as the molecules strike against the rough ground. The continual loss of momentum to the earth, an effect we have called "friction," means that the air near the ground will necessarily move more slowly than the air above, thereby maintaining the wind shear. The gain in momentum by the earth is not noticeable, because its large mass prohibits it from being appreciably accelerated in the direction of the wind. If there were no other influences, the winds would eventually cease as the air loses all of its momentum to the earth. In fact this does not occur, because there are other influences—for example, solar radiation—which indirectly stimulate horizontal motion. The solar energy is thereby converted into kinetic energy of the winds, sustaining the wind at higher altitudes. And so long as these winds continue, there will be a continual downward flow of momentum through the atmosphere to the earth. As we have seen in Section 5.4, the Ekman spiral with its deviation of the wind direction from geostrophic balance near the earth's surface is dictated by the effect of friction with the ground and the upward transmission of this retarding effect by the viscosity of air. Viscous effects are so important that it is worthwhile estimating their magnitude by considering the molecular nature of air.

We first recall that we were previously able to derive the ideal gas law for the pressure of a gas by using molecular kinetics (Section 3.4). An analogous argument can be given for the viscosity by reference to Fig. 7.5: We need to find the net rate at which momentum is carried by molecules as they pass through a unit area of an imaginary horizontal surface, say at $z = z_0$. First we note that the rate at which the molecules pass in the downward direction is $nv/6$, where n is the number of air molecules per cubic meter, or the number density, and v is their average speed. Molecules pass upward through a unit area at the same rate, so there is no net vertical migration. Of course, once again we have been cavalier with the statistics by using the factor of $\frac{1}{6}$ to correspond to the approximation that only $\frac{1}{6}$ of the molecules in any volume are traveling along one of the six principal coordinate directions.

But the average momentum of the molecules' movement in the x-direction varies with altitude. This is possible because molecules are continually striking each other as they move about and, during the innumerable collisions, they take on the characteristics of the ambient air at the altitude where they momentarily reside. This is in fact an important assumption, for it implies that locally an equilibrium is reached. Molecules moving downward lose some of their higher momentum to the other molecules in the lower regions. They travel only a short distance before this is accomplished. The distance is on the order of the mean distance which they travel between collisions, the *mean free path*. The mean free path is such an important parameter that it is given a special symbol l. Molecules which pass downward through a level $z = z_0$ have a net velocity in the x-direction which we shall denote by $\bar{u}(z + l)$. The net momentum in the x-direction for each molecule is therefore $m\bar{u}(z + l)$, where m is the mass of a molecule. Those which pass upward have an average velocity corresponding to the wind speed at $z = z - l$, and thus have a momentum $m\bar{u}(z - l)$. The net rate at which molecules convey momentum through the level $z = z_0$ is given by

$$\frac{\Delta P_x}{\Delta t} = \frac{1}{6} nmv \left[\bar{u}(z_0 - l) - \bar{u}(z_0 + l) \right], \tag{7.1}$$

where P_x represents x-directed momentum. The signs in this expression are chosen so that the quantity on the right is the rate at which x-directed momentum is transmitted upward, toward larger values of z. The quantity in brackets can be expressed in terms of the gradient of the wind speed $\Delta\bar{u}/\Delta z$:

$$\bar{u}(z_0 + l) - \bar{u}(z_0 - l) = 2l(\frac{\Delta\bar{u}}{\Delta z}), \tag{7.2}$$

where the gradient is to be evaluated at $z = z_0$. This equation is true for any value of z_0, so we can drop the subscript zero. Thus the rate at which momentum is transmitted upward is

(a)

$$\frac{\Delta P_x}{\Delta t} = -\frac{1}{3} \rho v l \left(\frac{\Delta\bar{u}}{\Delta z}\right) \tag{7.3}$$

(b)

$$= -\eta(\frac{\Delta\bar{u}}{\Delta z}).$$

The negative sign appears because the direction in which momentum flows is opposite to the gradient $\Delta\bar{u}/\Delta z$ of the wind velocity. That is, if the wind is stronger at higher altitudes, momentum is transmitted downward. The symbol ρ is the air density, $\rho = nm$. The constant of proportionality $\eta = \rho v l/3$ is called the *viscosity coefficient*, and for air at atmospheric pressure and 18°C it has the value 18×10^{-6} kg/m-sec (180×10^{-6} poises in the cgs system). As the smallness of this parameter suggests, the viscosity of air is not very great. For glycerin, by contrast, the coefficient is empirically found to be 2 kg/m-sec (20 poises). The large coefficient for this liquid describes the well-known fact that it cannot easily be poured; much of the momentum gained from the force of gravity is quickly transferred to the container.

A significant conclusion we draw from the above derivation is that the molecular theory, when the statistical aspects are correctly taken into account, predicts the value of the viscosity coefficient. The coefficient is given in terms of well-known parameters of molecules and gases—the density, mean speed, and mean free path. Unfortunately, however, the magnitude of this coefficient does not in fact describe what is observed to happen in the atmosphere. The viscosity of air is considerably greater than we have predicted, perhaps 10^3 times greater on a sunny day. Before we discuss this discrepancy in more detail, let us take up one more aspect of molecular motion.

Diffusion

We now consider the important case of the dispersal of gaseous pollutants. The trend toward uniformity applies also to the concentration of pollutant molecules within the atmosphere. This is illustrated in Fig. 7.6, which shows a condition of the atmosphere in which there is a horizontal gradient of the concentration χ of a pollutant. We suppose that the concentration varies only along the x-direction, with a negligible variation along the transverse y- and z-directions. The number of pollutant molecules per cubic meter is $n\chi(x)$, where n is the total number of air molecules per cubic meter. We suppose also that $\chi(x)$ is indepen-

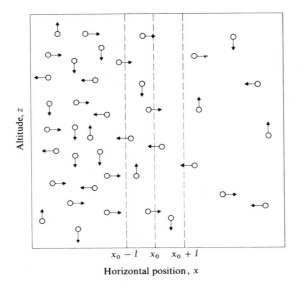

Horizontal position, x

Fig. 7.6   Molecular diffusion of pollutant molecules from a region of high concentration to one of low concentration. The density of circles represents the concentration of pollutant molecules which, in this case, decreases uniformly with increasing distance x.

dent of time, or approximately so, just as in the previous example for viscosity we had assumed that $\bar{u}(z)$ was independent of time.

The rate at which molecules of any kind pass toward the right through a unit area of an imaginary surface at $x = x_0$ is equal to the rate at which they pass toward the left. However, because a greater proportion of those headed toward the right are pollutant molecules, there will be a net flow of pollutant molecules in that direction. The net rate of flow can be estimated from our molecular model. The rate at which they pass toward the right is $nv/6$, and therefore the rate at which pollutant molecules pass toward the right is $nv\chi(x_0 - l)/6$. Similarly the rate at which pollutant molecules pass toward the left is $nv\chi(x_0 + l)/6$. Thus the net rate at which they pass toward the positive x-direction is given by

$$q_x = \frac{1}{6} nv [\chi(x_0 - l) - \chi(x_0 + l)]$$

(a)
$$q_x = -\frac{1}{3} nvl \left(\frac{\Delta \chi}{\Delta x}\right)$$

(7.4)

(b)
$$= -nD \left(\frac{\Delta \chi}{\Delta x}\right).$$

This diffusion equation was first developed by Albert Fick, a German physiologist of the mid 1800's, and the process it describes is known as *Fickian diffusion* or *molecular diffusion*. A negative sign appears because the direction of net motion is counter to the gradient $\Delta \chi / \Delta x$ of the relative concentration. The molecules are hindered in their progress by collisions with other molecules, as indicated by the appearance of the mean free path l in Eq. (7.4a), so the process is fairly slow.

The rate at which molecules diffuse is thus seen to be proportional to the gradient of the concentration. The proportionality constant D is called the *molecular diffusivity*. For air at 20°C the diffusivity has a value of roughly $1 \times 10^{-5} \mathrm{m}^2/$ sec. The fact that diffusion by Eq. (7.4) proceeds in a direction determined solely by the direction of the concentration gradient implies that the actual motion of the molecules is random, and that one direction of motion is not favored over another. Only spatial variations in the concentration are important in determining the direction of the net diffusion of pollutant molecules.

This derivation shows once again a process by which the atmosphere attempts to establish uniformity. The rate at which it does is proportional to the gradient of the nonuniform parameter. A similar analysis of the statistics of molecular motion would predict that the rate at which gas molecules transfer sensible heat from a warmer to a colder portion of the atmosphere is proportional to the temperature gradient. The resulting formula is similar to Eq. (5.16), and the constant of proportionality is called the *molecular thermal conductivity*, with a value of 2.5×10^{-2} J/m-sec-°K. In this as well as the foregoing examples, the constant of proportionality is predicted uniquely by the statistics of molecular dynamics.

A deficiency

The molecular theory of gases as indicated above yields expressions for the thermal conductivity, viscosity coefficient, and molecular diffusivity in terms of the known parameters of the atmospheric gases—the density, average molecular speed, mean free path, and molecular mass. However, it is found that the evaluation of the expressions fails to give agreement with the coefficients determined by empirical measurements of the conductivity, viscosity, and diffusivity in the lower atmosphere. The observed parameters are several orders of magnitude greater than is predicted by the molecular theory. Therefore, although the foregoing molecular processes do occur to some extent within regions of the atmosphere, there must be another more important mechanism that serves to transport energy, momentum, or pollutants from one place to another.

It is found that molecular theory fails on two counts: (1) In most situations within the atmosphere there is a violation of the steady-state assumption as a result of turbulence. All of the preceding derivations assume, as a property of the air, a well-defined gradient which is essentially constant in time; but this assumption is not valid for a turbulent atmosphere. (2) We have also assumed that the molecules move in a random fashion, with no preferred direction of motion (except for an average velocity when there is a wind). This condition also may be violated in a turbulent atmosphere, for there is an eddy-like motion of the air in which a correlation exists between the movement of molecules in different locations.

Let us illustrate these two effects by an oversimplified example. In Fig. 7.7 is shown the lower edge of an exhaust plume, where the concentration of a pollutant has a large gradient. (The concentration decreases sharply beneath the plume.) Thus, according to Fickian diffusion, pollutant molecules will mix downward at a rate given by Eq. (7.4), and this motion is indicated by the thin arrows in Fig. 7.7. However, the air also has a turbulent motion, and the eddies associated with this behavior can affect sizable volumes of the plume. We show one such eddy that has

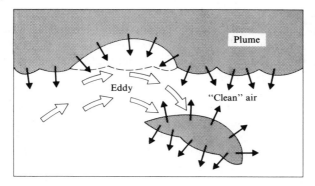

Fig. 7.7 Grossly simplified view of how a turbulent eddy breaks up a plume and causes much more rapid mixing of pollutants with ambient "clean" air.

carried away a portion of air, radically changing not only the pattern of the concentration of the pollutant but also the gradient of the concentration in the process. Hence our molecular kinetics model cannot be applied to this situation, for the concentration gradient changes rapidly with time. In addition, we see that a large number of air molecules in the eddy have a similar trajectory; therefore their movements are correlated when the portion of the plume is carried away. This violates the assumption of random motion, which is basic to our molecular diffusion model.

Turbulence in this example has the effect of hastening the mixing of pollutants with ambient air. So too, it enhances the thermal conductivity and viscosity. It is obvious that turbulence is a key factor in air pollution meteorology.

7.3 TURBULENCE

As might be imagined, the details of turbulence in the lower atmosphere are so complicated that very few situations have been described successfully by mathematical theories. Some attempts have focused on dimensional arguments to predict a coefficient for thermal conductivity, viscosity, and diffusivity assuming that the flow of sensible heat, of momentum, or of pollutant molecules is proportional to the appropriate gradient. Except in a few instances the resulting so-called *eddy viscosity*, *eddy thermal conductivity*, and *eddy diffusivity* have been applied successfully only in a limited region within a meter or so of a rough surface, such as the surface of the earth. This theory, called the *mixing length theory*, has been developed notably by L. Prandtl in 1925 and is useful for dealing with transport phenomena associated with fluid flow. Unfortunately, it is not applicable to unstable situations in which buoyancy of air parcels is an important feature, and for this reason has not been especially useful for problems of large-scale dispersal of pollutants in ambient air.

Another mathematical formulation of the effects of turbulence was developed from a statistical basis by G. I. Taylor in 1921.[3] Application of his theory requires knowledge of the details of the air turbulence which are not usually known. Consequently, the statistical theory has not proven to be especially useful for assessing the effects of pollution downwind from a source. However, as we shall see later, the statistical theory has yielded some pertinent information about the temporal dependence of average levels of pollution.

In view of the lack of a general theory of turbulence that can be applied to the prediction of the eddy diffusion of atmospheric contaminants, it may not be surprising that simple, empirical methods have been sought. The idea that Eqs. (5.16), (7.3b), and (7.4b) can describe in an empirical fashion the transport of energy, momentum, or pollutants within a turbulent atmosphere is taken for granted by many researchers even though there is no proof that this is generally justifiable. Whether this approximation is appropriate can be judged only on the basis of the particular application.

The definition of turbulence remains a matter of debate.[4] For our purpose, it can be described as an almost random character of the velocity of a fluid as viewed at a point in space, in contrast to the constancy in steady-state streamline flow or the periodicity of wave motion. It may be surprising, but even the definition of average wind speed has been the subject of debate. It is generally agreed that the mean wind speed cannot be defined without specifying the time interval over which measurements are made. In a practical sense, fluctuations about the mean are then called turbulence. Such variations are associated with complicated eddy patterns of flow in both horizontal and vertical directions. Therefore the characteristics of turbulence can depend upon whether the wind velocity component of interest is directed along the average wind direction, which we shall assume is horizontal, or the horizontal transverse direction, or the vertical direction. It will be useful to refer to a coordinate system in which the x-axis is along the wind direction and the z-axis is vertical. The average wind speed will be denoted by $\bar{u}$ and is along the x-axis. The fluctuating components of the wind along the x-, y-, and z-directions will be denoted by u', v', and w', respectively.

Turbulence spectra

Turbulence can be characterized by two average quantities, the *relative intensity* and the *length scale*. The *relative intensity* is a measure of the strength of the turbulent disturbances relative to the strength of the average wind. A convenient measure is the quantity $(v'^2/\bar{u}^2)^{1/2}$ for the y-component of turbulence, where v'^2 is the average value of v'^2. Analogous formulas apply to the x- and z-components. The *length scale* refers to the extent of correlation between local wind velocities at two different points in space. Perfect correlation means that the velocity at one point is uniquely related to the velocity at another, either by a fixed time delay or by a fixed change in magnitude or direction. No correlation means that the velocities have no relationship with each other at the two different points.

   For a more detailed description of turbulence, we might consider the *velocity spectrum* of the fluctuations which are observed at a particular point in space. If we monitor each of the fluctuating components of the wind, u', v', and w', for a period of time, we will see the effects of numerous eddies, many of them superimposed, as they parade by.

   A recording of the wind velocity along the x-axis might appear as in part (a) of Fig. 7.8. The value of u is seen to fluctuate about its average value $\bar{u}$, so that the difference $u - \bar{u} = u'$ at any instant is attributed to turbulence. We can imagine that the fluctuating portion u' arises from a superposition of eddies, each of which has a different frequency and intensity. For example the contributions from each of four separate eddies is illustrated in (b), (c), (d), and (e) of Fig. 7.8. When added together, these would describe fairly accurately the curve in Fig. 7.8(a). The intensity of each eddy component could be uniquely determined by an analysis of our recording, and we have labeled the intensity by $u(f_1)$ for the turbulent component at frequency $f_1 = T_1^{-1}$; similarly we let $u(f_2)$ be the intensity of the component at frequency $f_2 = T_2^{-1}$, and so on as indicated in the figure.

   Figure 7.8 illustrates an unrealistically simple case. In practice, an analysis of

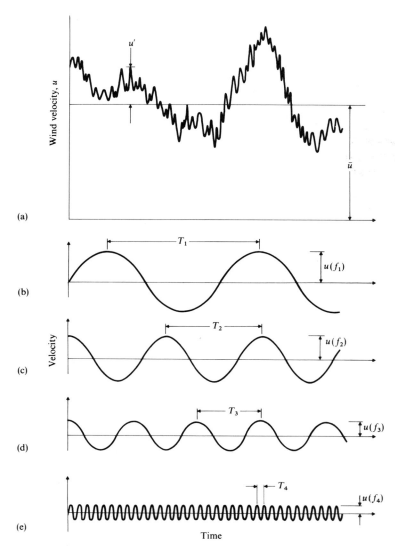

Fig. 7.8   (a) Fluctuations in the wind velocity; (b) (c) (d) (e), the turbulent eddies that cause them.

the data shows there are eddy components of the velocity over a broad range of frequencies; it is informative to analyze such data by asking what the intensity of the wind-velocity fluctuations observed at each possible frequency is. This is analogous to asking in the case of solar radiation what the intensity of the electromagnetic radiation at each frequency is. Just as we previously labeled the curve giving the solar radiation intensity *versus* frequency as the "emission spectrum," so we call the curve giving the intensity of velocity fluctuations *versus* frequency the "velocity spectrum."

An example of such a velocity spectrum is shown in Fig. 7.9 for the vertical component w' over fairly even terrain during superadiabatic conditions, as measured by H. A. Panofsky and R. A. McCormick at Brookhaven for various heights above ground.[5] The spectrum is actually displayed in terms of a horizontal axis which gives $fz/\bar{u}$, where f is the frequency. The reason why this parameter is chosen for reference is that it eliminates the simple translational effect of the wind; if we did not divide f by $\bar{u}$, a given disturbance carried with the wind would appear to have a higher frequency if the wind speed were higher. Thus the normalization by $\bar{u}$ removes much of this effect when the curves are plotted as in Fig. 7.9. To convey to the reader an appreciation of the frequency scale, we note that the value of $fz/\bar{u} = 0.1$ on the horizontal axis corresponds to a frequency of 1.6 cycles per minute for the upper curve and 0.54 for the lower.

The vertical axis of this figure gives the fluctuation intensity which is observed at each frequency, and the data clearly shows a decrease in intensity with increasing frequency. Similar measurements at a height of 100 m indicate that the intensity of the high-frequency components in the vertical direction (those with a frequency greater than 2 cycles per minute) are insensitive to the intensity of the incident solar radiation. It is therefore reasonable to conclude that high-frequency

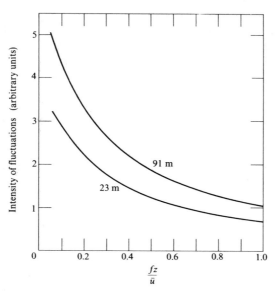

Fig. 7.9   Velocity spectrum for the vertical component w' of wind turbulence, measured over fairly even terrain during superadiabatic conditions, at heights of 23 m and 91 m. The intensity of the observed fluctuations at each frequency f is given by the curves. [Data from H. A. Panofsky and R. A. McCormick, *Quarterly Journal of the Royal Meteorological Society*, **86**, 495 (1960).]

components are insensitive to the instability of the atmosphere, suggesting that they are generally *mechanical* in origin.

Mechanical turbulence, we recall, is initiated when wind passes over land and is affected by friction against surface features. Viscosity transmits this effect upward, with the result that instability and turbulence develop. In a sense this is much like a man tripping over his own feet. The intensity of the high-frequency velocity components is observed to increase with increasing wind speed.

The behavior of the lower-frequency components of turbulence indicates that in some cases they may have a different origin. Panofsky and McCormick[5] found that the vertical components increased in intensity with increasing solar radiation, suggesting that they have a *thermal* cause. Thus turbulence can exist even in the absence of a wind, when it is produced by rising thermals.

The relationship between the frequency of a turbulent eddy and its length scale has not yet been discovered, perhaps because of the absence of a regularity in the phenomenon. There is general belief that large-scale turbulence occurs only at low frequencies, whereas high-frequency components have a much shorter length scale, in comparison with water waves and sound waves. It is commonly maintained—without proof in most cases—that the scale of mechanical turbulence may be determined by the size of perturbing obstructions such as buildings or terrain. Thus mechanical turbulence caused by large obstacles may have a length scale sufficiently long to be commensurate with that of thermal turbulence.

Generally it is found that turbulence is more isotropic (direction independent) the greater the height above ground, but that the intensity of turbulence diminishes with altitude.[4] At altitudes exceeding 700–1500 m, frictional effects from level terrain have little effect on the motion of air masses. Studies of the turbulence spectrum at 23 m and 91 m at Brookhaven showed that the intensity of high-frequency vertical components decreased with increasing altitude relative to the intensity of the low-frequency components.[6] This would be reasonable in view of the commonly accepted relationship between frequency and length scale and the fact that long wavelength vertical disturbances cannot be fully developed very close to the ground. Another example of this is the fact that at low frequencies the horizontal turbulence at a height of only 100 m is generally found to be much greater than the vertical components. A simple rule, often found to apply, is that the altitude at which a given frequency component has an isotropic intensity is approximately equal to the length scale of the disturbance.

A model

An insight into the nature of turbulence has been developed in a theory by A. N. Kolmogorov.[7] He hypothesizes that energy enters the spectrum at relatively low frequencies, through thermal or mechanical effects. In the former case, it is the direct conversion of sensible heat into the energy of turbulent motion. In the latter, it is the conversion of the kinetic energy of steady-state motion (associated with the average wind speed) into turbulent energy, thus reducing the average wind speed. In either case, most of the energy of turbulence is associated with the low-frequency, long length scale components. These components transfer part of their energy to higher-frequency components, which in turn pass it on to still higher-frequency ones. Eventually the length scale of the turbulence is so small that the transfer of momentum through the effect of viscosity is a highly chaotic process, and the energy of the turbulent components is converted into the kinetic energy of random motion of the molecules. Turbulent motion is locally isotropic in the high-frequency end of the spectrum where it is dominated by the effects of viscosity.

The details of turbulence in most urban areas are not known, nor is it clear whether the theoretical concepts such as introduced by Kolmogorov and others will prove of value in determining the influence turbulence will have on the diffusion and dispersal of pollutants. The variation of turbulence with horizontal position and altitude is imperfectly understood except in the simplest situations, and is a matter for urgent investigation.

7.4 POINT SOURCES AND TIME AVERAGES

Pollutants may be released into the air by various means: Burning fields, backyard incinerators, foundry operations, and pulp wood processing are some examples. The traditional situation which has aroused public complaints is the effect of direct downwind pollution from a single source. The origin of the pollution may be clearly evident from the behavior of an exhaust plume, or by an identifying odor. In most cases, the pollutant producing the adverse effects is a primary pollutant emitted by a particular source. This type of situation is the simplest to consider, because the responsibility for the pollution is clearly evident.

The ambient concentration of a pollutant at ground level monitored downwind from an exhaust stack will fluctuate with time as turbulence disturbs the position of the plume. At any instant of time, a view from above the source toward the ground will show the plume as seeming to meander over the ground, as illustrated in Fig. 7.10(a). At this instant, the variation of the ambient pollutant concentration along the y-axis (which is transverse to the average wind direction) may appear as shown in Fig. 7.10(b). As the source continues to emit, the plume will wander to either side of the average direction of the wind velocity, with the result that a one-hour average of the ambient concentration at each point along the y-axis will appear quite different from the profile the plume has at any one instant. A hypothetical one-hour average for each point on the y-axis is illustrated in Fig. 7.10(c). The hourly average has a wider spatial extent but a lower maximum concentration. Thus, as a result of atmospheric turbulence, any one location downwind from the source will be subject to an ambient concentration which strongly fluctuates with time, but the average concentration over a long period of time will be considerably less than the peak values measured during short time intervals.

In an ideal situation a knowledge of the turbulence velocity spectra and the mechanisms of atmospheric diffusion would allow us to calculate the expected

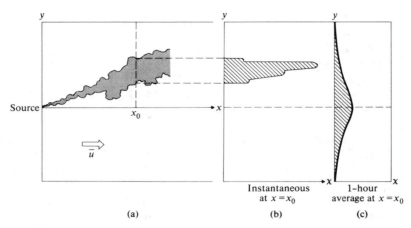

Fig. 7.10 (a) Instantaneous top view of a plume; (b) instantaneous horizontal profile of the plume concentration x along a transverse direction at some distance downwind from the source; (c) one-hour average profile for the same downwind distance.

time variation of the ambient concentration at any point downwind from a source. However, the theory of atmospheric diffusion and our knowledge of turbulence have not advanced so far as to permit such a *tour de force*. No theory has yet been developed and tested which will permit the temporal variation of downwind ambient concentrations to be calculated from meteorological parameters. We must therefore settle for a more restricted goal.

In fact, the temporal variations of the ambient concentration cannot even be precisely measured. The reason is that the ambient concentration is not measured for an infinitely long time interval nor are any existing instruments capable of responding instantaneously to changes in the ambient concentration. Both practical limitations affect the faithfulness of the data which characterize measurements of ambient concentrations, sometimes in a significant way. Measuring instruments, because of their inherent finite response time—a result of the time constants of the electronic circuits, the need to process the pollutant chemically before measuring, or merely the inertia of the indicating meter—always average their measurements over a period of time called the *averaging time*. In chemical analyzers, this may be several minutes; in other instruments, only a matter of a few seconds. Thus variations of the ambient concentration which are more rapid than the averaging time are missed; that is, the components of the high-frequency spectrum are overlooked.

On the other hand, because the total time interval over which some measurements are carried out has a limited duration—perhaps dictated by the stamina of the operator—variations with periods much longer than this *measuring interval* or *sampling time* will also be missed. The sampling time must be sufficiently long to encompass the periods for the most important components of the fluctuations. In practice, this is generally a minimum of three minutes for "steady winds" and considerably longer in very unstable conditions. Measurements at a particular location of the angular variation of wind direction, or of the turbulent intensity transverse to the average wind, usually show a rapid increase during the initial few minutes and a less rapid increase thereafter. Measurements of the ambient concentration at greater distances from a source should extend over a correspondingly longer time interval, since long-period components of the turbulence with a longer length scale play a more important role in conveying and dispersing pollutants.

Static models

Lacking the ability to predict the temporal variation of the plume from an exhaust stack accurately, we must turn to alternative methods. The current fashion in diffusion analysis is to characterize downwind pollution by long-term averages, with the anticipation that the effects of atmospheric fluctuations can be characterized by empirical parameters. Figure 7.11 illustrates this with three examples showing the instantaneous and the respective long-term average shapes of plumes, given for different classes of atmospheric stability. The long-term average plume behavior is more regular and therefore more simply described in mathematical terms.

In some limited circumstances, connections between parameters describing atmospheric turbulence and the observed plume width downwind have been established quantitatively. One study found a direct relationship between the angular fluctuations of the wind direction (measured on the horizontal plane with respect to the mean wind direction) and the average angular spread of the plume a distance 100 m downwind, both angles measured from the position of the source. Figure 7.12 shows the correlation for a source 2 m above the ground, as measured at two different experimental field stations, one in Nebraska and the other in Massachusetts. Some such direct relationship between the horizontal spreading of a plume and the horizontal turbulence must be expected from any logical treatment of the statistics of turbulent diffusion for, as we have noted, it is the turbulence that is primarily responsible for diffusion. The plume width of the experiments cited in Fig. 7.12 did not consistently correlate with other parameters such as the temperature lapse rate, so indeed horizontal turbulence is evidently the important factor in this experiment.

In one respect, the need to analyze pollution dispersal in terms of average characteristics is unfortunate, because a prediction of the magnitude and frequency of short-term fluctuation of the ambient concentration is consequently not available. Many experimental studies indicate that the ambient concentration, when averaged over an interval of only three minutes, often leads to peak values somewhere downwind which are 2–3 times higher than the peak values for one-hour periods at the same location.[8] The difference may be even greater for less stable atmospheres. Data obtained for averaging times τ of between 10 minutes and 5 hours suggest that the peak concentration will diminish as $\tau^{-1/2}$, in agreement with several statistical theories for the effect of atmospheric turbulence.[9]

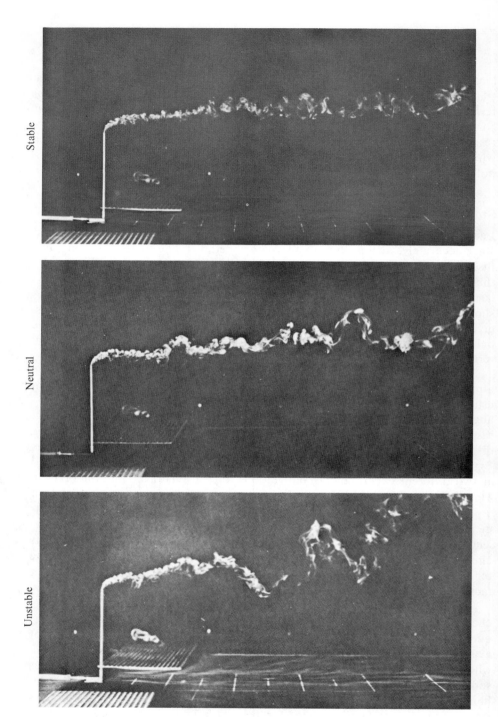

Fig. 7.11 Instantaneous (left) and long-term average (right) behavior of plumes from a model exhaust stack in a wind tunnel. The long-term averages were obtained by taking a number of photographs in succession of the scene. (Photographs taken in the air pollution wind tunnel

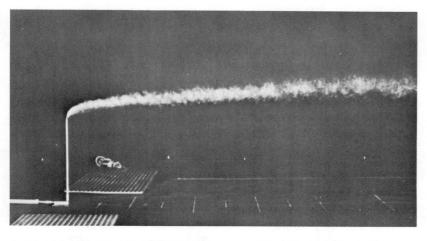

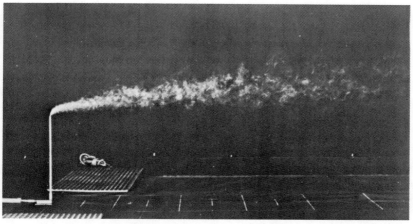

are reprinted by courtesy of Environmental Engineering Research Laboratories, New York University.)

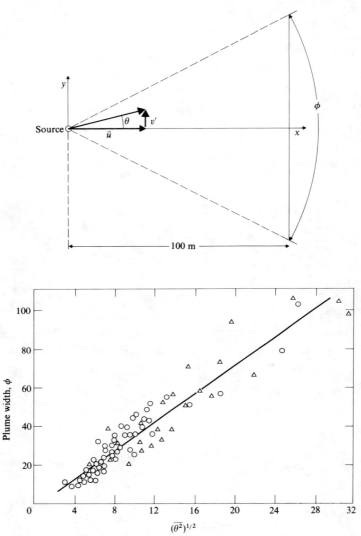

Fig. 7.12 (a) Top view of an experiment indicating the angle θ between the wind at any instant and its average direction (the x-axis), and the angle ϕ defining the plume width observed 100 m downwind from the source; (b) observed relation between the root mean square value of θ and the resulting plume width ϕ. (H. E. Cramer, Proceedings of the First National Conference on Applied Meteorology, American Meteorological Society, 1957.)

Thus a one-hour average concentration would be $6^{-1/2} = 0.41$ times the concentration obtained for a 10-minute average. For intervals of less than 10 minutes, the experimental data for some sources indicate that a $\tau^{-1/5}$ dependence seems to be valid.[10,11] The departure of the data from a $\tau^{-1/2}$ dependence is illustrated in Fig. 7.13. The dependence on sampling period τ for short periods should be influenced

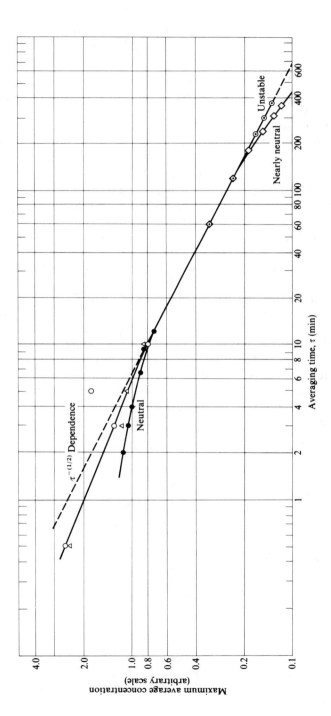

Fig. 7.13 Dependence of the maximum average value of the downwind concentration on the averaging time. Data from several researchers is shown. [From M. Hino, *Atmos. Environ.*, **2**, 149 (1968)]

by local conditions and the turbulence spectrum, in a way which has not yet been determined for most localities, so we shall not pursue the subject further.

How a plume spreads

A crucial factor in our ability to deal rationally with air pollution is the availability of models by which pollution downwind from a source can be predicted on a quantitative basis. One approach has been to argue that the turbulent diffusion of pollutants can be treated by an average diffusivity constant and an equation similar to that in Eq. (7.4b); such an average "eddy diffusivity constant" would be evaluated numerically on the basis of empirical studies. The meandering of a plume is accounted for in such models by assuming straight line downwind travel of the plume; the average lateral spreading is governed by a diffusion-like equation, analogous to Eq. (7.4b), and the details of the meandering motion are averaged out. It is generally agreed that such an approach envisioning an average diffusive-like behavior must incorporate eddy diffusivity constants which have no precise significance in the physical sense. Hence sole justification for their use must be based on empirical success.

Now we shall focus our attention on the lateral diffusion of a pollutant from within a plume, and using average eddy diffusivity constants, we shall derive the equation which governs the downwind spreading of a plume. Let X again denote the concentration of pollutant molecules, so that nX is the number of pollutant molecules found in a cubic meter of air. The equation governing the diffusion process can be deduced by reference to Fig. 7.14. This illustrates a hypothetical volume, stationary above ground, through which air and pollutants move with the wind direction parallel to the x-axis. The dimensions of this volume are Δx, Δy, and

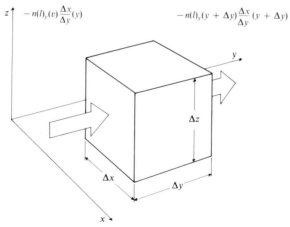

Fig. 7.14 Diffusion of pollutant molecules parallel to the y-axis. The rate at which pollutants accumulate in the volume element is the difference between the rate at which they enter and the rate at which they leave. The large arrows show the sense of flow of molecules if $\Delta X/\Delta y$ is a negative quantity in the region of the volume element.

Δz, with sides parallel to the axes of the coordinate system. The rate at which the concentration of a pollutant builds up within the volume is given by the rate at which pollutant molecules enter minus the rate at which they leave.

For the moment, let us consider only the movement of molecules along the y-direction. By assuming a Fickian type of diffusion, we know that the rate at which pollutant molecules accumulate within the volume is given solely by the difference in the gradients of the concentrations at the two faces of the volume, one face at y and the other at $y + \Delta y$, with the gradient at each face weighted by the eddy diffusivity constant at the respective face. The diffusion of pollutant molecules will be in the direction of the arrows shown in Fig. 7.14 if X decreases with increasing y. There will be a net build-up of the average value of X within the volume if more pollutant molecules enter from the left face than leave through the right. The rate at which pollutant molecules accumulate in a region of the atmosphere of volume $\Delta x \Delta y \Delta z$ is therefore governed by the equation:

$$(\Delta x \Delta y \Delta z)\,(n\frac{\Delta X}{\Delta t}) = (\Delta x \Delta z)\left\{ -nD_y(y)\left[\frac{\Delta X(y)}{\Delta y}\right]_{x,z} + nD_y(y + \Delta y)\left[\frac{\Delta X(y + \Delta y)}{\Delta y}\right]_{x,z} \right\}.$$

(7.5)

The prefactor on the right-hand side of this equation is the area of the face through which the molecules pass. The subscripts x and z indicate that the gradient of the concentration with respect to y must be evaluated for a constant value of x and of z. A subscript y is attached to the diffusivity D because, in a turbulent atmosphere, we must anticipate that the diffusivity along the x- and z-directions would differ from that along the y-direction. Dividing both sides of Eq. (7.5) by $\Delta x \Delta y \Delta z$ and converting to an obvious notation, we find

$$n\frac{\Delta X}{\Delta t} = n\left\{ \frac{\Delta\left[D_y\left(\frac{\Delta X}{\Delta y}\right)_{x,z}\right]}{\Delta y} \right\}_{x,z}.$$

(7.6)

The implication of this equation can be seen more easily if we assume that the eddy diffusivity is independent of position and for the moment drop the subscript notation:

$$\frac{\Delta X}{\Delta t} = D_y\left\{ \frac{\Delta\left(\frac{\Delta X}{\Delta y}\right)}{\Delta y} \right\}.$$

(7.7)

The sign of the quantity in curly brackets on the right-hand side indicates whether the variation of pollutant concentration X with y has positive curvature or negative curvature. In extreme instances, this corresponds to whether the concentration has a local minimum or local maximum, respectively, as illustrated in Fig. 7.15. Equation (7.7) predicts that if the concentration has a local minimum, then the right-hand side of the equation has a positive value, and the concentration will increase with time. If it has a local maximum, the right-hand side has a negative

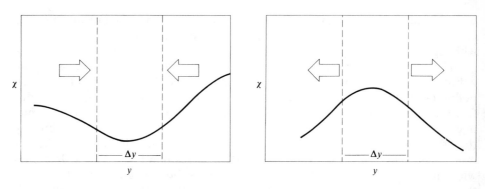

Fig. 7.15 An illustration that Eq. (7.7) describes a trend toward a uniform distribution of pollutant concentration in X in the atmosphere. The arrows show the direction of diffusion of pollutant molecules for (a) a local minimum in X and (b) a local maximum.

value and X will decrease with time. Thus Eq. (7.7) properly predicts the tendency of the pollutants within a plume to spread uniformly throughout the atmosphere along the y-direction, transverse to the wind.

Now when we turn to consider the diffusion process in the vertical z-direction, we conclude that it is governed by an equation analogous to Eq. (7.6), but containing an eddy diffusivity D_z which may differ in magnitude from the horizontal diffusivity D_y.

This leaves only the movement of molecules along the x-direction still to be described. The introduction and elimination of pollutant molecules from the volume for this direction is different, because the major cause of the movement is the average wind, not the diffusivity. The rate at which pollutant molecules enter the volume is approximately $\bar{u}nX(x)\Delta y\Delta z$, where $\Delta y\Delta z$ is the area of the face which they cross. The wind speed $\bar{u}$ is so much less than the molecular speeds that, when the pollutants enter the volume, their concentration readily adjusts to the local value $X(x)$. Similarly, as the wind carries air along the x-direction from the face of the volume at x to the face at $x + \Delta x$, the relative concentration X will be affected by the number of pollutant molecules which have entered or left the volume by diffusion in the y- and z-directions. Thus the concentration at $x + \Delta x$ will in general differ from that at x, and the rate at which pollutant molecules leave must be written as $\bar{u}nX(x + \Delta x)\Delta y\Delta z$. The rate at which they accumulate per unit volume is therefore the difference between the rate at which they enter and leave:

$$n\frac{\Delta X}{\Delta t} = n\bar{u}\left[\frac{X(x) - X(x + \Delta x)}{\Delta x}\right]. \tag{7.8}$$

Now if the accumulation of the pollutant molecules moving along all three principal directions is added together, we find that the rate of increase in the relative concentration is given by

$$\frac{\Delta X}{\Delta t} = -\overline{u}\frac{\Delta X}{\Delta x} + \left\{\frac{\Delta\left[D_y\left(\frac{\Delta X}{\Delta y}\right)_{x,z}\right]}{\Delta y}\right\}_{x,z} + \left\{\frac{\Delta\left[D_z\left(\frac{\Delta X}{\Delta z}\right)_{x,y}\right]}{\Delta z}\right\}_{x,y}. \tag{7.9}$$

This equation describes the accumulation of pollutants at any position in the atmosphere.

The important point to realize is that a source emitting pollutants at a constant rate has a plume whose average shape does not change with time. In other words, a steady-state condition prevails; so the left-hand side of Eq. (7.9) is zero. Pollutants are neither accumulated nor lost at any position although they continually diffuse outward from the center line of the plume (the x-axis); the average concentration at any point in space remains constant in time.

The equation describing the downwind dispersal can be more accurately written if we take the limiting form of Eq. (7.9) when the size of the volume element is considered to be infinitesimally small. In the notation of calculus, we have

$$\overline{u}\frac{\partial X}{\partial x} = \left\{\frac{\partial}{\partial y}\left[D_y\left(\frac{\partial X}{\partial y}\right)_{x,z}\right]\right\}_{x,z} + \left\{\frac{\partial}{\partial z}\left[D_z\left(\frac{\partial X}{\partial z}\right)_{x,y}\right]\right\}_{x,y}, \tag{7.10}$$

where the signs of the type $\partial/\partial y$ indicate partial derivatives; that is, derivatives with respect to one variable (in this instance, y) when all other independent variables (for example, x and z) are considered as constants. This equation, together with the stipulation that $X = 0$ an infinite distance from the source, governs the shape of the plume as pollutants are carried downwind. The eddy diffusion constants D_y and D_z need not be equal, nor need they be independent of positions x, y, and z.

A solution

For the special case in which the diffusion constants D_y and D_z in Eq. (7.10) are independent of position, the type of diffusion described by this equation is said to be *Fickian*. In this instance, the equation can be solved by a simple expression for the concentration X of pollutant if we also assume that the average wind speed $\overline{u}$ does not vary with position. Such conditions can be satisfied only approximately by actual atmospheric behavior. If pollutants are emitted from a point source at ground level, the solution for Eq. (7.10) is then

$$X(x,y,z) = \frac{Q}{\pi\overline{u}\sigma_y\sigma_z}e^{-1/2[(y^2/\sigma_y^2) + (z^2/\sigma_z^2)]}. \tag{7.11}$$

This equation gives the concentration which would exist in the ambient air at any position downwind from the source. We recall that the source is located at the origin of our coordinate system ($x = 0$, $y = 0$, $z = 0$), and that x is the distance directly downwind from the source along the plume line, y is the horizontal distance from the plume line, and z is the vertical height above the ground.

If we are interested in what concentration is experienced by a person or object (hereafter called the *receptor*), we must evaluate the formula by inserting the x-, y-, and z-coordinates of the receptor.

Let us examine the significance of this result by first turning our attention to the prefactor on the right. We note that X is directly proportional to the parameter Q, which we use to represent the rate at which the source emits the pollutant in question. The value of Q must be expressed in units commensurate with X. That is, if X is in kilograms per cubic meter, then Q must be what is called the *mass emission rate* $Q = Q_M$, which is expressed as the number of kilograms per second of the pollutant issuing from the exhaust stack. On the other hand, if X is to be the fraction of air molecules which are pollutant molecules, then the emission rate must be the *volume emission rate* $Q = Q_V$, which is the volume of the pollutant gas emitted each second, evaluated for standard pressure (1000 mb) and ambient Kelvin temperature T. The volume emission rate Q_V (in cubic meters per second) is related to the mass emission rate Q_M (in kilograms per second) by the equation:

$$Q_V = 22.4 \times 10^{-3}(Q_M/M_p). \ (T/273). \tag{7.12}$$

The quantity M_p is the molecular weight of the pollutant expressed in kilograms. Either Q_V or Q_M should be used in place of Q in Eq. (7.11), as appropriate.

Equation 7.11 gives us an important prediction: The profile of the plume transverse to the wind is governed by what is called a *binormal distribution*. If for any value of x the receptor is a distance y from the plume line, the decrease in X with increase in y follows a normal distribution, as described by the following

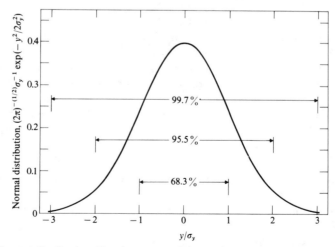

Fig. 7.16 Normal distribution. The percentage of the total area under the curve within one, two, and three standard deviations of the center is indicated.

factor contained in the solution:

$$\frac{1}{(2\pi)^{1/2}\sigma_y}e^{-(y^2/2\sigma_y^2)}.$$

This is an exponential function, but the exponent contains the square of the independent variable (y) and not the first power that we have previously encountered in this book. Thus it has a qualitatively different appearance, as we illustrate in Fig. 7.16. The bell-shaped curve described by this factor has a width proportional to the parameter σ_y, called the *standard deviation* of the distribution. As indicated in the figure, about 68% of the pollutant molecules are found within 1 standard deviation of the plume line, and 96% within 2 standard deviations. The concentration at 1 standard deviation is about 60% of the concentration on the plume line; at 2 standard deviations, it is 14%.

As we said previously, Eq. (7.11) is a *binormal* distribution, which means that the decrease of X with vertical distance from the plume line also follows a normal distribution. In general, the diffusion of pollutants vertically does not proceed at the same rate as it does in the horizontal direction, so the standard deviation σ_z usually differs from σ_y. Since the source is located at ground level, it must also be implicitly understood that $X = 0$ for negative values of z.

Equation (7.11) as written contains no explicit mention of the downwind distance x of the receptor, although we certainly expect such a dependence. In fact the x-dependence is contained in the standard deviations σ_y and σ_z. In order for Eq. (7.11) to be a solution of Eq. (7.10), the standard deviations must satisfy the conditions

$$\sigma_y^2 = \left(\frac{2D_y}{\bar{u}}\right)x, \quad \sigma_z^2 = \left(\frac{2D_z}{\bar{u}}\right)x. \tag{7.13}$$

The square of the standard deviations, known as the *variances* (σ_y^2 and σ_z^2), are thus required to be proportional to the respective diffusivities and to the downwind distance. They are inversely proportional to the wind speed. From these conditions it is evident that the larger the diffusivity for the transverse direction, the more rapidly the plume spreads with downwind distance. The solution to Eq. (7.10) thus logically accounts for a more rapid diffusion of a plume when there is a greater intensity of turbulence. The shape of the spreading plume as σ_y increases downwind is illustrated in Fig. 7.17 for the concentration at ground level.

Our preceding discussion assumes that the pollutants in the effluent are gaseous, because they respond to turbulence and mix just as air molecules do. The results are not valid for large particles which may be in the effluent, because for them the sedimentation owing to gravity is an important effect. In vigorously turbulent areas, particles up to 30 microns radius do not settle at an appreciable rate since they are more responsive to the eddy motions and effectively follow the surrounding gas, at least for a few hours. Thus the diffusion of such small particles can also be described by Eq. (7.11).[12] Larger particles will settle and will produce higher concentrations near the source than is predicted by this formula.

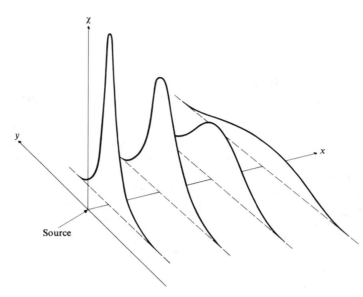

Fig. 7.17 Representative ground-level concentration for several transverse profiles downwind from a point source.

One additional point about Eq. (7.11) deserves a comment. The formula predicts a finite concentration of pollutant at large values of y or z, even when the downwind distance x is very small. To accomplish this, some molecules must move outward at nearly infinite speeds! In fact this prediction does not correspond with reality. The anomalous prediction is a result of our simple physical model which assumes that the rate of diffusion of a pollutant is determined solely by the local gradient of the concentration. The model neglects some important physical aspects of the molecular nature of air. This does not pose a practical difficulty, because—from a practical standpoint—the concentration of pollutant at large distances is negligibly small.

Random or correlated?

If the plume width is considered to be the standard deviation σ in either transverse direction, then Eq. (7.13) indicates that the width increases with the square root of the distance traveled downwind. This feature is characteristic of Fickian diffusion from a point source. Readers who are familiar with Albert Einstein's solution of the random walk problem of Brownian motion may recognize that this dependence of the plume width on distance is equivalent to a square root dependence on the time t after release of the pollutant. The root mean squared distance traveled by a randomly moving particle was found by Einstein to be proportional to $t^{1/2}$.

If, on the average, there is a random migration of pollutant molecules, then Eq. (7.11) has a basis in established theory. If not, then the problem must be approached with a much higher degree of theoretical sophistication. Any correlation between the behavior of different portions of a plume implies that the relationships between the variances and distance x in Eq. (7.13) are invalid.

Now our discussion on the characteristics of atmospheric turbulence shows that in some instances there will be a degree of correlation with atmospheric diffusion, because a few components of the turbulence spectrum are more important than others. Statistical theories based upon the work of G. I. Taylor can relate the parameters σ_y and σ_z to expressions for the correlation of wind velocities in different portions of the plume.[13] However, in most cases we do not know what the correlation expressions are for the actual turbulence in a given geographical locality. Thus we are at a standoff so far as their applications are concerned. But nevertheless we are warned that for most practical cases the Fickian diffusion model is to a degree inadequate.

It is still instructive to consider the general features of plume dispersal that the Fickian model predicts. We might anticipate that these are at least qualitatively accurate for describing the dispersal in turbulent atmospheres. The important features of Eq. (7.11) are:

1. The downwind concentration at any location is directly proportional to the emission rate of the source. This is because Fickian diffusion is a linear process, dependent only upon the gradient of the concentration.

2. The more turbulent the atmosphere, the more rapid the spread of the plume in the transverse direction. Turbulence increases the eddy diffusivity in Eq. (7.13).

3. The maximum concentration at ground level is found directly downwind, on the *plume line*, and is inversely proportional to the downwind distance from the source.

4. The maximum concentration decreases for higher wind speeds $\bar{u}$. Even on the plume line, where at ground level there is no explicit dependence on $\bar{u}$ (because $\sigma_y \sigma_z$ is inversely proportional to $\bar{u}$), the ground-level concentration will actually decrease with increasing wind. This is because the eddy diffusivity D_z in Eq. (7.13) increases with wind speed due to increased mechanical turbulence.

These are four key features of most models which describe the dispersal of emissions from a point source at ground level.

7.5 SMOKESTACK PLUMES

Dispersion from an elevated source is a different matter. Pollutants have a longer time to diffuse laterally before the high concentration region of the plume touches ground. Therefore the maximum ambient concentration at ground level is not found at the source but at some distance from the stack. The magnitude of the maximum concentration and where it is expected to occur are two of the most important questions involved in the effectiveness of an exhaust stack.

The simplest way to adapt Eq. (7.11) to describe this case is by substituting

(z-H) for z, where H represents the height of the stack. We must also adjust the multiplying factor to ensure that the right-hand side is properly normalized. However, instead of doing that, let us first consider a more pressing question: What happens when pollutants strike the ground? Do the molecules stick, or do they in effect bounce off? The answer is not simple. The proportion that stick depends upon the type of surface (vegetation, building materials, etc.).[14] Due to our lack of knowledge, this proportion is generally not predictable in advance. It is simpler from a mathematical standpoint to avoid this question and to assume that the pollutants are reflected when they hit the ground. In this case, the generalization of Eq. (7.11) would be

$$\chi = \frac{Q}{2\pi \bar{u}\, \sigma_y \sigma_z}\, e^{-(y^2/2\sigma_y^2)} \left[e^{-[(z-H)^2/2\sigma_z^2]} + e^{-[(z+H)^2/2\sigma_z^2]} \right]. \qquad (7.14)$$

It is possible to visualize how this describes the situation by referring to Fig. 7.18. We suppose that the ambient concentration at any position above the ground level ($z \geq 0$) is given by the contributions from two sources: the real source at $z = H$ and an imaginary source of equal emission rate at $z = -H$. The symmetry of the geometry suggests the name of *image source* for the source underneath. The ambient concentration predicted by Eq. (7.14) is the same as it would be if pollutants were reflected from the ground when they hit. If a fraction of the pollutant was actually absorbed on impact, the image source should be weaker. Equation (7.14) can therefore be regarded as giving the maximum expected ambient concentration at any point, and its use would thus provide a conservative estimate.

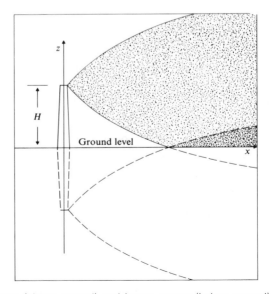

Fig. 7.18   Real source (above ground) and image source (below ground). The total ambient concentration for the spatial region $z > 0$ is the same as it would be if the pollutant molecules were reflected upward as they struck the ground.

Empirical methods

The difficulties facing theoretical efforts to describe diffusion in turbulent atmospheres, and the lack of data for their application to individual geographical localities, have led to an abandonment in most practical applications of theoretical formulas in favor of empirical ones. A generalization of the binormal distribution Eq. (7.14) is in contemporary favor with most diffusion experts. The generalization is performed by assuming that the variances are not given by Eq. (7.13) as required by a solution of Eq. (7.12), but are to be determined by observing the behavior of a number of plumes under various conditions of source operation, wind speed, and atmospheric stability. An example is the Pasquill-Gifford method in which σ_y and σ_z are given for each value of x by empirical graphs.[15] These empirical values for the variances could be interpreted according to Eq. (7.13) as corresponding to empirical diffusivities which depend upon x although it must be kept in mind that diffusivities determined in this way have no direct physical interpretation.

Maximum ground-level concentration

If the ambient concentration downwind from an elevated point source is given by Eq. (7.14) with empirical values for the x-dependence of σ_y and σ_z, then several statements can be made concerning the magnitude and distribution of the ground-level concentration. The location which received the maximum ground-level concentration will be of course on the plume line ($y = z = 0$). For the special case in which the ratio σ_y/σ_z is found to be independent of x, it can be shown that the maximum concentration X_{max} will occur where

$$\sigma_z = 2^{-1/2}H. \tag{7.15}$$

This location will of course depend upon the empirically determined dependence of the standard deviation σ_z on the value of x. For lack of simpler or more accurate alternatives, a power law dependence is often assumed as introduced by O. G. Sutton:

$$\sigma_y^2 = x^{2-n}\, C_y^2/2$$
$$\sigma_z^2 = x^{2-m}\, C_z^2/2, \tag{7.16}$$

where n and m are appropriate numbers; these numbers, together with the constants C_y and C_z are called "diffusion parameters." Thus for the case when $n = m$, so that σ_y/σ_z is independent of x, the point of maximum ground-level concentration is given by

$$x^{2-n} = H^2/C_z^2. \tag{7.17}$$

Under the same conditions, the ambient concentration at this location can be obtained by evaluating Eq. (7.14), with the result

$$X_{max} = \frac{2Q}{\pi \bar{u}\, e\, H^2}\left(\frac{C_z}{C_y}\right). \tag{7.18}$$

This formula illustrates some important features concerning the effectiveness of exhaust stacks. The symbol e represents as usual the base of the natural logarithms $e = 2.718$. Near the ground, the empirical parameter C_z tends to be smaller than C_y because of suppression of large-scale vertical turbulence. Thus a value of 0.7 for C_z/C_y under a neutrally stable lapse rate may be appropriate. For altitudes in excess of 25 m, the turbulence becomes more isotropic; field studies indicate that C_y approximately equals C_z above this level.[16] An example of how diffusion parameters vary with atmospheric stability and elevation is given in Table 7.1. Of course, the values of the constants applied in the evaluation of Eq. (7.14) must depend upon individual circumstances. Generally one chooses values which correspond to an elevation midway between ground level and the height of the exhaust stack. This is true for the wind speed, which also appears in the formulas. Wind speeds are frequently found to increase by 30% to 50% from a height of 10 m to 300 m. We shall return to a discussion of the appropriate numerical values of diffusion constants after we first examine some additional important features of Eq. (7.18).

TABLE 7.1

Sutton's diffusion parameters for $n = m$ in Eq. (7.16), as evaluated on the basis of a 3-minute averaging period*

Stability condition		C_y and C_z $(m^{n/2})$			
			Elevation (meters)		
	n	25	50	75	100
High lapse rate	0.20	0.21	0.17	0.16	0.12
Zero lapse rate	0.25	0.12	0.10	0.09	0.07
Moderate inversion	0.33	0.08	0.06	0.05	0.04
Strong inversion	0.50	0.06	0.05	0.04	0.03

*From G. H. Strom, in *Air Pollution*, A. C. Stern, Ed., New York: Academic Press, 1968, Vol. I. Copyright © 1968 by Academic Press and reprinted by permission.

The maximum ground-level concentration from a plume released from a height H is seen to be inversely proportional to H^2, a feature that was first enunciated by C. H. Bosanquet and J. L. Pearson in 1936.[17] The second power of H enters because the plume is diffusing in two dimensions as it travels downwind; thus the same mass of pollutant is spread over a wider area, an area which is effectively proportional to H^2. However, the fact that X_{max} is also proportional to C_z implies that the more vigorous the vertical diffusion, the sooner the lower edge of the plume touches ground [a fact apparent from Eq. (7.17).] If C_y for some reason were to be small, relatively little diffusion would have occurred in the horizontal direction and a high ground-level concentration would result. It is fortunate that in most cases C_z/C_y is on the order of unity.

The maximum concentration according to Eq. (7.18) is also inversely pro-

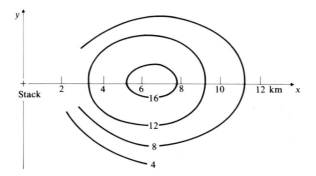

Fig. 7.19 Downwind concentration of a pollutant emitted from a 300-MW power plant with a 100-m stack. Contours give relative values of the ground level concentration measured for a three-hour average. [M. Hino, *Atmospheric Environment,* **2,** 149 (1968)]

portional to the wind speed. Strong winds draw out the plume and dilute it more thoroughly because more air passes by the exhaust stack during a given length of time. Strong winds also affect the degree of turbulence, owing to an enhancement of mechanical turbulence, a factor which should be incorporated into selecting the proper values for the diffusion parameters.

A feature of Eq. (7.18) worth special emphasis is the fact that the maximum ground-level concentration depends upon the *mass emission rate;* for a given release height H it is independent of the *concentration* of the pollutant in the exhaust gas. Thus simply emitting a more dilute mixture without decreasing the mass emission rate will not reduce the ground-level concentration at the receptor, other factors being equal. The reason is that our model implicitly assumes that only pollutant—and no air—is emitted from the stack, and that dilution of the effluent in ambient air following after its emission is the primary means by which the concentration is reduced.

An aspect of concern in the analysis of exhaust plumes is the extent of the area affected by a pollutant concentration exceeding a certain value. The affected area will, of course, depend upon how the diffusion parameters vary with distance x, and so a detailed analysis of Eq. (7.14) is required. One empirical example, obtained by monitoring tracer pollutants which were intentionally placed in the exhaust gas from a power plant, is shown in Fig. 7.19. The contours trace locations of uniform ambient concentrations.[9] It is important to realize that the distance downwind from large sources at which the maximum ground-level concentration is observed may be impressively great. In this example, the maximum was observed a distance of slightly more than 6 km from the exhaust stack.

The patterns of ground-level concentrations shift with changing stability. As an example, Fig. 7.20 illustrates the fact that a much more extensive length along the plume line will experience high concentrations during periods of high stability. However away from the plume line, the concentrations are much lower.

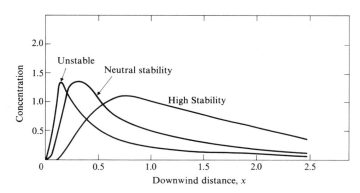

Fig. 7.20   Effect of atmospheric stability on the ground-level concentration under the plume line. The source is assumed to have a volume emission rate of $Q_v = 4.7$ liters/sec (10 ft^3/min) from a 30-m stack; and the wind speed is $\overline{u} = 0.5$ m/sec (100 ft/min). The distance x is in kilometers. (After C. A. Gosline, Jr., L. L. Falk, and E. N. Helmers, *Air Pollution Handbook*, P. L. Magill *et al.*, Eds., New York: McGraw-Hill 1956, by permission.)

Diffusion parameters

We return now to discuss in more detail some important aspects of the diffusion parameters which determine the spreading of an exhaust stack plume. The determination of σ_y and σ_z for Eq. (7.14) is a crucial step in predicting the ground-level ambient concentration, for the predicted concentration is very sensitive to their values. Small inaccuracies in their estimation will usually produce large inaccuracies in the predicted concentration. It is unfortunate that, for most applications, the diffusion parameters cannot be determined with high reliability. One reason is that parameters such as those of the Sutton method and the Pasquill-Gifford method were measured for diffusion over smooth terrain. In application, however, features of uneven terrain, such as hills, tall trees, or buildings are frequently important, because they alter the spectrum of mechanically produced turbulence.

To avoid much of this difficulty, one could try to relate the diffusion parameters to the characteristics of the turbulence measured at the location of interest. For example, these could be the turbulence relative intensities $((v')^2)^{1/2}/\overline{u}$ and $((w')^2)^{1/2}/\overline{u}$ in the horizontal and vertical directions, respectively. Or we might instead choose a more graphically descriptive quantity and measure the turbulence by the wind's *angular deviation* from the mean direction, letting $\tan \theta = v'/\overline{u}$ give the angle θ of horizontal deviation and $\tan \alpha = w'/\overline{u}$ give the vertical angle α. Averaging over an appropriate period of time, one could empirically determine the standard deviations σ_θ and σ_α which are defined by the mean square value of the angles: $\sigma_\theta^2 = \overline{\theta^2}$ and $\sigma_\alpha^2 = \overline{\alpha^2}$. Thus σ_y and σ_z, respectively, in Eq. (7.14) could be related to the standard deviations σ_θ and σ_α at the source or at some point near the source.[18,19] This correspondence is suggested by experiments such as the one whose results are summarized in Fig. 7.12.

The diffusion parameters will evidently depend upon the atmospheric stability. As an illustration, for one-hour sampling periods in stable air, M. E. Smith[18] suggests that

$$\sigma_y = 0.15 \, \sigma_\theta x^{0.71}$$
$$\sigma_z = 0.15 \, \sigma_\alpha x^{0.71};$$

(7.19)

for unstable air,

$$\sigma_y = 0.045 \, \sigma_\theta x^{0.86}$$
$$\sigma_z = 0.045 \, \sigma_\alpha x^{0.86}.$$

(7.20)

For these equations, the distance x should be expressed in meters, and σ_θ and σ_α in degrees. The drastic change in the magnitudes of the coefficients in Eq. (7.20) compared with Eq. (7.19) emphasizes how sensitive the diffusion of a plume is to the degree of atmospheric stability. Before applying these and similar equations to local situations, one should always consult the original references in which the formulas are given. Only then can one have some assurance that the conditions under which the coefficients and exponents were evaluated have any bearing on a particular local condition. All too often prepackaged formulas are applied to circumstances which do not justify the application. Undoubtedly equations such as (7.19) and (7.20) will be refined in the future to yield better agreement with a larger body of empirical data as it is accumulated.

Before we proceed to the next subject, let us emphasize what equations such as Eq. (7.18) mean. These equations do *not* yield the maximum possible ambient concentrations. They give only the maximum *average* concentration for an averaging period which is appropriate for the diffusion parameters. Furthermore, the binormal form for the profile of the plume has no firm theoretical basis, since it has not yet been established that the average eddy diffusion in an unstable atmosphere proceeds at a rate proportional to the average gradient of the concentration of pollutant. Strictly speaking, these results are only empirical and have no validity beyond the range of parameters and conditions for which they have been empirically verified.

As we stated at the beginning, the results are also invalid when the plume enters a region where the temperature lapse rate suffers a marked change. For example, a plume released within a mixing layer capped by an elevated inversion will cease its upward dispersal at the inversion base. From there on, Eq. (7.14) cannot be applied, and another dispersal formula must be sought. Often suggested is the following gross approximation: Once the plume has spread to an extent whereby the inversion base is less than, say, two standard deviations from the plume center line, it is assumed that for distances further downwind the pollutant is uniformly distributed in the vertical direction from ground level to the inversion base; the only subsequent mixing can occur in the horizontal direction. Whether this is a reasonably accurate prescription has unfortunately not yet been established, for there is little data on pollutant concentrations which are experienced by a receptor at ground level when vertical dispersal is limited by an elevated inversion.

7.6 PLUME RISE

In the preceding section, it was assumed that the conditions of release of an effluent could be neglected except for consideration of a constant H, the release height. In many situations such an assumption is not warranted. The effluent from a power plant is warm and possesses buoyancy, so the plume will continue to rise after it leaves the stack; this happens in the case of smoke from brush fires as well. In addition, the effluent from large sources may be propelled along by fans, which in the case of power plants draw air into the combustion chamber; the action of the fans and the expansion of gas in the chamber force the effluent to leave by the exhaust stack. Thus the effluent may have a high upward velocity and momentum as it exits from the stack, and this encourages plume rise. As a result, the physical characteristics of a plume can be described by various parameters. The most common are the exit speed V and the heat content or heat emission rate Q_h. The first is a measure of the initial upward momentum of the plume, and the second of its buoyancy.

Let us examine what Q_h characterizes. The production of exhaust gases through combustion in a steam-producing combustion chamber of a power plant, or by processes associated with foundry operations, or by burning logs in a home fireplace imparts kinetic energy to the gas molecules and thus raises their temperature. Not all of this thermal energy can be removed for useful work, so it is exhausted with the waste gases. The rate at which it is exhausted is called the heat emission rate Q_h which is an important parameter because the difference in temperature between the exhaust gas and ambient air is proportional to this excess thermal energy. Therefore, the buoyancy should be directly related to Q_h; that is, the hotter the plume, the higher it will rise.

Large steam-generating electrical power plants often are designed to emit their plumes with an exhaust speed of $V = 100$ km/hr and a temperature of 140°C. Modern coal-burning power plants have an overall efficiency of about 40%, implying that about 60% of the fuel energy is lost either in the gaseous exhaust or in the cooling water used to condense the steam from the turbine. Thus a station rated for an electrical output of 1 gigawatt (abbreviated 1 GWe, for 1 gigawatt electrical) must convert chemical energy to thermal energy at a rate of 2.5 gigawatts (abbreviated 2.5 GWt for 2.5 gigawatt thermal). In fact only about 5–10% of the calorific value of the fuel is lost with the waste gases, so that this station would emit a plume with $Q_h = 125$ to 250 MW. It is not uncommon in smaller, less efficient operations to have 20% of the calorific value of the fuel exhausted up the stack. Hence the momentum and buoyant rise of the plume after release are not negligible factors when determining the effective release height H for application of the plume diffusion equations.

The plume rise above the stack, h_r, must be added to the geometrical height of the stack h_g to obtain the *effective release height* H. Both V and Q_h will influence the value of h_r. Atmospheric stability and the wind speed will also play important roles. For small sources, such as the home chimney, the exit speed and heat emis-

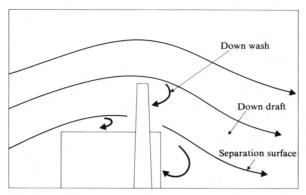

Fig. 7.21   Air trajectories as wind blows from left to right.

sion rate are often so small as to be negligible factors. The effective release height is then essentially identical to the geometric stack height.

Turbulent effects

Before we discuss techniques by which plume rise may be estimated, let us look at the conditions under which it is anomalously affected. Most complications arise for small sources, those which have a small exit speed and negligible buoyancy. The reason is the physical presence of an exhaust stack or building causes mechanical turbulence in the air if the wind is sufficiently strong. One situation is illustrated in Fig. 7.21. Experience indicates that V should exceed the average wind speed $\bar{u}$ by 20–30% to prevent entrapment of the effluent in the *downwash* of the exhaust stack. The downwash condition occurs as air tries to "fill in" behind the stack as the wind drives it by.

Perhaps the reason for this can be seen more easily if we view the situation from the top of the stack, as in Fig. 7.22. As a gross simplification, we imagine that one of two types of air flow occurs. At one extreme is nonviscous flow in which no momentum or energy would be transferred from the air to the stack. Therefore the air would not lose energy as it moved around this obstacle. In Fig. 7.22(a) work done by pressure forces increases the kinetic energy of the air traveling from A to B. This must be the case, since the air near the stack must travel a greater distance in going around than air further to the side, and so the nearby air must have a higher average speed to maintain smooth flow. The reverse situation occurs as the air passes from B to C, the nearby air slowing as it moves into a region of higher pressure until it reaches its initial speed and continues its smooth, uniform motion. But for a viscous fluid such as air, this ideal type of flow does not occur at high wind speeds. In viscous flow, as illustrated in Fig. 7.22(b), friction with the stack tends to slow the air during its passage from A to B, so that on continuing from B to C there is insufficient kinetic energy to overcome the slowing pressure effects. Some of the air becomes trapped in an eddy to the lee of the stack. As more air accumulates, an interesting response is observed. At some point

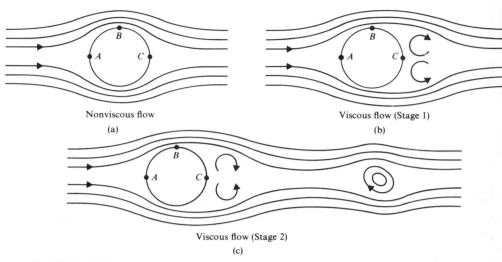

Nonviscous flow
(a)

Viscous flow (Stage 1)
(b)

Viscous flow (Stage 2)
(c)

Fig. 7.22   (a) Top view of streamline flow past an upright cylindrical object such as an exhaust stack. (b) Turbulence develops behind the stack. (c) Formation of a Kármán vortex street.

a *vortex* will be shed, as shown in Fig. 7.22(c), and thereafter a regular pattern of vortices will leave, in a sequence of alternating rotation. The parade of vortices in the wake of an obstruction is known as a *Kármán vortex street*, named after Th. Von Kármán, a pioneer in the study of turbulence.

We have not been precise in distinguishing between the conditions for nonturbulent, or streamline, motion in Fig. 7.22(a) and turbulent motion in Fig. 7.22(c). The transition between streamline flow and turbulent flow is found to depend on the ratio of inertial forces to viscous dissipative forces in the fluid motion. The relative importance of these two effects can be appreciated from a dimensional argument. The inertial force is the centrifugal force experienced by the air molecules as they are deflected back into the lee of the stack. This is proportional to the density of air ρ multiplied by its acceleration for circular motion $\bar{u}^2/L$, where L is a characteristic distance such as the diameter of the stack D_s. The viscous force per unit area on the stack is very approximately equal to $\eta\bar{u}/D_s$, where $\bar{u}/D_s$ represents the magnitude of the gradient of the wind speed, assuming that the wind flow patterns are disturbed out to a distance about equal to D_s from the wall of the stack. By Newton's third law this must also be the magnitude of the force experienced by the air, per unit area of the stack wall. On a unit *volume* of the affected air, the total volume including all of the air within a distance D_s from the stack, the viscous force is approximately equal to $\eta\bar{u}/D_s^2$. The ratio of the centrifugal force per unit volume to the viscous force per unit volume gives a parameter which is known as the Reynolds number $\mathbf{R}_e$, named after Sir Osborn Reynolds who discovered its importance:

$$\mathbf{R}_e = \frac{\rho\bar{u}D_s}{\eta}.$$

Evidently there is some degree of arbitrariness in the choice of the parameter L for all but the simplest of situations. But the value of the Reynolds number gives a rough indication whether streamline flow or turbulent flow predominates.

For small values of $\mathbf{R}_e$ (on the order of unity), there is streamline flow. For large values (100 or greater), the streamline flow *separates* from the obstruction, leaving a turbulent region in the wake. Eddies and vortex patterns form. The bulk of the air behaves as though it were frictionless, and only very close to the geometrical boundaries of an obstruction are frictional effects pronounced. If friction is present, eddies may develop even for flow over smooth, flat surfaces, provided the Reynolds number is sufficiently large. This happens when wind flows over the ground. In the atmosphere, $\mathbf{R}_e$ is always much greater than 100, so mechanically produced turbulence of this type is generally found. The characteristic length in this case is the height of the boundary layer of air in which the wind speed varies appreciably.

Effluent from a stack which has insufficient exit momentum may be drawn into the vortices and the eddies behind the stack and conducted downward. If it descends sufficiently low, the effluent may then be entrapped in the *downdraft* of air on the lee side of a nearby building and from there be carried quickly to the ground. The strength and location of a downdraft will of course depend upon the size and location of adjacent structures. One example is illustrated in Fig. 7.21. The downdraft is streamline flow, in contradistinction to the turbulent downwash.

Turbulence will also be formed on the lee side of a building for sufficiently high wind speeds (or Reynolds numbers). Streamline flow still occurs far from the building, delineated from the region of turbulence by a *separation surface*, as shown in Fig. 7.21. Smokestacks improperly located close to a building in the turbulent region may see their effluent caught and mixed with the air in the wake, if the exit momentum is insufficient. This may be especially common near incorrectly designed house chimneys, in which case eddies on the lee side of the house may actually trap the effluent and permit it to reenter the interior through an open window. Momentum rise is necessary to clear plumes from the wake of a stack or building. Many illustrations of unusual plume behavior may be found in R. S. Scorer's book *Air Pollution*.[20]

As a plume exits from a stack, vertical and horizontal expansion quickly cause it to cool and perhaps condense water if the plume humidity is high. If the plume has sufficient thermal energy to be warmer than the ambient air, it rises as it drifts downwind. In an unstable atmosphere, the plume would rise without limit. If it encounters an inversion layer, it may quickly level off, depending upon its temperature and the lapse rate within the inversion. It is not always possible to define unambiguously a plume "height" and a plume rise h_r, illustrated in Fig. 7.23; a rule of thumb has often been to choose the plume height as the point where the plume has leveled off to an angle of, say, $10°$ with respect to the horizontal plane.

Many physical models have been devised in attempts to describe plume rise under various conditions, but the predictions often disagree because of the different assumptions of basic turbulent processes. The current fashion is to establish empirical formulas in terms of $\bar{u}$, V and Q_h.[21]

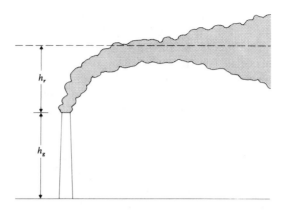

Fig. 7.23 The effective stack height is the sum of the geometrical stack height h_g and the plume rise h_r.

Since we expect from physical reasoning that plume rise h_r will increase with Q_h and decrease with $\bar{u}$, it is reasonable to attempt to describe observed plume rises by a formula which involves a ratio of the two parameters. One such empirical relation is known as the simplified CONCAWE formula (Conservation of Clean Air and Water, Western Europe):

$$h_r = k\frac{Q_h^{1/2}}{\bar{u}^{3/4}}. \tag{7.21}$$

A moderately good fit to data from some European sources is obtained with $k = 0.086$ when Q_h is expressed in watts, h_r in meters, and $\bar{u}$ in meters per second.[22]

Moses and Carson have proposed another formula based upon 700 observations of various stacks under different conditions.[23] When applied to operations in different stability classes, it takes the form:

$$h_r = a\frac{VD_s}{\bar{u}} + b\frac{Q_h^{1/2}}{\bar{u}}\ , \tag{7.22}$$

TABLE 7.2

Parameters for Eq. (7.22).[*]

Stability class	Negative lapse rate, in degrees Celsius per kilometer	a	b
Stable	$\Delta T/\Delta z > 1.5$	-1.04	$+0.071$
Neutral	$1.5 > \Delta T/\Delta z > -12.2$	$+0.35$	$+0.084$
Unstable	$-12.2 > \Delta T/\Delta z$	$+3.47$	$+0.163$

[*]J. E. Carson and H. Moses, *Journal of the Air Pollution Control Association,* **19**, 862, 1969.

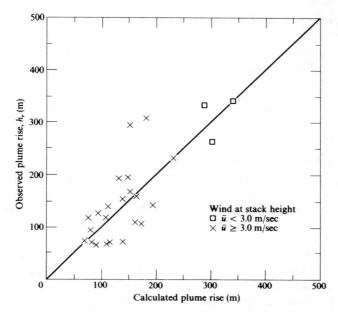

Fig. 7.24   Observed versus calculated plume rise for a large power plant during neutral stability conditions. The calculated rise is based on Eq. (7.22). Data points falling on the straight line are in agreement with the predicted rise. [J. E. Carson and H. Moses, *Journal of the Air Pollution Control Association*, **20**, 477 (1970); courtesy of Argonne National Laboratory.]

where D_s is the stack diameter in meters, and the other parameters are in MKS units. The constants a and b depend upon atmospheric stability. Somewhat arbitrarily, stability is characterized by the three classes given in Table 7.2. As expected, the plume rise is sensitive to the degree of atmospheric stability. The momentum rise term, proportional to V, is small in practical applications. The negative value of a for the stable conditions implies that the second term, proportional to Q_h, must also include some effect of the exit momentum, since upward momentum alone ($Q_h = 0$) could never lead to a lowering (or negative) plume rise.

An appreciation of the accuracy which may be expected from Eq. (7.22) can be gained from a comparison with data of the observed plume emitted by the Paradise Power Plant of the Tennessee Valley Authority during near-neutral conditions, shown in Fig. 7.24. Deviations from the predicted rise are seen to be about $\pm 50\%$ at most, with general agreement between the trends of the predictions and the observations. The data show that the buoyancy of large plumes may carry them upward for many hundreds of meters above the stack.

Comparisons by Carson and Moses[23] of predictions from several proposed formulas and observations of plume rise have not succeeded in determining the power to which the heat emission rate Q_h should be raised, although formulas proportional to $Q_h^{1/2}$ were slightly superior. Neither has it been possible to determine whether a separate term for momentum rise is needed. What has been shown is that the quality of presently available plume rise data is simply inadequate.

Plume rise formulas determined by fits to many observations are thus accurate only in a statistical sense, and have not been designed to predict plume height on an hourly basis.[24] This limitation should be recognized when plume rise formulas are used in conjunction with downwind dispersion equations to calculate predictions for the ground-level concentrations of pollutants.

An example

To reinforce the essential features of the preceding chapters, let us take an example. Suppose it is desired to calculate the location of the maximum ground-level concentration of SO_2 from a fossil fuel power generating plant rated for an electrical output of 1×10^9 W. The fuel used is coal with a 2% sulfur content. It is also desired that the value of the maximum ground-level concentration be known. Meteorological conditions for the time in question are characterized by overcast skies and a neutral atmospheric stability, with a breeze of 4.4 m/sec (10 mi/hr) at the top of the 100 m stack. The exit velocity of the effluent is 100 km/hr (60 mi/hr) and the rate at which thermal energy is exhausted up the stack is 6% of the rate at which energy is liberated by combustion. For simplicity, we assume that the neighboring terrain is flat and featureless.

Exhaust parameters: If the efficiency of the station is 40%, the rate of production of thermal energy from the combustion of coal must be 2.5×10^9 W. To produce this, coal must be burned at a rate of 83 kg/sec, according to the calorific value for the fuel characterized in Table 8.3. Therefore the rate at which sulfur is exhausted is 1.7 kg/sec, and the rate at which SO_2 is exhausted is $Q_M = 3.4$ kg/sec (or 290 Tonnes/day). The heat emission rate is 0.06 of 2.5×10^9 W or $Q_h = 150 \times 10^6$ W.

Plume rise: We shall use Eq. (7.22) to estimate the plume rise h_r. A quick computation indicates that the momentum term for neutral stability is less than 10% of the buoyancy term for stack diameters D_s less than 10 m, so we shall neglect it. Plume rise from buoyancy alone is then found to be $\Delta h_r = 230$ m. Therefore the effective plume height is about $H = 330$ m.

Pollutant diffusion: Suppose that on-site measurements give $\sigma_\theta = 12.7°$ and $\sigma_\alpha = 8°$ for one-hour averages. As a first estimate, we shall assume that σ_y and σ_z are related to the measured parameters by equations similar to Eq. (7.19) and Eq. (7.20), but with a multiplicative factor and exponent of x which are interpolations between the two stability classes. We have no experimental evidence that this is a correct procedure, so for a more reliable estimate one could resort to tabulated coefficients or results in graphical form, such as appear in the handbook by D. B. Turner.[17] However from a pedagogical viewpoint, it is simpler to assume $\sigma_y = 0.10\ \sigma_\theta x^{0.8}$ and $\sigma_z = 0.10\ \sigma_\alpha x^{0.8}$.

Since σ_y/σ_z is independent of x, we can use Eq. (7.17) to obtain the location where the maximum concentration will be found, and Eq. (7.18) to estimate the magnitude of concentration. The solutions are $x = 1.2$ km for the distance and

$\chi_{max} = 1.0$ mg/m^3 for the maximum mass concentration of SO_2 for a one-hour average at ground level. This is equivalent to 0.4 ppm concentration by volume.

Under the best of conditions—level terrain, even heating of the ground, and minimal wind shear—the greatest accuracy we might expect from this calculation allows us to anticipate that the actual maximum concentration will be within a factor of 2 of our prediction. This is true even if we had used handbook values for the diffusion coefficients. It is also important to keep in mind that a prediction of this nature has no relevance unless preceded by a detailed meteorological study of local conditions. Only then might one have some confidence that the appropriate assumptions are satisfied.

7.7 LINE SOURCES

The point source is the simplest one with which to deal. Only local terrain and meteorological conditions influence the downwind diffusion of pollutants, provided the exhaust plume is emitted with sufficiently high velocity. This is not true for sources with other geometrical shapes such as a line source. The line source often can represent a road along which pollutants are continually emitted; it is thus spatially extended, and the direction of the wind with respect to the geometry of the source will influence the downwind distribution of pollution.

In Fig. 7.25 we give an example: The highway in this illustration makes two sharp bends near an observer at point P. The ambient concentration at P will depend upon the direction of the wind, since the length of road, and therefore the strength of the source which effectively contributes to pollution at P, also varies. The ambient concentration for case (a) will generally be greater than for (b). How much greater will depend upon the magnitudes of the diffusion coefficients. This contrast has an important implication because motorists who are on a long segment of straight road, perhaps in slowly moving traffic, will be subjected to much higher concentrations of carbon monoxide from other cars when the wind parallels the road than when it is transverse. High concentrations of CO affect the human central nervous system, slowing response and distorting perception. Thus we might anticipate a higher rate of traffic accidents for the parallel wind condition. However, this remains a point of speculation, for studies have so far failed to uncover any relationship between CO levels and accident rates.

Line sources are of concern not only for the CO emitted from motor vehicles. Hydrocarbons and nitrogen oxides are two other contaminants that come from automobile tailpipes and they are the chief ingredients for photochemical smog. Actually the major irritants of this smog are not these primary pollutants, but are secondary pollutants formed as the effluent drifts downwind. We shall see in Section 10.2 that the nature of the ambient pollutant evolves on sunny days with the passage of time. Thus a prediction for the ground-level concentration at downwind locations must be based not only on how the pollutants diffuse but also on how they react with each other. This difficult dispersal problem is highly nonlinear, and it has not yet been solved.

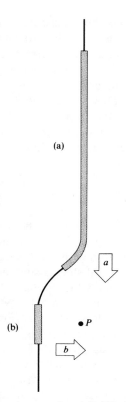

(a)

(b)

Fig. 7.25   A line source curves around a receptor at *P*. Wind parallel to the long stretch of road (a) will produce a higher ambient concentration than transverse (b) provided that atmospheric mixing is minimal.

We can make headway with understanding aspects of diffusion from a line source by taking the diffusion of CO as an example. This compound appears not to participate as actively in photochemical smog as do the hydrocarbons and nitrogen oxides, and so is nearly a conserved quantity. Consequently, its dispersal may be described by diffusion formulas which are elaborations of those which we have used to treat point sources. More complicated situations involving local micrometeorological details are now being developed by several researchers.[25]

Unfortunately, the diffusion of pollutants from line sources is more difficult to handle mathematically than it is for point sources. Only in very special circumstances of high symmetry can the shape of the plume be predicted by techniques of algebra and calculus. For most practical applications, useful results can be obtained solely through numerical techniques on high-speed computers. Therefore we will restrict our attention to a single aspect—the downwind pollution in cross winds.

Before we derive an equation for the plume, let us see qualitatively what

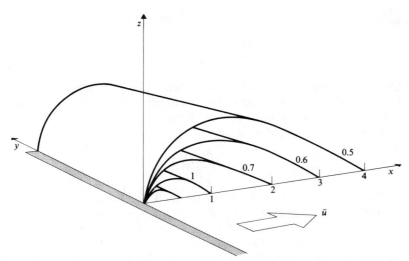

Fig. 7.26 Cross section revealing surface of constant ambient concentration for an infinitely long line source on the y-axis. Only relative concentrations are indicated.

happens. The situation is illustrated in Fig. 7.26. Each segment of a line of length L emits a pollutant at a rate Q. The natural parameter for characterizing the strength of the source is therefore Q/L, the emission rate per unit length. Having a negligible buoyancy, automobile exhaust can be considered as being released at ground level. As the effluent drifts downwind along the x-direction transverse to the road, molecules diffuse laterally and upward. However, for the simplest model of an infinitely long road, symmetry dictates that there can be no variation in the concentration of pollutant in the direction parallel to the road, so long as the emission rate is uniform. Therefore, because $\Delta X/\Delta y = 0$, there can be no diffusion parallel to the road. We can anticipate that X will decrease less rapidly downwind than for the point source, since there is diffusion only in the vertical direction. The resulting equation for diffusion is simpler than Eq. (7.10), for it involves only a variation of X with x and z:

$$\bar{u}\,\frac{\partial X}{\partial x} = D_z\,\frac{\partial^2 X}{\partial z^2}. \tag{7.23}$$

This equation is of course theoretically justified only if diffusion according to Eq. (7.4b) is the dominant process by which mixing occurs in the atmosphere. This means that it is assumed that the process of average eddy diffusion can be completely characterized by one constant, D_z. In general, this assumption has not been proven to be valid; and we have mentioned instances in which it is certainly not true. Nonetheless, on the optimistic assumption that the qualitative features of Fickian diffusion do not vary greatly from those of average turbulent mixing, we shall proceed to examine what Eq. (7.23) predicts. The solution of this equation is

$$X = \frac{2Q/L}{(2\pi)^{1/2}\,\bar{u}\,\sigma_z}\,e^{-z^2/2\sigma_z^2},$$

(7.24)

where Q/L is the source strength per unit length of road. If X is desired in grams per cubic meter, then Q/L must be expressed in grams per meter per second. The comparable volume emission rate can be obtained from Eq. (7.12). The standard deviation σ_z must satisfy the condition

$$\sigma_z^2 = \left(\frac{2D_z}{\bar{u}}\right)x$$

(7.25)

for Eq. (7.24) to satisfy Eq. (7.23). This is identical with Eq. (7.13) for the standard deviation of the vertical profile of the plume from a point source.

A significant feature of the prediction in Eq. (7.24) is the slow decrease of X with downwind distance x. This is due to suppression of horizontal diffusion by the symmetry of the source. Equation (7.24) indicates that $X(x)$ is proportional to $x^{-1/2}$ at ground level ($z = 0$), whereas it decreases much more rapidly, as x^{-1}, for the point source at ground level. Contours of constant ambient concentration are illustrated for the line source in Fig. 7.26. The slow decrease in ambient concentration downwind is one reason why road systems are important contributors to community-wide pollution.

7.8 AREA SOURCES

A multitude of point and line sources compose the pattern of pollutant sources in an urban region. Any receptor in the community receives pollutants not only from sources directly upwind but from those slightly to the side as well. This is illustrated in Fig. 7.27 which shows several plumes overlapping as they drift. Horizontal diffusion and the geometry of the sources makes difficult the task of assigning responsibility for adverse effects to any one source. Further complications arise from temporal variations that accompany patterns of commuter traffic or work shifts at factories. No simple algebraic expressions can adequately include all the various factors, so numerical techniques are used exclusively.

Historically, the main value derived from such studies has been in predicting the changing patterns of pollution which might be introduced by new sources. For example, suppose that it is proposed that a nitric acid plant be built, but it is known that the plant will contribute nitrogen oxides to the atmosphere. Can the emissions be tolerated? In polluted urban environments, it is not sufficient to consider the effect of one source alone, but how it contributes to the overall problem. For example, locating a large source of nitrogen oxides adjacent to a petroleum refinery emitting large quantities of hydrocarbons is—on the face of it—an invitation for photochemical smog. Whether in fact there is adequate cause for prohibiting this can only be based on the results of area modeling studies that assess the total consequences. Such modeling studies, utilizing numerical tech-

Fig. 7.27 Area source formed by overlapping plumes, June 1966. Current emission standards now make such plumes illegal. (Courtesy of Mr. A. Proudfit, Sign X Labs, and the U.S. Public Health Service)

niques, provide information essential for making rational decisions. But the validity of the results will necessarily be limited by the applicability of the basic assumptions concerning turbulent mixing and the reliability of micrometeorology data.

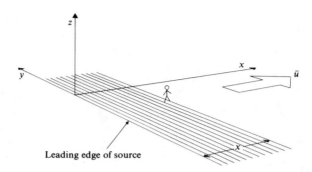

Fig. 7.28   An area source with a receptor downwind a distance x from the edge. The source up-wind from the receptor can be imagined as a series of closely-spaced line sources of infinite extent along the y-direction.

Some features of the area source problem can be extracted from a simple calculation. We imagine that we are in a large urban region, a distance x downwind from one boundary. The boundary is aligned transverse to the wind, as illustrated in Fig. 7.28, and the sources are assumed to be uniformly distributed throughout the urbanized area. Therefore the rate of pollutant emission per unit area Q/L^2 is a constant. Here Q is the rate of emission of a pollutant from the region within a square of surface area whose side has a length L. We suppose that the urbanized region is so extensive that net diffusion of pollutants to the side can be neglected. This is equivalent to assuming that the city has an infinite extent along the y-axis.

The ground-level concentration ($z = 0$) can be estimated by supposing the area source to be a series of closely spaced line sources, each parallel to the x-axis. Then by summing the contributions according to Eq. (7.24), we obtain:

$$X(x) = \frac{2(Q/L^2)}{(\pi\bar{u}D_z)^{1/2}}x^{1/2}. \tag{7.26}$$

What we all know is predicted: the farther the receptor is from the windward edge of the city, the higher the concentration. However, because upward mixing is possible, the concentration increases only as the square root of x, hence less rapidly than the area of the source upwind.

The increase is faster than $x^{1/2}$ if an elevated inversion prohibits upward diffusion above a certain height W. In this case, contaminants continue to build up in the mixing layer as the wind carries them through the city. If we imagine the pollutant to be uniformly mixed, the rate at which molecules pass a given point x downwind—to either side and above the receptor—is $\bar{u}nXW$ (per unit length in the y-direction). This quantity must equal the rate at which they enter the air upwind: $(Q/L^2)x$ (per unit length in the y-direction). From this equality, we deduce that

$$X = \frac{(Q/L^2)}{\bar{u}nW}x. \tag{7.27}$$

Thus with a low inversion the concentration increases linearly with x far from the windward edge of a city. The consequences of this marked increase are well known to inhabitants of the downwind regions of a large urban area.

Equations (7.26) and (7.27) should not be taken too seriously. They are only qualitatively correct. The assumptions on which they are based are usually not satisfied; few cities have a uniform distribution of pollution sources. But the calculations have a value, for they underscore qualitatively the way pollutant levels are augmented by spreading urbanization. The extensive nature of the source discourages horizontal eddy diffusion, and the continual addition of pollutants to air which has already been polluted by upwind sources leads to a steady increase of the ambient concentrations as one proceeds toward the down-wind edge of an urban region.

It is significant that, as the population of an urbanized region increases, there is an even greater increase in the area of downwind terrain affected by a given level of pollution. This occurs for two reasons:

1. The rate at which pollutants are emitted increases more rapidly than the population.
2. The region affected by a given level of pollution from an area source increases more rapidly than the total rate at which pollutants are emitted from the source.

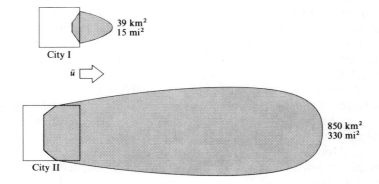

Fig. 7.29 Effect of doubling the area and population of a city on the area covered by a given pollutant concentration. (P. A. Leighton, Statewide Conference on Man and Air in California, 1964; figure courtesy of Metronics Associates, Palo Alto, California.)

In regard to the first point, we note that in Los Angeles County in the period from 1940 to 1960 the population increased by slightly more than a factor of 2, but the estimated rate of emission of nitrogen oxides increased by more than a factor of 3, and the hydrocarbon emissions from automobiles increased by a factor of 4. The more rapid increase of emissions illustrates the growing dependence upon the energy which is released from the combustion of fuels such as natural gas, gasoline, and fuel oil.

The effect downwind from an area source whose population and area have doubled is illustrated in Fig. 7.29 for a city which has the shape of a square. The downwind area subjected to a given concentration of a pollutant, or to a higher concentration, is shown by the shaded regions. The assumptions on which the calculation was based are fairly restrictive, and include the approximation that the rate of emission of the pollutant from the entire source increased by a factor of 3.2, as well as the condition that upon release the pollutants were confined to a region below an elevated inversion layer. Despite such restrictions, the calculation illustrates the important fact that the area affected by a given level of a contaminant can increase more rapidly than the increase in the population. The affected area in this example is estimated to increase by a factor of 22.

SUMMARY

Gaseous and aerosol contaminants mix with ambient air mainly through the action of atmospheric turbulence. Mechanical and thermal effects cause turbulence, so the rapidity with which pollutants are mixed with ambient air depends upon many factors: the roughness of the terrain, the average wind speed and shear, the height above ground for pollution release, and the intensity of solar radiation. A useful quantitative theory for permitting the exact calculation of the downwind pollution from a point source, which incorporates measurable features of atmospheric turbulence in the intervening distance, has not yet been achieved. In fact, the characteristics of turbulence in most localities are not known in sufficient detail to even enable the completion of such a calculation were a theory available. A practical alternative has been the use of empirical formulas which utilize average characteristics of the local turbulence; and for this are developed diffusion parameters which can be applied to standard binormal dispersion formulas. Empirical formulas are also used to estimate the plume rise due to buoyancy and upward momentum. However, the accuracy of such formulas is limited, giving an uncertainty of perhaps a factor of 2 for the downwind ground-level concentration of a pollutant in the simplest of cases. The geometry of the pollution source also influences the region which is affected downwind. Ground-level concentrations of a contaminant decrease less rapidly downwind from a line or area source than from a point source at low levels, because horizontal dispersal is not as rapid for the extensive sources.

NOTES

1. P. L. Leighton, "Geographical Aspects of Air Pollution," *Geographical Review,* **61**, 151 (1966).

2. R. S. Dean, R. E. Swain, E. W. Hewson, and G. C. Gill, "Report Submitted to the Trail Smelter Arbitral Tribunal," *U.S. Bureau of Mines Bull.*, **453** (1944).

3. G. I. Taylor, reprinted in the *Scientific Papers of Sir Geoffrey Ingram Taylor*, G. K. Batchelor, Ed., London: Cambridge Univ. Press, 1960.

4. F. Pasquill, *Atmospheric Diffusion*. London: D. Van Nostrand, 1962.

5. H. A. Panofsky and R. A. McCormick, "Properties of Spectra of Atmospheric Turbulence at 100 Meters," *Quart. J. Roy. Meteorol. Soc.*, **80**, 546 (1954).

6. H. A. Panofsky, "Variation of the Turbulence Spectrum with Height under Superadiabatic Conditions," *Quart. J. Roy. Meteorol. Soc.*, **79**, 150 (1953).

7. A. N. Kolmogorov, "The Local Structure of Turbulence," *Compt. Rend. Acad. Sci. USSR*, **30**, 301 (1941). In Russian.

8. G. H. Strom, "Atmospheric Dispersion of Stack Effluents," Vol. **I**, 2nd ed, in *Air Pollution*, A. C. Stern, Ed., New York: Academic Press, 1968, p. 227.

9. M. Hino, "Maximum Ground-level Concentration and Sampling Time," *Atmos. Environ.*, **2**, 149 (1968).

10. G. Nonhebel, "Recommendations on Heights for New Industrial Chimneys," *J. Inst. Fuel*, **33**, 479 (1960).

11. F. Wippermann, "Der Effeckt der Messdauer bei der Ermittlung von Maximal-Konzentrationen eines sich in turbulenter Strömung aus breitenden Gases," *Air and Water Poll. Intern. J.*, **4**, 1 (1961).

12. G. T. Csanady, "Turbulent Diffusion of Heavy Particles in the Atmosphere," *J. Atmos. Sci.*, **20**, 201 (1963).

13. Y. Ogura, "Diffusion from a Continuous Source in Relation to a Finite Observation Interval," *Adv. Geophys.*, **6**, 149 (1959), and references quoted in this article.

14. A. C. Chamberlain, "Aspects of the Deposition of Radioactive and other Gases and Particles," *Intern. J. Air Poll.* (London), **3**, 63 (1960).

15. H. A. Panofsky, "Air Pollution Meteorology," *American Scientist*, **57**, 269 (1969).

16. O. G. Sutton, "The Theoretical Distribution of Airborne Pollution from Factory Chimneys," *Quart. J. Roy. Meteorol. Soc.*, **73**, 426 (1947).

17. C. H. Bosanquet and J. L. Pearson, "The Spread of Smoke and Gases from Chimneys," *Trans. Faraday Soc.*, **32**, 1249 (1936).

18. M. E. Smith, Ed., "Recommended Guide for the Prediction of the Dispersion of Airborne Effluents," American Society of Mechanical Engineers, New York (1968).

19. D. B. Turner, *Workbook of Atmospheric Dispersion Estimates*, U.S. Dept. of Health, Education, and Welfare, Public Health Service, National Air Pollution Control Administration, Cincinnati, Ohio: P.H.S. Publication No. 999-AP-26, 1969.

20. R. S. Scorer, *Air Pollution*, Oxford: Pergamon Press, 1968.

21. For some comparisons of theory and model studies under special atmospheric

conditions see T. A. Hewett, J. A. Fay, and D. P. Holt, "Laboratory Experiments of Smokestack Plumes in a Stable Atmosphere," *Atmos. Environ.*, **5**, 767 (1971).

22. F. W. Thomas, S. B. Carpenter, and W. C. Colbaugh, "Plume Rise Estimates for Electric Generating Stations," *J. Air Poll. Control Assoc.*, **20**, 170 (1970).

23. H. Moses and J. E. Carson, "Stack Design Parameters Influencing Plume Rise," *J. Air Poll. Control Assoc.*, **18**, 456 (1968); J. E. Carson and H. Moses, "The Validity of Several Plume Rise Formulas," *J. Air Poll. Control Assoc.*, **19**, 862 (1969); also **20**, 476 (1970).

24. T. L. Montgomery and M. Corn, "Adherence of Sulfur Dioxide Concentrations in the Vicinity of a Steam Plant to Plume Dispersion Models," *J. Air Poll. Control Assoc.*, **17**, 512 (1967).

25. For example, see R. Lamb and M. Neiburger, "An Interim Version of a Generalized Urban Air Pollution Model," *Atmos. Environ.*, **5**, 239 (1971).

FOR FURTHER READING

E. KENNARD	*Kinetic Theory of Gases*, New York: McGraw-Hill, 1938.
R. S. SCORER	*Air Pollution*, Oxford: Pergamon Press, 1968.
F. PASQUILL	*Atmospheric Diffusion*, London: D. Van Nostrand, 1962.
J. HINZE	*Turbulence*, New York: McGraw-Hill, 1959.
J. L. LUMLEY and H. A. PANOFSKY	*The Structure of Atmospheric Turbulence*, New York: John Wiley and Sons, 1964.
H. TENNEKES and J. L. LUMLEY	*A First Course in Turbulence*, Cambridge, Massachusetts: M.I.T. Press, 1972.
O. G. SUTTON	*Micrometeorology*, New York: McGraw-Hill, 1953.
C. H. B. PRIESTLEY	*Turbulent Transfer in the Lower Atmosphere*, Chicago: Univ. of Chicago Press, 1959.
G. A. BRIGGS	*Plume Rise*, Air Resources Laboratory, ESSA for USAEC Division of Technical Information, 1969.

QUESTIONS

1. What are two reasons why high wind speeds cause more rapid dilution of contaminants?

2. What causes turbulence in the lower atmosphere? How does turbulence depend upon altitude in the troposphere? What causes a variation in the vertical and horizontal spectrum with altitude?

3. Which factors of source operation and meteorology enhance the rise of a plume from an exhaust stack?

4. What is the difference between a power rating expressed in MWe and MWt for an electrical power plant? What is meant by the efficiency of a plant?

5. Why is the vertical component of turbulence usually less than the horizontal component near the ground?

6. Which meteorological conditions invalidate the types of diffusion calculations and empirical formulas for point sources that we have discussed?

7. How may elevated or ground-based inversion layers affect the plume from an exhaust stack or the effluent from motor vehicles on a straight road?

8. What are downwash and downdraft? How may such effects on the effluent of a plant be minimized?

9. What approximations are involved when eddy diffusion parameters defined by Eq. (7.13) are used to describe diffusion of contaminants in the atmosphere?

10. Why does the ground-level concentration of a pollutant decrease less rapidly with distance downwind from a line source (transverse to the wind) than from a point source?

11. It has been said that a typical fossil fuel power plant runs at about 60% of its rated capacity as an average over its 30-year lifetime, and therefore that the mass emission rate used in a pollutant diffusion equation for a smokestack plume should be 60% of the capacity rate. Comment on the significance of this procedure.

PROBLEMS

1. If the maximum ground-level concentration over a 10-minute averaging time downwind from a stack is 0.3 ppm of a given pollutant, what might be expected for a one-hour averaging time? Upon what assumptions is your answer based?

2. At what wind speed should vortices be formed in the lee of a smokestack whose diameter is 3 meters?

3. A 500 MWe power station burns natural gas without any controls to minimize the emission of nitrogen oxides. What maximum concentration of NO may be expected at ground level during stable meteorological conditions if it emits NO at a rate of 50 Tonnes per day? The plant operates at an efficiency of 40%, with a heat emission rate in the gaseous effluent which is 8% of the energy released from the combustion of the fuel. The stack has a diameter of 4 m, is 70 m high, and the plume is emitted with a velocity of 80 km/hour. The wind speed is 2 m/sec.

4. Two coal-burning power plants are to be constructed in different localities having similar meteorological conditions. If one has a geometrical stack height at 100 m with plume rise of an additional 100 m for a heat emission rate of 1×10^8 W and the other, which produces twice as much power, has twice the heat emission rate, how much higher or lower should the geometrical stack height of the larger plant be to yield the same maximum ground-level concentration of SO_2? Assume that the rate of emission of SO_2 is proportional to the rate at which power is generated.

5. How much higher should an exhaust stack be in order not to permit the maximum ground-level concentration of SO_2 to increase for a given wind speed if a power generating station switches to a fuel which releases twice as much SO_2 for the same rate of power production? Consider the cases when (a) the plume has negligible

buoyancy and (b) the plume rises 100 m above a geometrical stack height of 50 m. Meteorological conditions, the heat emission rate, and the plume exit velocity are assumed to be unchanged when the type of fuel is changed.

6. Show that Eq. (7.11) is a solution of Eq. (7.10) provided that Eq. (7.13) applies.

7. Prove that the rate at which pollutant molecules move downward and cross an imaginary plane of infinite extent transverse to the plume at a distance $x = x_0$ is equal to the rate at which they are emitted by the source. Use Eq. (7.11) to describe the plume shape.

8. Show that Eq. (7.15) gives the location of the maximum downwind concentration at ground level from an elevated point source if σ_y/σ_z is independent of x.

CHAPTER 8

SULFUROUS SMOG

In this chapter we shall take up the subject of the most widespread type of community air pollution: sulfurous smog. By this term we mean air which is predominantly polluted by sulfur oxides and other sulfur-bearing compounds, usually in association with particulate matter. Our attention will be directed to the source of both the gaseous and the particulate components of this form of smog. There are three reasons why we do this:

1. In many instances, sulfur-bearing compounds and particulate matter have a common source, the most important being the fuels, coal and oil. These fuels contain: *sulfur compounds* which when oxidized produce sulfur oxides, mainly SO_2; they also contain minerals which remain as ash or may be entrained in the waste gas and removed as *fly ash*; and they release volatile compounds when heated, which may escape combustion and later agglomerate to form a loose-knit, solid particle known as *soot* or an amorphous particle called *tar*.

2. Both SO_2 and aerosols actively influence the evolution of sulfurous smog through their participation in atmospheric reactions, and the SO_2 is partially or completely converted into an aerosol form.

3. Sulfur dioxide and aerosols have an adverse effect on humans which is synergistic, as we noted in Chapter 2. Exposure to both simultaneously, as from community air, produces a much greater effect than the sum of the effects either would cause separately.

Historically, the first important source of sulfur dioxide pollution was not the combustion of sulfur-bearing fuels. It probably originated with the development of smelting techniques for processing mineral ores. Copper, lead, and zinc are generally found in ore deposits as sulfides, and the common method for separating out the desirable metallic component was by roasting the pulverized ore at sufficiently high temperatures to oxidize the sulfur and cause it to be released as SO_2. As early as Roman times, smelters on the Iberian peninsula were equipped with chimneys 7 to 8 meters high to provide a good draft and safely vent the gaseous waste to protect the health of the workers. Unfortunately such a low stack provided little benefit to the neighbors some distance downwind. Even today sulfurous pollution from smelters is a serious problem in some communities, but its relative importance on a global scale has declined, for only 10% of the global SO_2 emissions from anthropogenic sources is now from smelters.

Considerably more widespread is the problem which arises from the use of sulfur-bearing fuels, especially coal. The popular use of coal began in the early Middle Ages when man's urbanization led to depletion of timber supplies near large cities. His search for a concentrated fuel which is economical to transport was satisfied by the utilization of soft coal, a material relatively easy to mine and consume. In the thirteenth century, inhabitants and industries in the city of London, for example, grew to depend upon coal brought in from Newcastle by barge. Unfortunately this fuel released great quantities of tar, soot, and sulfurous fumes; and the resulting air pollution became so intolerable that in 1273 the king induced Parliament to pass an act prohibiting the burning of such "sea coal" within the city.[1] This did not end the problem, for a decade later workers were still complaining that they could not work at night on account of the unhealthy air. Little wonder that medieval man believed that sickness was caused by a corrupt atmosphere!

But at the turn of the century the enforcement of air pollution controls in London appears to have been taken seriously, for in 1306 a man was reportedly condemned and executed for using sea coal. Then with the passage of centuries and the growing dependence of the nobility on the economic benefits of industrial expansion, vigilance relaxed and atmospheric conditions worsened. Simple prohibitions against the use of coal were no longer possible. The changing situation was eloquently dramatized when Queen Elizabeth I removed her court from the city, apparently in order to escape the smog. It required a severe episode of air pollution in 1952, as a result of which 4000 persons died, to shock the government into a response which has resulted in effective measures to abate smoke and alleviate conditions.

It should be appreciated at the outset that the cause, evolution, and effects of sulfurous smog involve many interrelated processes. The key factor is the amount of sulfur in the fuel commonly used in a community, but also important is the ambient concentration and composition of particulate matter, much of which comes from the fuel itself. Thus an examination of fuel chemistry will be our first goal. We shall then discuss how the combustion effluent undergoes changes within the atmosphere which markedly increase its potential for causing adverse effects, and we shall examine some of the more significant effects. How the emission of gaseous and particulate matter can be controlled will be discussed in Chapter 9, which reviews the general principles of control devices. We shall conclude our discussion of sulfurous smog and tie together numerous components of the air pollution syndrome introduced in earlier chapters by examining the contributing factors of fuel usage, effluent dispersal, and meteorology which led to the tragic 1952 episode of air pollution in London.

8.1 FOSSIL FUELS

In the United States over 95% of the utilized energy comes from the combustion of fossil fuels—coal, oil, and gas. On a global scale the figure is closer to 99%.

Despite increasing dependence on nuclear reactors it is estimated that by the year 2000 some 75% of our energy will still be derived from fossil fuels, and such fuels will be consumed at twice the present rate. These energy sources consist of decayed organic compounds which were formed under varying degrees of temperature and pressure within the crust of the earth. The original compounds are typified by long-chain molecules such as cellulose $(C_6H_{10}O_5)_n$, where the integer n is on the order of a thousand or more, but the action of bacteria and subsequent treatment by prolonged heating and compression eliminates some of the oxygen, thereby making it a better fuel. The reason for this improvement is that the energy released by the combustion of such organic compounds is roughly the same as the energy which would be released by complete oxidation of the carbon atoms alone. Thus it is advantageous to have fuel initially in a reduced state, with a high ratio of carbon to oxygen atoms. For this reason charcoal is a good fuel, for it is nearly pure carbon, produced by heating wood and driving off oxygen and hydrogen.

Natural gas

Marsh gas, consisting mainly of methane CH_4, is ideal from the standpoint of its carbon to oxygen ratio. *Natural gas* is its more common name and it is often found in association with oil deposits, containing from 50 to 99% methane by volume. Natural gas is usually processed before combustion to remove most of the sulfur-bearing compounds. As a result, its contribution to sulfurous pollution is usually negligible.

Measurements of the emissions from various gas-burning sources provide us with quantitative data from which the relative importance of such pollution sources may be evaluated. The quantity of each pollutant emitted by a source without special controls over its pollutants is found to be approximately proportional to the quantity of fuel consumed. Therefore a useful parameter called the *emission factor* is commonly used to indicate how much pollutant is emitted for a given amount of fuel. Table 8.1 summarizes emission factors for several pollutants emitted when natural gas is burned.

It should be appreciated that the values listed in Table 8.1 are representative factors; that is, the factor for a particular boiler or space heater will depend upon the design of the apparatus and the conditions under which it operates. What it emits may differ from the listed values by 20% or more, depending upon conditions. This variability is also illustrated by a comparison of emission factors in the three columns. Electric generating stations usually operate under careful controls which ensure complete oxidation of the fuel. Therefore most of the carbon monoxide which might be produced in the initial stages of a combustion chamber is oxidized to carbon dioxide before it leaves. This is not true for less efficient space heaters in the home. As a consequence, CO emission factors for power stations are substantially lower than they are for domestic heating units.

One pollutant indicated in the table merits a special comment. This is the entry labeled NO_x, which actually represents two pollutants, nitric oxide, NO,

TABLE 8.1

Emission factors for the combustion of natural gas. The first column of each category indicates the mass in grams of the respective pollutant emitted when one cubic meter of gas is burned. The second column gives the weight of pollutant in pounds per million cubic feet of gas consumed. The density of natural gas is assumed to be 0.83 kg/m^3 (0.052 lb/ft^3). It is assumed also that no control is exercised to remove pollutants from the effluent. Almost all emitted particulates have a size of less than 5 microns. If an emission factor in terms of the thermal energy is desired, the conversion factor of 3.70×10^7 J of liberated heat energy/m^3 of gas (1000 BTU/ft^3) may be applied.

Pollutant	Type of unit					
	Power plant		Industrial process boiler		Domestic and commercial heating units	
	g/m^3	lb/10^{6}ft^3	g/m^3	lb/10^{6}ft^3	g/m^3	lb/10^{6}ft^3
CO	negligible		0.64	0.4	0.64	0.4
Organic gases	0.06	4	0.11	7	negligible	
NO$_x$ (as NO$_2$)	6.2	390	3.4	214	1.9	116
SO$_x$ (as SO$_2$)	0.006	0.4	0.006	0.4	0.006	0.4
Particulate matter	0.24	15	0.29	18	0.31	19

*R. L. Duprey, "Compilation of Air Pollutant Emission Factors," National Air Pollution Control Administration, 1968.

and nitrogen dioxide, NO$_2$. These gases contribute to the formation of photo-chemical smog and are listed in the table for completeness. In Chapter 10 we shall discuss in detail how they are produced; for the moment, we shall mention only that the quantity produced is dependent mainly on the flame temperature at which material burns in the combustion chamber. Power plants operate with high flame temperatures to maximize efficiency, so the NO$_x$ production is correspondingly higher than for domestic heating units which burn gas at lower temperatures. The amount produced is relatively insensitive to the type of fuel used, so long as the flame temperature is the same.

There has been some confusion in the past about how to represent the emissions of NO$_x$ as a numerical value since both NO and NO$_2$ are emitted. It is now conventional to indicate the mass which is emitted assuming that all of the NO has been oxidized to NO$_2$. (In fact, however, only about 10% of the emitted nitrogen oxides may be nitrogen dioxide.) The motivation for the convention is that the dioxide is a toxic gas, whereas nitric oxide is not; for pollutant dispersal calculations of the maximum possible NO$_2$ concentration at ground level (as discussed in the preceding chapters), it is convenient to have the emissions stated for this gas.

A similar convention applies when representing the emission factor for the sulfur oxides SO$_x$, including SO$_2$ and SO$_3$. It is usually assumed that all of the gaseous sulfur oxides appear as sulfur dioxide. In this case, the convention is also a good approximation, since 98 to 99% of the SO$_x$ is indeed sulfur dioxide.

From the factors given in Table 8.1 reasonable estimates can be made for the emissions from almost any type of gas combustion. For example, if the reader knows the rate at which his home furnace uses gas, he can calculate the rate at which it emits pollutants. In fact, available data on fuel consumption usually do provide the basis for estimates of the rate at which pollutants are emitted from various sources within a community. Emission factors are thus the basic data from which an inventory of emitted pollutants can be established.

Petroleum

Oil constitutes another form of fossil fuel. It can be grouped into two classes: shale oil and petroleum. They have essentially identical uses and there are only slight chemical differences. Shale oil is distilled from certain shales which may yield up to 10 to 20% of their weight as oil. Petroleum may be found impregnating porous rock which is capped by a dome or other obstruction of impervious rock. Crude petroleum as it comes from the ground is a mixture of thousands of hydrocarbons and other organic molecules. A representative sample contains 84% carbon, 12% hydrogen, and 4% oxygen, nitrogen, and sulfur by weight.

Components of petroleum are classified according to their range of boiling temperatures at atmospheric pressure, each group being called a "broad fraction." Lighter molecules, usually containing fewer carbon atoms, have lower boiling points. Heavier and more complex molecules have higher boiling points partly because they have more difficulty disentangling from a liquid. Table 8.2 shows the list of broad fractions and their characteristics for a representative sample of petroleum.[2] The subscripts on the carbon symbols indicate the number of carbon atoms in the molecules contained in each broad fraction; for example, 33.2% by volume of petroleum consists of molecules with from 6 to 10 carbons. Residual oils with the highest boiling points are commonly used by industries and power generating utilities. This fuel is highly viscous because the large, complex molecules of which it is composed entangle and oppose shear; such residual oils require heating before they will flow. Lighter oils, such as distillate oil, are used for domestic or commercial fuel.

The sulfur content of petroleum depends upon the location of the oil field and many run as high as 2% or more by weight. Petroleum from Indonesia is noted for its low sulfur content of 0.5% or less; by contrast, California residual oil has a content of about 1.7%. It is believed that oil from the north slope of Alaska is fairly low in sulfur.

Similarly the ash found in oil varies from one field to another. Virtually nil in light oils, it may constitute as much as 0.1% of residual oil. It is interesting that the ash content of Indonesian oil is quite low, amounting to roughly 0.01%; whereas California residual oil has about 0.06%.

Particulate matter released during the combustion of fossil fuels is not limited to inorganic material. Gaseous hydrocarbons in the combustion flame form acetylene ($H-C\equiv C-H$), which is believed to participate in reactions with free radicals to produce polyacetylene-free radicals; and these in turn react with

TABLE 8.2

Broad fractions distilled from a representative sample of petroleum*

Property		Broad fraction					
	Gas	Gasoline	Kerosene	Light gas oil	Heavy gas oil and light lubricating distillate	Lubricant	Residue
Boiling range at atmospheric pressure (°C)	below 40°	40–180°	180–230°	230–305°	305–405°	405–515°	—
Range of normal paraffin component	C_1-C_5	C_6-C_{10}	C_{11}-C_{12}	C_{13}-C_{17}	C_{18}-C_{25}	C_{26}-C_{38}	—
Estimated percentage of the original petroleum constituted by the given fraction	4	33.2	12.7	18.6	14.5	10.0	7
Number of compounds isolated in the sample	7	101	37	12	10	8	—
Estimated percentage of the fraction accounted for by the compounds isolated	100	82.3	38.4	30.2	20.5	9.9	—

*F. D. Rossini, *Journal of Chemical Education,* **37**, 554, 1960.

TABLE 8.3

Emission factors of the combustion of fuel oil.* The first column of each category indicates the mass of emitted pollutant in grams per liter of oil burned. The second column gives the weight in lb/1000 gal. It is assumed that no control is exercised to reduce emissions; it is assumed also that power plants burn residual oil and that domestic consumption uses the lighter distillate oil. If an emission factor in terms of the thermal energy is desired, the conversion factor of 4.2×10^7 J of liberated heat energy/liter of distillate oil $(1.5 \times 10^5$ BTU per gallon) may be applied. The symbol s represents the percentage of sulfur in the oil by weight; that is, 2% oil would yield $2 \times 19 = 38$ g/liter of SO_x.

| | Type of unit | | | | | | | |
| | Power plant | | Industrial residual | | Commercial distillate | | Domestic | |
Pollutant	g/liter	lb/10³gal	g/liter	1lb/10³gal	g/liter	1lb/10³gal	g/liter	lb/10³gal
CO	.005	0.04	0.24	2	0.24	2	0.24	2
Organic gases	0.46	3.8	0.48	4	0.48	4	0.6	5
NO_x(as NO_2)	12.5	104	8.6	72	8.6	72	1.4	12
SO_x(as SO_2)	19s	160s	19s	160s	19s	160s	19s	160s
Particulate	1.2	10	2.8	23	1.8	15	0.1	8

*R. L. Duprey, "Compilation of Air Pollutant Emission Factors," National Air Pollution Control Administration, 1968.

acetylene and polyacetylene to nucleate particle formation through the chemical agglomeration of the molecules. Chain-breaking and ring-closure processes yield a wide spectrum of chemical compounds, including the carcinogenic polycyclic compound benzo(a)pyrene.

Typical emission factors are listed in Table 8.3 for several broad classifications of sources. As we shall see later, particulate emissions are generally not as serious as they are for the combustion of coal; but nitrogen oxides may be emitted in quantity, especially by power stations with high flame temperatures. Sulfur dioxide emissions are high unless low sulfur oil is used.

Coal

The most common fossil fuel is coal. This material is fairly dense, because it is formed by compression and heating, and historically it has been economically attractive for use at sites distant from the mine on account of the low cost of transportation. The distinction between "soft" coal (*bituminous*) and "hard" coal (*anthracite*) is believed to depend not only upon the higher temperature and pressures to which anthracite was subjected, but also on differences in the original vegetative constitution of each.[1] Bituminous (derived from the Sanskrit for "pitch-producing") describes coals containing more than about 20% volatile matter; the name describes the tarry substances and soot emitted on heating. Unless the burning is well controlled, much of this material may not be combusted and hence escapes with the waste gas. Many of the compounds found in coal tar and soot are polycyclic aromatics, especially naphthalene, which may constitute 10% of the tar. More than half of the earth's fossil fuel reserves consist of bituminous coal. The composition of bituminous coal, as well as that of other members of the coal series, is given in Table 8.4.

TABLE 8.4

Characteristics of the coal series*

Material	Calorific value kJ/g	Calorific value BTU/lb	Volatile matter %	Perfectly dry fuel Hydrogen %	Perfectly dry fuel Oxygen %	Perfectly dry fuel Carbon %
Wood	16.0	7,000	—	6.5	43	50
Peat	18.5	8,000	50	6.0	32	60
Lignite (brown)	23.2	10,000	47	5.5	26	67
Lignite (black)	23.5	10,200	41	5.4	19	74
Coal, bituminous	25.0–33.0	10,700– 14,400	35–30	5–5.3	16–4.7	77–87
Coal, carbonaceous	35.0	15,000	11	4.0	2	92
Coal, anthracite	35.0	15,000	8	3.0	2	94

*From A. R. Meetham, D. W. Bottom, and S. Cayton, *Atmospheric Pollution*, Oxford: Pergamon Press, 1964. Copyright © 1964 by Pergamon Press and reprinted by permission.

The coal used as a fuel in Donora, Pennsylvania, during the smog episode described in Chapter 6 came from a bituminous deposit called the Pittsburgh bed. It was also a common fuel for industries in the city after which the bed was named, and before strict emission controls were implemented, Pittsburgh had achieved some degree of notoriety for its dirty air. The composition of the Pittsburgh bed breaks down into 31–41% volatile matter, about half being carbon, and an additional 56–66% in nonvolatile carbon or "fixed carbon." Hydrogen constitutes 5–5.6%, nitrogen, 1.4–1.6%, oxygen 6.0–8.2%, and sulfur 1.3–1.8%.[3]

Anthracite (from the Greek *anthrax* for coal) contains less than 10% volatile matter, with the dry fuel consisting of about 94% carbon, 3% hydrogen, and 2% oxygen. The heat released from the complete combustion of 1 kg of anthracite is about 35×10^6 J (15,000 BTU/lb fuel). The analogous calorific values of bituminous are inferior to this by as much as 30%. The attractiveness of anthracite as a cleaner, and more economical fuel for transporting substantial distances is apparent, although it was not used in great quantities until recently, because it is difficult to ignite.

Coal contains mineral constituents which have been introduced both through the composition of the original vegetation and through the interdiffusing of organic matter with the soil which bore the original plants. The mineral content of coal is variable, but is at least 1–2% by weight. The Pittsburgh bed, mentioned above, has an ash content of about 7–8%. When coal is burned in bulk form, most of the mineral content remains as ash. This is mainly aluminum silicate, the constituent of clay. However, it is now common in large operations such as power, generating stations to use finely pulverized coal. The particles have a size of less than 80 microns, with about half of the fuel's mass consisting of particles whose radius is less than 10 microns. This fuel can be transported as a slurry when mixed with water and consequently can be piped from the mine to where it is to be burned. After drying, it is combusted with efficient utilization, much like a gas, and most of the volatile matter is burned as it evaporates in the intense heat; but as much as 80% of the ash may be fly ash which is carried away with the exhaust gases.

Table 8.5 summarizes the composition of ash collected from the effluent of 30 power stations in the United Kingdom. This material is in a highly oxidized state, with silicon dioxide (SiO_2) and alumina (Al_2O_3) as major constituents. Calcium oxide (CaO) and hematite (Fe_2O_3) also occur in substantial concentrations. The presence of the latter is significant, for we shall see that ferrous oxide may be a catalyst which encourages the oxidation of SO_2 to sulfuric acid in a humid atmosphere.

Sulfur occurs in coal as a constituent of several types of compounds; they include mineral sulfates, iron sulfide (and other pyrites), and organic compounds. The total sulfur content can run from 0.2 to 7% by weight. The organic component is chemically bound to the coal and cannot be removed without decomposing the fuel; it yields H_2S on distillation or decay, and all of the sulfur is converted into SO_2 or SO_3 on combustion. Trace amounts of chlorine (0.1–0.7%),

TABLE 8.5

Constitution of pulverized-fuel ash ignited at 800°C*

| Oxide | Percentage weight | |
	Normal range encountered	Typical value
SiO_2	40–50	45
Al_2O_3	20–30	25
Fe_2O_3	5–15	10
CaO	7–20	8
$Na_2O + K_2O$	4–6	5
MgO	0.5–3	2
SO_3	0.5–1.5	1
P_2O_5	0.1–0.5	0.3
Mn_3O_4	0–0.4	<0.2

*Reprinted with permission from P. M. Foster, *Atmospheric Environment*, **3**, 157, 1969, Pergamon Press.

fluorine (up to 0.01%), phosphates, lead, mercury, and arsenic are also commonly found.[1] Emission factors for the predominant gaseous and particulate pollutants which result from the combustion of coal are listed in Table 8.6.

The main concern with respect to community air pollution is focused on the emissions of sulfur oxides and particulate matter. Emission of nitrogen oxides may also constitute a problem, but is generally considered of secondary importance. The emission of CO by inefficient domestic or commercial furnaces is a health threat if the area is not well ventilated. Table 8.6 shows the size distribution of the fly ash from three types of combustion furnaces, to emphasize the fact that the predominant particle size depends very much upon how the fuel is processed and burned.

Although the emission factors in Table 8.6 feature only the principal pollutants, it may be that emission of trace amounts of certain pollutants can be signficant on a global scale, simply because great quantities of coal are consumed each year. In 1970, global consumption of coal amounted to 2.18×10^9 Tonnes, and of lignite, 0.77×10^9 Tonnes. Thus, for example, mercury, which is found in concentrations of 0.01 to 30 ppm by weight in coal, would be emitted at an annual rate of 3,000 Tonnes if a mean concentration of 1 ppm by weight were assumed.[4] This is comparable to the 10,000 Tonnes per year in industrially discharged waste, the major anthropogenic source of mercury found in the environment. Crude oil production in 1969 amounted to 2.13×10^9 Tonnes, but the mercury content of various broad fractions from different wells has not yet been determined. Although fossil fuel combustion is a major contribution compared with other anthropogenic sources, recent estimates indicate that natural sources, such as outgassing from the earth's crust, release mercury at an annual rate of about 100,000 Tonnes and thus are dominant on a global basis.

TABLE 8.6

Emission factors for the combustion of coal.* The first column of each category indicates the mass of emitted pollutant in kilograms per metric ton of coal which is burned. The second column gives the weight in pounds per U.S.A. short ton. It is assumed that no control is exercised to reduce emissions. Conversion to an energy basis can be accomplished by assuming that 30×10^9 J of thermal energy are released by combustion of 1 metric ton of coal (26×10^6 BTU/U.S.A. short ton). The thermal capacity of power stations usually exceeds 30 MW (100×10^6 BTU/hr), and that of domestic and commercial units is usually less than 3 MW (10×10^6 BTU/hr). The symbol s equals the percentage of sulfur in the coal by weight; that is, 2% sulfur coal would yield $2 \times 19 = 38$ kg/T of SO_x. The emission factor for particulates is for a 1% ash content. Coal with a higher content produces correspondingly higher emissions.

Pollutant	Power station kg/T	Power station lb/t	Type of unit Industrial kg/T	Type of unit Industrial lb/t	Domestic and commercial kg/T	Domestic and commercial lb/t
CO	0.25	0.5	1.5	3	25	50
Organic gases (mainly CH_4)	0.1	0.2	0.5	1	5	10
NO_x (as NO_2)	10	20	10	20	4	8
SO_x (as SO_2)	19s	38s	19s	38s	19s	38s

Particulates	Pulverized	Type of unit Spreader stoker	Cyclone
(R = particle radius in microns)			
Total	8.0 kg/T 16 lb/t	6.5 kg/T 13 lb/t	1 kg/T 2 lb/t
$R > 22$	25%	61%	10%
$22 > R > 10$	23	18	7
$10 > R > 5$	20	11	8
$5 > R > 2.5$	17	6	10
$2.5 > R$	15	4	65

*R. L. Duprey, "Compilation of Air Pollutant Emission Factors," National Air Pollution Control Administration 1968.

8.2 OXIDATION OF SULFUR DIOXIDE

The combustion of fossil fuels involves a number of complex reactions but the overall result is a conversion of the predominant hydrocarbon constituents into carbon dioxide and water vapor through oxidation at high temperatures with the oxygen of ambient air. A simple example would be the oxidation of methane:

$$CH_4 + 2O_2 \rightarrow CO_2 + 2H_2O. \tag{8.1}$$

Many other reactions will simultaneously take place for other hydrocarbons in a more complex fuel. As oxidation occurs in different parts of the combustion chamber—be it power plant or home fireplace—reactions proceed under a wide variety of temperatures and concentrations of oxygen. For this reason, innumerable side reactions accompany the main processes; but the details and relative importance of these side reactions have not been explained completely.[5] One product, which results from incomplete oxidation of the fuel in regions of insufficient oxygen, is carbon monoxide, CO. Another we have seen is caused by the oxidation of sulfur-bearing molecules; these are the sulfur oxides, labeled as a class by the symbol SO_x. About 98% by weight of the emitted sulfur oxides consists of sulfur dioxide, and only 2% or so of sulfur trioxide.

The sulfur oxide component of the combustion effluent is not in chemical equilibrium, and as the exhaust plume disperses in the atmosphere, oxidation of SO_2 to SO_3 continues. We have previously mentioned when discussing visibility reduction in Section 2.3 that SO_3 is hygroscopic and, by encouraging the condensation of water, forms droplets of sulfuric acid. A dense, slightly blue plume is frequently seen to issue from the exhaust stack of a source whose effluent is nearly saturated. The bluish tint is due to light scattering from the many small droplets formed about SO_3 condensation nuclei, as well as particulate nuclei.

Much of the remaining gaseous SO_2 may then be oxidized as it diffuses to the droplets and is dissolved. The oxidizing agent is molecular oxygen which has also been dissolved in the droplets. The reaction proceeds according to the following overall equation:

$$2\,SO_2 + 2\,H_2O + O_2 = 2\,H_2SO_4. \tag{8.2}$$

Johnston and Coughanowr[6] have shown that this process is catalyzed by iron and manganese salts, commonly found in the fly ash of burned coal (cf. Table 8.5). These particles serve as nucleation sites for droplet formation and, at the high concentrations of SO_2 often emitted in industrial plumes (250 ppm or more), molecules of SO_2 can readily diffuse into the interior of a droplet where they may come into contact with the nucleating particle. Oxidation will then proceed throughout the droplet. This production of H_2SO_4 lowers the vapor pressure of the droplet, and encourages more water molecules to diffuse through the air to the liquid surface and condense.

In addition, metal oxides such as MgO, Fe_2O_3, ZnO, and Mn_2O_3 react with the surrounding sulfuric acid to form metal sulfates, with the result that the liquid of the droplet is maintained as a relatively neutral solution. This makes possible the absorption of still more SO_2 from the air. An analysis of nucleation and oxidation on $MnSO_4$ nuclei suggests that once the metal oxides in the droplets are depleted, or once the relative humidity has decreased to the point where no further condensation takes place, absorption of SO_2 is terminated when the H_2SO_4 concentration is about 1 mole per liter of water.[7] At this stage, oxidation within the droplets essentially ceases.

The importance of high relative humidity as a condition for appreciable oxidation of SO_2 within a plume is underscored by the results of a study by helicopter of the plume from a power plant burning pulverized coal.[8] The effluent as released had a concentration of about 2000 ppm of SO_2. Below 70% relative humidity, oxidation of SO_2 at distances as far as 15 km from the stack amounted to 1–3%, corresponding to the original percentage of SO_3 or H_2SO_4 in the hot, undiluted flue gas.

Thus little oxidation had occurred. But for higher relative humidities, oxidation was rapid. At about 1.5 km downwind from the stack, over 20% was oxidized and, at 13 km, over 30%. The highest total oxidation observed during these studies occurred when there was a slight mist in the ambient air; in this instance, about 55% of the SO_2 was oxidized by the time the plume was 15 km from the stack, corresponding to a time of about 108 minutes after the effluent had been emitted. It is noteworthy that the most severe cases of sulfurous smogs in Donora and London have also occurred during humid conditions, with a dense fog at ground level.

Considerably less oxidation would be expected under the same conditions within the plume from an oil-fired combustion chamber, because the resulting fly ash in the effluent is diminished by a factor of 10 or more from the ash emitted from uncontrolled burning of coal (assuming the same amount of power is produced). Unfortunately, owing to the numerous constituents within the ash of a plume and the highly variable meteorological conditions of dispersal, very little has been established quantitatively as regards the evolution of the various gaseous, liquid, and solid constituents in the effluent.

Sulfur dioxide can also be oxidized on the surface of certain solids, such as Fe_2O_3 and Al_2O_3, and this may play an important role in atmospheres that carry a heavy particulate burden. However, the details and rates of such reactions have yet to be explored.

A third mechanism may also be responsible for some oxidation of SO_2 in the atmosphere although it is not as important as catalyzed oxidation. The first step is the photoexcitation of the SO_2 molecule through its absorption of solar radiation. Schematically, this can be written as

$$SO_2 + E = SO_2^*, \tag{8.3}$$

where E is the absorbed solar energy, and SO_2^* represents an excited sulfur dioxide molecule whose electrons describe orbits with a higher total energy than that of the unexcited molecule. One property of these new electron orbits is that the chemistry of the molecule is affected, and it more readily reacts with certain atmospheric constituents such as molecular oxygen.[9] After this reaction, several others follow to complete the oxidation to sulfuric acid:

$$\begin{aligned} SO_2^* + O_2 &= SO_4, \\ SO_4 + O_2 &= SO_3 + O_3, \\ SO_3 + H_2O &= H_2SO_4. \end{aligned} \tag{8.4}$$

Both SO_3 and SO_4 are intermediate products of this series of reactions. The rate-limiting step is Eq. (8.3), so the oxidation rate of SO_2 by this mechanism depends upon the intensity of solar radiation. But experiments indicate that the reaction rate for bright sunlight is still fairly slow; less than 2% of the SO_2 is oxidized during an hour. It is therefore not nearly so effective as the catalyzed oxidation in a humid atmosphere, which we previously described.

However important the above-mentioned catalyzed reactions may be in urban pollution, they do not account for the relatively short lifetime of 2–4 days that SO_2 enjoys in the atmosphere when emitted from regions of the world where particulate mass concentrations are low. It is believed that naturally occurring ammonia NH_3 is a key factor which determines the mean lifetime in the troposphere. The source of ammonia is not known with certainty. Some is emitted during industrial processes, but this appears to be only a minor contribution to the mean NH_3 concentration of about 0.01 ppm in the troposphere. The greatest fraction appears to be released when microorganisms in the soil and oceans oxidize amino acids from decayed protein to yield CO_2 and H_2O. When tropospheric NH_3 is absorbed in water droplets together with SO_2, it will neutralize the solution as the SO_2 oxidizes and will thus encourage more absorption of SO_2 and accelerate its removal from the atmosphere. Ammonia is highly soluble in water, and SO_2 moderately so. Solubility increases with decreasing temperature, and the rate of SO_2 oxidation is accelerated appreciably at lower temperatures, such as those commonly found at a height of several kilometers above ground.[10] The fact that ammonium sulfate $(NH_4)_2SO_4$ is a common constituent of aerosols suggests that this mechanism may be one of the more important in scavenging SO_2 from the atmosphere.

Photochemical aerosol

So far, we have mentioned four means by which SO_2 is converted into sulfuric acid or sulfate ion: oxidation in water droplets, perhaps catalyzed by metal ions; oxidation in droplets containing NH_3; catalyzed oxidation on solid surfaces; and photo-oxidation. But perhaps the most rapid means of oxidizing SO_2 is a fifth process: the photo-oxidation of SO_2 in the presence of nitrogen dioxide and unsaturated hydrocarbons. This condition exists in photochemical smog. As we shall find in Chapter 10, photochemical smog is an oxidizing atmosphere since pollutants which constitute this smog—such as the secondary pollutants, ozone and the peroxyacyl nitrates—are potent oxidizing agents. Thus we might anticipate that they would quickly react with sulfur dioxide. In fact, the emission of SO_2 into photochemical smog quickly produces a sulfate aerosol.[11] Most of the particles are less than 1 micron in size, and they are consequently very effective in scattering light. Consequently, the development of photochemical smog in a city can result in a reduction of the SO_2 ambient concentration, but at the expense of increasing the sulfate concentration and reducing visibility. Very little is known about the detailed chemistry of the aerosol formation or the composition of the particles, but it appears that one of the major constituents is H_2SO_4.

8.3 THE SULFUR CYCLE

Anthropogenic sources account for about one-third of the total emissions of sulfur compounds into the atmosphere. Robinson and Robbins[12] have estimated that the annual output of SO_2 from anthropogenic sources amounts to about 63×10^6

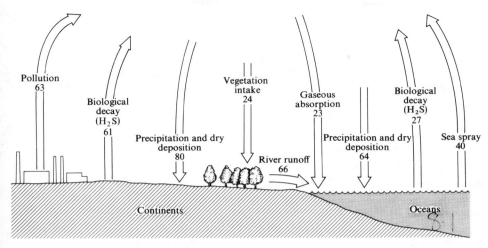

Fig. 8.1 Circulation of sulfur through the environment. The numbers give the amount of sulfur in each process, expressed in 10^6 Tonnes per year. (E. Robinson and R. C. Robbins, *Journal of the Air Pollution Control Association*, **20**, 233, 1970.)

Tonnes of sulfur each year, some 93% of it being emitted in the Northern Hemisphere. The total from all natural sources is approximately 128 Tonnes per year. Most of this is in the form of H_2S released from the decay of terrestrial and marine organic matter. Another source of atmospheric sulfur is the sulfate released as an aerosol in sea spray. There is a continual cycling of sulfur in gaseous and aerosol form through the atmosphere, with an exchange between that of marine and terrestrial origin, as summarized in Fig. 8.1.

Within the troposhere, H_2S is oxidized in a day or two to SO_2, apparently by the small concentration of ozone which is normally present. The reaction may be greatly accelerated to completion in a couple of hours in the oxidizing atmosphere of photochemical smog. Further oxidation will convert it into sulfuric acid or $(NH_4)_2SO_4$. The relative importance of various scavenging mechanisms for removing these sulfates is not yet determined. Precipitation of SO_2 and sulfates in rain is certainly a major factor, and dry sedimentation to the ground is another. The average amount of SO_2 that remains in the troposphere on a global basis amounts to a concentration of roughly 0.2 ppb.

The SO_2 residence time of 2–4 days in the troposphere is not so short that sulfurous smog can be considered a local phenomenon. Recently the rain in Scandinavia has been found to be noticeably acidic, having a pH as low as 3 on occasion. This has commonly been attributed to SO_x released in the United Kingdom, the Netherlands, and West Germany which then drifts with the prevailing winds as it is oxidized. But it is also possible that some of the sulfate could be oxidized H_2S released from sewage now being dumped into the North Sea.

8.4 ADVERSE EFFECTS

From our discussion in the preceding chapters we see that the nature of sulfurous smog continuously evolves, depending upon the ambient conditions of (1) the concentration of SO_2, (2) the concentration and composition of suspended particulate matter, (3) the relative humidity, (4) the intensity of sunlight, and (5) the chemistry of other gaseous pollutants in the ambient air. These affect the relative concentrations of SO_2, SO_3, H_2SO_4, and the solid sulfates that constitute the common variety of sulfurous smog. Thus it is difficult, if not impossible, in epidemiologic studies to separate out the effects of each of these pollutants. Some human physiological responses from exposure to high concentration of SO_2 are well established, and we shall first consider these. More subtle effects from chronic exposure to lower levels of sulfurous smog in community air are not so clearly defined, but there is strong statistical evidence of an association between pollution levels and the incidence of acute and chronic pulmonary disease. After we consider this evidence, we shall take up the adverse effects of sulfurous smog on plants and materials.

Toxicology

Sulfur dioxide is a pungent respiratory irritant, but when experienced at concentrations below about 25 ppm it affects only the upper respiratory system. It does not cause eye irritation at concentrations below about 10 ppm, but the threshold is reduced when it is experienced in combination with aerosols. The threshold for taste of about 0.3 ppm appears to be below the odor recognition threshold of 0.5 ppm. Exposure to 1.5 ppm or more for a few minutes may produce a reversible bronchiolar constriction in healthy individuals, with an increase in pulmonary flow resistance. This may be accompanied by shallow breathing (a low tidal volume) and an increased respiratory rate. Many of these effects may diminish during continuous exposure, but the reason for this is not understood. Individuals generally exhibit a reaction to SO_2 at concentrations of 5 ppm and above, although some sensitive people have responded at 1 ppm.[13]

Sulfur dioxide becomes an irritant of the lower respiratory system when experienced in association with aerosols. As we have previously mentioned in Chapter 2, SO_2 can be adsorbed onto the surface of aerosols and thereby be carried deep into the lung. The moist air in the bronchioles and alveoli establishes conditions appropriate for the oxidation of SO_2 and formation of sulfuric acid droplets where particles become lodged. An aggravated (or potentiated) adverse response has been demonstrated by experiments in which animals were exposed to SO_2 and to high concentrations of aerosols (at about 1 mg/m^3).[14] Salts of ferrous iron, manganese, and vanadium enhanced the potency of levels of SO_2, as indicated by measurements of the pulmonary flow resistance. The potentiated response was equivalent to a pure SO_2 exposure at concentrations which need to be *higher* by a factor of 3 or 4 to produce the same flow resistance. Salt was also found to enhance the response. We have noted that the constituents of some of these aerosols are precisely the substances which catalyze the oxidation of SO_2 into sulfuric acid. Insoluble particles such as manganese dioxide were found to have little synergistic effect. Thus it would appear that particles act in synergy with SO_2 not only because they carry it deep into the lung but because their chemical composition can enhance the rate of its oxidation to sulfuric acid.

Studies of the alteration in lung function in humans and lower animals indicate that sulfuric acid droplets and metal sulfates are also much more potent irritants than SO_2 at the same ambient concentration of sulfur. The potency increases as the particle size decreases, from about 2 microns to 0.3 micron. This is consistent with our previous discussion in Chapter 2 of the size dependence of the effectiveness with which the lung retains aerosol particles.

Epidemiology

The air pollution episodes in the Meuse Valley, Donora, and London have demonstrated that prolonged exposure to high levels of contaminants found in sulfurous smog can cause morbidity and mortality in persons with chronic lung disease. No specific pollutant has been identified as the causative agent, but sulfur oxides, H_2SO_4, and particulate levels were all high. These extreme cases have promoted considerable interest in identifying lower levels at which harmful effects occur.

It is reasonable to expect that members of a population with *preexisting lung disease* may more readily exhibit adverse reactions to pollutants, and some epidemiologic studies have focused on this possibility in order to have a sensitive indicator. For example, evidence has been uncovered that chronic lung diseases such as asthma may be aggravated by prolonged exposure to moderately high levels of pollution.[15] The incidence of hospitalization of children under 15 years of age with asthma was found in the geographical areas near Buffalo, New York, to correlate with the average level of suspended particulate matter. Between 1956 and 1961 an average annual incidence rate of 32.4 cases per 100,000 exposed population was found in the cleanest neighborhoods with less than 80 $\mu g/m^3$ average particulate concentration, but the rate was up to 50.7 in neighborhoods where average levels exceeded 135 $\mu g/m^3$. Similar trends are obtained if the population within each neighborhood is divided into social classes and the incidence for a given class is compared with the mean levels of pollution. Although particulate matter was used as a measure of pollution levels, it would be incorrect to conclude that it alone is the causative agent, because the region suffers from the complex mixture of pollutants typically found in sulfurous smog.

Several studies have linked sulfurous smog and the incidence of *acute respiratory disease*. Douglas and Waller, for example, have followed a group of 3900 British school children from their birth in 1946 until they left school in 1961.[16] The frequency and severity of acute illness in the lower respiratory tract were found to increase directly with increasing levels of air pollution. No dependence upon social class was found. Another survey of 800 school children in Sheffield, England, has also established an association between the incidence of illness in both upper and lower respiratory systems with levels of sulfur dioxide and particulate matter.[17] These findings are further supported by a study of the effects of air pollution on Japanese school children in Tokyo and Osaka.[18,19]

A clear statistical association has been established between the levels of SO_2 and particulates in sulfurous smog and the incidence of *chronic bronchitis* in Great Britain. In reviewing the published studies, Lave and Seskin[20] conclude that if the air in all of the boroughs were improved to the quality of the air in the cleanest one, the death rate attributed to bronchitis would decrease by about 40%, or perhaps more.

Some studies have sought correlations between specific components of sulfurous smog and disease. From a study of the incidence of chronic bronchitis among steelworkers in Sheffield, there is evidence that smoke levels (particulate concentration) correlated with disease, but that SO_2 levels did not.[21]

No relationship has been established between levels of the constituents of sulfurous smog and *cancers of the respiratory tract*. Although coal and tar, at very high concentrations, have been associated with cancer of the lung and larynx, as soot has with lung cancer, it has not been established that their presence at concentrations normally found in community air is detrimental.[22] Nevertheless it would be prudent to consider their presence in the air a hazard until there is definite proof to the contrary.

In Section 2.5 we discussed evidence for an *urban factor* in the mortality from lung cancer, by which we mean that mortality is higher among members of an urban population than a rural one, whether or not the populations consist of smokers or nonsmokers. However, we emphasized that a causal relationship between air pollution and this factor has been neither established nor refuted. Nevertheless there is a strong possibility that air pollution is at least a contributary cause. And sulfurous smog is the form of pollution which predominates in the urban areas included in the study. This is also the form of pollution experienced by the residents of St. Louis; and we recall that a study by Ishikawa *et al.*[23] found an enhanced incidence of advanced emphysema in this city compared with the incidence in less-polluted Winnipeg. The New York City study by Hodgson also shows a relationship between mortality from respiratory and heart disease and daily levels of particulate matter.[24] Thus although direct causal relationships have not been established, there is statistical evidence that levels of sulfurous smog currently being experienced in metropolitan areas are causing increased mortality. There is strong evidence that prolonged exposure can cause chronic bronchitis and suggestive evidence for causation of emphysema and lung cancer.

The time of exposure to sulfurous pollutants is a poorly understood variable determining the severity of adverse effects. In Fig. 8.2 we show a summary of established and suspected responses for a given level and length of exposure to SO_2. (Unfortunately some of the data in this figure are influenced by the presence of aerosols in unknown amounts or composition, so the circumstances of exposure have not been identified precisely.) In some cases, threshold concentra-

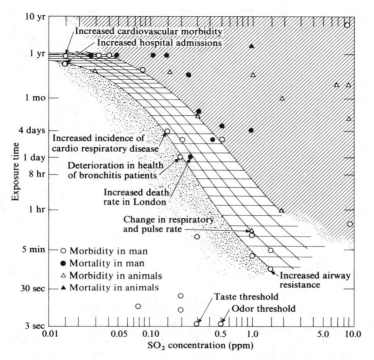

Fig. 8.2   Effects of sulfur oxides pollution on health. Shaded area represents range of concentrations and exposure times in which deaths have been reported in excess of normal expectation. Grid area represents range of concentrations and exposure times in which significant health effects have been reported. Speckled area represents ranges of concentrations and exposure times in which health effects are suspected. ("Air Quality Criteria for Sulfur Oxides," National Center for Air Pollution Control, 1967, U.S. Department of Health, Education and Welfare, Public Health Service.)

tions are the significant parameters, in others perhaps the dosage, or the dosage sustained for a given time interval. Very little is known concerning the distribution of sensitivities, and therefore response threshold, among the general public.

Plant damage

Sulfur dioxide damages the leaves of some plants exposed to a sufficiently high concentration and dosage. Apparently a threshold does exist because plants exposed to only low concentrations are able to take in small amounts of SO_2 through the stomata and dispose of it without damage. The sulfur dioxide tends to be oxidized to sulfates in the cells of the leaf. Concentrations above the threshold for a particular plant cause deterioration of the internal cells and leaf discoloration. Spinach, lettuce, and some other leafy vegetables are very sensitive, as are cotton and alfalfa. Many other plants are moderately sensitive. Some

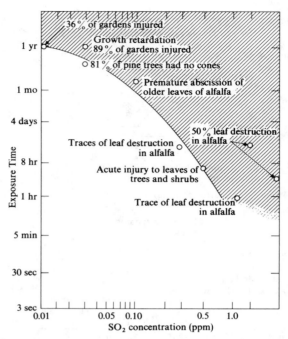

Fig. 8.3 Effects of sulfur oxides pollution on vegetation. Shaded area represents the range of concentrations and exposure times in which injury to vegetation has been reported; white area, the range of concentrations and exposure times of undetermined significance to vegetation. ("Air Quality Criteria for Sulfur Oxides," U.S. Department of Health, Education, and Welfare, Public Health Service, National Center for Air Pollution Control, 1967.)

examples of SO_2 levels that produce a response are summarized in Fig. 8.3. Most plants display maximum sensitivity during the growing season in sunlight when the stomata are open and the plants are engaged in strong physiological activity.

In comparison with SO_2, aerosols of sulfuric acid appear to be less toxic to plants. Effects are noted only in special circumstances—for example, when large droplets may strike the leaves. About 50 years ago near Ducktown, Tennessee, sulfurous mists from two copper smelters denuded trees and killed plants for several miles around. This was followed by erosion and such loss of topsoil that even today the area can support only a few hardy weeds.

Material damage

Prolonged exposure to sulfurous smogs is known to have caused serious damage to building marble, limestone, and mortar because the carbonates in these materials are replaced by sulfates (as $CaSO_4$ or gypsum, $CaSO_4 \cdot 2H_2O$), and these sulfates are soluble in water. Consequently, over a period of time, rain removes the altered minerals. The damage may be accompanied by discoloration from aerosols that adhere to the surface and deposit their burden of chemicals. Historical monuments in Venice, Italy, have suffered extensive damage from exposure to the sulfurous pollutants from a nearby industrial community.

In sufficient concentrations, the sulfurous pollutants can discolor paint, corrode metals, and cause organic fibers, nylon hose, and leather to weaken. The severity of the effect is influenced by the prevailing relative humidity. For example, the oxide surface on most metals protects the underlying material from corrosion if the relative humidity is less than about 80%. However, at higher relative humidities, the oxide covering apparently can be broken down when exposed to SO_2, and corrosion from the acidic pollution commences. Also there can be a synergistic effect between SO_2 and aerosols when the gas is adsorbed onto the particle's surface and conveyed to a material on which the particle becomes lodged. The highly concentrated pollutant then readily attacks the very localized area to which it is exposed. Soot is ideally suited for this because of its loose, porous structure and large surface area for adsorbing gases.

8.5 AN EPISODE: LONDON, 1952

Toward the end of the first week of December, 1952, meteorological conditions near London had all of the characteristics for an onset of pronounced stability. Gusty winds subsided as evening fell on December 4; a cold anticyclone had approached slowly from the northwest and showed signs of stagnating over the suburbs. Several layers of clouds, advected across the high-pressure area at high altitudes, reflected the incident solar radiation and, without the sunshine at ground level, the temperature fell precipitously. The relative humidity stood at 80%, which was sufficiently high to form a dense fog within the polluted city air during the night. Developing upward, the fog was stopped only at the base of the elevated inversion layer of the anticyclone, which rode at a height of 100–150 m. Some instability developed within the fog as the uppermost layer cooled by emitting thermal radiation and subsequently subsided as it lost buoyancy. Surface temperatures were close to freezing as the next day commenced.

But the fog did not dissipate as the sun rose. Because of the high albedo of the fog layer, sunlight could only weakly penetrate to the ground; and consequently there was no appreciable warming of the lower layers of air. The calm center of the anticyclone stagnated over Westminster, and winds were exceptionally light. In some places there was a dead calm; in others, the breeze amounted to no more than 5 km/hr (about 3 mph). Lacking advection, the fog remained within the shallow land depression which defines the Greater London Basin. By December 6 the noontime temperature was down to $-2°C$ (28°F) and the relative humidity stood at 100%.

Within the pea-soup fog thousands of sources continued to emit their wastes. The cold, damp conditions were met with increased use of fuel in homes and factories, and since the effluent from the multitudes of chimney pots had little

Fig. 8.4 Reduced visibility during the London sulfurous smog on December 8, 1952. Copyright © by the Radio Times Hulton Picture Library; reprinted by permission.

buoyancy, it accumulated at low levels in the mixing layer. And there was greater demand for electricity in the darkness. The heights of the exhaust stacks of the power plants and the buoyancy of their plumes, with few exceptions, weren't enough to permit the effluent to penetrate the inversion, so their pollutants added to the community problem. The chief domestic and commercial fuel was bituminous, and consequently emissions of soot, tar, ash, and sulfur oxides steadily mounted. Ventilation, confined by the inversion and unaided by advection, was minimal. The ambient concentrations of particulates and sulfur dioxide increased day by day as the stagnant conditions persisted. The fog had quickly turned into smog. Tar and soot clung to everything and gave the atmosphere a "taste."

Condensation onto the aerosols further reduced visibility, and oxidation of SO_2 proceeded apace, catalyzed by the high ash content and humidity. The smog was so thick that the visual range was reduced to only a few meters. Bus conductors walked in front of their vehicles to guide them through the all-but-stalled traffic (Fig. 8.4). The smog penetrated every accessible area. Even the concert at Sadler's Wells was affected, for the singers could no longer see the conductor.

The ambient concentration of SO_2, normally in the range of 0.1 ppm by volume, built up to about 0.7 ppm by December 8.[25] Carbon dioxide levels at nearby Greenwich were well over twice normal.[1] As SO_2 levels increased, so did the incidence of respiratory distress among the population. The death rate soared to over twice its normal value for that time of year. It closely correlated with the atmospheric burden of both SO_2 and particulates, as illustrated in Fig. 8.5. An increase in mortality among all age groups was observed, although the highest increase was for older people. Prevalent causes were chronic bronchitis, bronchopneumonia, and heart disease. The correlation exhibited in Fig. 8.5 does

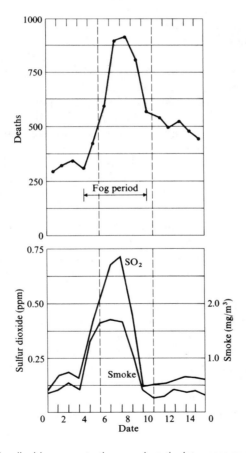

Fig. 8.5 Deaths, sulfur dioxide concentrations, and particulate mass concentrations during the December, 1952, episode in London. (After E. T. Wilkins, *Journal of the Royal Sanitation Institute,* **74**, 1, 1954; Crown copyright, reproduced by permission.)

not establish a causal relationship, for it may be that trace amounts of the contaminants which were not monitored—such as vanadium compounds or coal
distillates—were responsible for the increased mortality. This remains an open
question. The death rate remained high for several weeks after the episode. All
told, it is estimated that there were some 4000 excess deaths as a result of the
four-day episode.

It was not until December 9 that meteorological conditions improved as
a breeze developed. The following day a cold front passed, bringing with it
fresh air from the north Atlantic and an end to the episode.

It is apparent that the causes of the London episode were many. Meteorological factors set the stage but injurious conditions would not have resulted
without the emission of pollutants in large quantity and in such a way that they
remained within the mixing layer. The combination of the two factors intensified the problem, which could not be ended until one or the other was alleviated.

It would be wrong to presume that pollutants emitted during the episode
remained in the air during the entire time that the high-pressure region was
stagnant. The London smog was in fact a dynamic situation, with pollutants
being continually emitted and scavenged. That this is the case can be seen from
the following argument. The volume of air into which pollutants were diluted,
or the mixing volume, can be estimated from the known area of the city and the
height of the base of the inversion. The height is variously estimated as $W = 100–$
150 m, and the area is about $A = 1300$ km^2. Taking the greater height, we have
a mixing volume of about 200 km^3. Into this each day went an estimated 2000 tons
of SO_2.[1] If all of this had remained without loss, the ambient concentration after
24 hours would have risen by about 3.5 ppm (10 mg/m^3). Such a large increase
however was never observed, and the ambient concentration was never observed
to exceed 1.3 ppm anywhere in the city during the episode. Figure 8.5 shows that
the greatest daily increase in ambient concentration of SO_2 was only about 0.18
ppm (0.5 mg/m^3), only 5% of the expected increase. This discrepancy cannot be
attributed to advection away from the metropolitan area, for with the light wind
this was negligible. The fact that the observed daily increase of SO_2 concentration
was so small indicates that scavenging mechanisms quickly remove SO_2 from the
air.

Thus it is of interest to estimate the residence time for SO_2 in the London
smog. We can do this by first noting that the average SO_2 concentration at
six places in London during the four-day episode was about 0.73 ppm (2.1
mg/m^3).[1] If an emission rate equivalent to 10 mg/m^3 addition each day is to
yield the observed ambient concentration, the average residence time of SO_2
must be only about $(2.1/10) \times 24 = 5$ hours.

The scavenging mechanisms responsible for this residence time can easily
be imagined. Deposition onto building surfaces, the ground, or smoke particles
is an evident possibility. Perhaps more important is absorption by fog droplets.
Meetham has estimated that the total surface area of the fog droplets was more
than 100 times larger than the 1300 km^2 of the city's area.[1] This large area
presented by a high concentration of small droplets distributed throughout the

air, when compared with the smaller area of the ground underneath, indicates that SO_2 molecules struck fog droplets many times before reaching a point where deposition on the ground was possible. Thus the affinity of SO_2 for absorption and subsequent oxidation within a droplet appears to have been the important factor determining the residence time for SO_2 in the London air. In fact SO_2 may move in and out of various droplets during its five-hour residence time, until eventually it is oxidized inside a droplet. This fixes the sulfur as sulfuric acid or sulfate within the droplet, and once in this form the sulfur would not have been counted as part of the ambient SO_2 concentration by monitoring instruments then in use.

We have introduced the preceding discussion to emphasize an important feature of some types of air pollution. A steady-state condition, one in which important parameters do not vary appreciably with time, does not necessarily imply that the situation is static. In the London smog, the ambient concentration of SO_2 was determined by a balance between a very high rate of emission of pollutant and a high rate of elimination of the pollutant. Unfortunately, because of the complexity of the many interrelated processes, the quantitative conditions for this balance are not well understood.

It is perhaps no surprise that the shock of the death toll of the London smog stimulated political agencies into action. After some debate, the initial emphasis for pollution control was placed on reducing the emission of particulate matter, and the control of SO_x was viewed as being of secondary importance. Ordinances were introduced to cut particulate emissions by regulating fuel quality and improving the efficiency of combustion. Since then, there has been a remarkable improvement in atmospheric conditions. Although SO_2 levels have, on occasion, reached those experienced during the episode, a correspondingly high mortality rate was not observed. This may illustrate the important role played by synergistic effects in 1952. Since 1955 when smoke controls were initiated in central London, the frequency and intensity of dense fogs have diminished, perhaps due to the decreased density of aerosol nucleation centers.[26] On the average, winter sunshine has increased by over 50% and the visual range by a factor of 3 in the last decade.[27]

SUMMARY

Sulfurous smog is a mixture of many pollutants—sulfur dioxide, sulfur trioxide, sulfuric acid, aerosols, and organic sulfur compounds. Most of the sulfur, and often the particulate matter, originates in the fossil fuels, coal and oil. In the atmosphere SO_2 is oxidized to SO_3 and sulfate by several possible mechanisms. One rapid way is by oxidation within a water droplet when catalyzed by perhaps iron or manganese oxides. Another is oxidation within a droplet in the presence of ammonia, which reduces the acidity of the solution and encourages further absorption of SO_2. Because of these and other mechanisms, the rate of oxidation of SO_2 may be affected by the type and concentration of other gaseous and particulate pollutants, the humidity, and the intensity of sunlight. The severity of adverse responses from exposed humans is also influenced by these factors, for SO_2 interacts synergistically with aerosols in affecting the lower respiratory system. The details as to how SO_2 or H_2SO_4 affect the pulmonary membranes are not known, but epidemiologic studies indicate that prolonged exposure to sulfurous smog may cause chronic bronchitis. There is strong statistical evidence that it increases the incidence of acute lower respiratory disease, at least among children, and there is suggestive evidence that it may cause other chronic lung diseases and increase mortality from respiratory and heart disease. Plants can tolerate low concentrations of SO_2 but have thresholds above which leaf damage occurs. Materials such as masonry, paints, and metals suffer from continual exposure to high levels of smog.

NOTES

1. A. R. Meetham, D. W. Bottom, and S. Cayton, *Atmospheric Pollution, Its Origins and Prevention*, Oxford: Pergamon Press, 1964.

2. F. D. Rossini, "Hydrocarbons in Petroleum," *J. Chem. Ed.*, **37**, 554 (1960).

3. H. H. Schrenk, H. Hieman, G. D. Clayton, W. M. Gafafer, and H. Wexler, "Air Pollution in Donora, Pa.," *Public Health Service Bull.*, 306, (1949).

4. O. I. Joensuu, "Fossil Fuels as a Source of Mercury Pollution," *Science*, **172**, 1027 (1971).

5. A. Levy, E. L. Merryman, and W. T. Reid, "Mechanisms of Formation of Sulfur Oxides in Combustion," *Environ. Sci. and Tech.*, **4**, 653 (1970).

6. H. F. Johnstone and D. R. Coughanowr, "Absorption of Sulfur Dioxide from Air," *Ind. Eng. Chem.*, **50**, 1169 (1958).

7. P. M. Foster, "The Oxidation of Sulfur Dioxide in Power Station Plumes," *Atmosph. Environ.*, **3**, 157 (1969).

8. F. E. Gartrell, F. W. Thomas, and S. B. Carpenter, "Atmospheric Oxidation of Sulfur Dioxide in Coal-burning Power Plant Plumes," *Am. Ind. Hyg. Assoc. J.*, **24**, 113 (1963).

9. P. A. Leighton, *Photochemistry of Air Pollution*, New York: Academic Press, 1961.

10. H. A. C. McKay, "The Atmospheric Oxidation of Sulfur Dioxide in Water Droplets in Presence of Ammonia," *Atmosph. Environ.* **5**, 7 (1971).

11. N. A. Renzetti and D. J. Doyle, "Photochemical Aerosol Formation in Sulfur Dioxide-Hydrocarbon Systems," *Intern. J. Air Poll.* (London), **2**, 327 (1960).

12. E. Robinson and R. C. Robbins, "Gaseous Sulfur Pollutants from Urban and Natural Sources," *J. Air Poll. Control Assoc.*, **20**, 233 (1970); W. W. Kellogg, R. D. Cadle, E. R. Allen, A. L. Lazrus, and E. A. Martell, "The Sulfur Cycle," *Science* **175**, 587 (1972).

13. M. O. Amdur, "Toxicologic Appraisal of Particulate Matter, Oxides of Sulfur, and Sulfuric Acid," *J. Air Poll. Control Assoc.*, **19**, 638, (1969).

14. M. O. Amdur, and D. Underhill, "The Effect of Various Aerosols on the Response of Guinea Pigs to Sulfur Dioxide," *Arch. Environ. Health*, **16**, 460 (1968).

15. H. A. Sultz, J. G. Feldman, E. R. Schlesinger, and W. E. Mosher, "An Effect of Continued Exposure to Air Pollution on the Incidence of Chronic Childhood Allergic Disease," *Am. J. Public Health*, **60**, 891 (1970).

16. J. W. B. Douglas and R. E. Waller, "Air Pollution and Respiratory Infection in Children," *Brit. J. Prev. Soc. Med.*, **20**, 1 (1966).

17. J. E. Lunn, *et al.*, "Patterns of Respiratory Illness in Sheffield Infant School Children," *Brit. J. Prev. Soc. Med.*, **21**, 7 (1967).

18. H. Watanabe, "Effect of Air Pollution on School Children," Presented at the United States-Japan Cooperative Science Group on Air Pollution Health Seminar, Boston, Mass., February, 1969.

19. T. Toyama, "Air Pollution and Its Health Effects in Japan," *Arch. Environ. Health*, **8**, 161 (1964).

20. L. B. Lave and E. P. Seskin, "Air Pollution and Human Health," *Science*, **169**, 723 (1970).

21. J. Gregory, "The Influence of Climate and Atmospheric Pollution on Exacerbations of Chronic Bronchitis, "*Atmos. Environ.*, **4**, 453, (1970); see also **5**, 435 (1971).

22. W. C. Hueper, "Carcinogens in the Human Environment," *Arch. Path.*, **71**, 237 (1961).

23. S. Ishikawa, D. H. Bowden, V. Fischer, and J. P. Wyatt, "The 'Emphysema Profile' in two Midwestern Cities in North America," *Arch. Environ. Health*, **18**, 660 (1969).

24. T. A. Hodgson, "Short-Term Effects of Air Pollution in New York City," *Environ. Sci. and Tech.*, **4**, 589 (1970).

25. E. T. Wilkins, "Air Pollution Aspects of the London Fog of December 1952," *Quart. J. Roy. Meteorol. Soc.*, **80**, 267 (1954).

26. J. H. Brazell, "Frequency of Dense and Thick Fog in Central London as Compared with Frequency in Outer London," *Meteorol. Mag.*, **93**, 129 (1964).

27. C. G. Collier, "Fog at Manchester," *Weather*, **25**, 25 (1970).

FOR FURTHER READING

P. A. LEIGHTON	*Photochemistry of Air Pollution*, New York: Academic Press, 1961.
A. R. MEETHAM, D. W. BOTTOM, and S. CAYTON	*Atmospheric Pollution, Its Origins and Prevention*, Oxford: Pergamon Press, 1964.
M. BUFALINI	"Oxidation of Sulfur Dioxide in Polluted Atmospheres—A Review," *Environ. Sci. and Tech.*, **5**, 685 (1971).
P. URONE and W. H. S. SCHROEDER	"Photochemical Reactions of SO_2—a Review," *Environ. Sci. and Tech,*, **3**, 436 (1969).
REVIEW:	"Toxicologic and Epidemiologic Bases for Air Quality Criteria," *J. Air Poll. Control Assoc.*, **19**, 629–732 (1969).
M. TRESHOW	*Environment and Plant Response*, New York: McGraw-Hill, 1970.
R. S. SCORER	*Air Pollution*, Oxford: Pergamon Press, 1968.

QUESTIONS

1. What is the difference between charcoal, coke, peat, bituminous, and anthracite?

2. What is natural gas? In what ways do the following fractions of petroleum differ: gasoline, light fuel oil, and residual fuel oil?

3. What are the essential chemical ingredients for sulfurous smog? Which pollutants affect its development?

4. In what ways may high humidity aggravate the adverse effects of sulfurous smog?

5. What components of sulfurous smog are reducing agents?

6. The overall reaction for photo-oxidation of SO_2 in the air can be written as $2SO_2 + O_2 \rightarrow 2SO_3$. How do you explain the fact that the rate of the reaction is not observed to vary as the square of the SO_2 concentration, but increases linearly with the concentration?

7. Coal-burning, power-generating stations are attractive for electrical utilities because of the simplicity and speed of their construction, low capital cost, economy of fuel, and reliability. Current estimates predict that substantial coal reserves are available for use in the future. From the standpoint of air pollution, which physical and chemical characteristics of these reserves are important?

8. Which meteorological features were influential directly or indirectly in the London smog episode of 1952? What were the similarities between this episode and the Donora episode in 1948?

9. List the possible mechanisms by which SO_2 is scavenged from the air.

10. If you had responsibility for assessing the contribution of SO_2 pollution to mortality in a city by an epidemiologic study, what additional contributing influences should you take into account to ensure that you had isolated SO_2 as the only reasonable cause?

PROBLEMS

1. A large steel mill emits 10^6 liters of exhaust gas each second, with an SO_2 concentration of 300 ppm. What is the daily tonnage output of SO_2 in the exhaust?

2. A smelter produces 320 Tonnes per day of particulate matter in the gaseous waste. If a particle collector (such as an electrostatic precipitator) of 96% efficiency is used to clean the effluent, what is the particulate output from the exhaust stack per day?

3. Compare the SO_x emission rate for a large industrial space heater rated at 1MW thermal when it uses 2% sulfur-bearing coal and when it uses 2% sulfur-bearing distillate fuel oil.

4. What is the rate at which particulates are emitted by a power station rated for 1 GW of electrical power production if it has an efficiency of 42% and burns pulverized coal? What is the daily amount emitted without pollution emission controls and what is the amount emitted if an electrostatic precipitator with 97% efficiency is used?

5. A small hospital relies on an oil-burning furnace to provide heat and electricity. If oil containing 2% sulfur is the fuel, and is burned at a rate sufficient to release 1.5×10^7 W of thermal energy, what is the maximum downwind concentration of SO_x that might be experienced at ground level for a 10 km/hr wind? The heat emission rate in the stack effluent is 1.5×10^6 W, the geometric stack height is 30 m, and the momentum rise of the plume is negligible. Select a few different classes of atmospheric stability to show how sensitive your prediction is to meteorological conditions.

CHAPTER 9

STATIONARY SOURCES AND THEIR CONTROL

In this chapter we broaden our view from that of the previous chapter to consider specific sources which contribute to sulfurous smog, including not only those which burn fossil fuels, but other types of major sources as well. Some emit only particulate matter in appreciable quantities, in which case it is not strictly correct to include them as contributors to sulfurous smog, unless other major sources of sulfur compounds are also polluting the same air. Nevertheless it is convenient to deal with them here. All of our examples are stationary sources; that is, they have a fixed location. We will postpone an examination of mobile sources such as the automobile, to Chapter 10 because in most communities they contribute mainly to photochemical smog.

It is useful to consider simultaneously both the sources of pollution and the means which have been implemented or suggested for reducing their emissions. To this end, we shall first examine the general principles upon which pollution control devices are based. Then we shall be in a position to discuss the application of these devices to specific sources, such as fossil fuel power plants, smelters, petroleum refineries, pulp mills, and incinerators.

9.1 PRINCIPLES OF CONTROL DEVICES

At the risk of stating the obvious, we note that there are two main sources of primary pollutants: (1) the fuel or raw materials employed in an energy-conversion or fabrication process, and (2) chemical reactions involved in the fuel or material processing. Thus to minimize pollution, it is logical to seek ways to prevent the formation of a pollutant in the first place by cleaning up the fuel and raw materials, e.g. by eliminating sulfur from fossil fuels. Similarly it is desirable to alter or introduce new processing operations, e.g. by better controlling the proportions of fuel and air in a combustion chamber to minimize the formation of carbon monoxide. In subsequent sections, we shall discuss several examples in which these basic principles of pollution control have been applied. But first we shall focus our attention on a more commonly adopted method of pollution control in which the basic materials and processes are unchanged and the effluent itself is cleansed of its burden of contaminants.

It is useful to distinguish between several types of devices that operate on different principles and hence are specific in controlling different pollutants effectively. These main classifications are: *filters*, *settling chambers*, *inertial collectors*, *electrostatic precipitators*, *scrubbers*, *adsorbers*, and *chemical reactors*. Combinations of these within a single device are possible, and often several devices may be operated in series to obtain more effective control.

One reason why effluent-cleansing devices are popularly used in industrial pollution control is because their installation often does not necessitate an expensive change in processing techniques; the device is mounted near the base of an exhaust stack and affects only the waste products. This is not to say that in every case the engineering design of the control device can be made independently of any consideration of the process; indeed, any obstruction to the free flow of the exhaust may adversely affect the preceding stages. In addition, the device must be matched to the characteristics of the effluent, for it often must withstand continual exposure to humid, high-temperature gases containing corrosives and particulate matter. Control devices must also be compatible with effective dispersal of the effluent. Thus it is desirable to keep the gases at high temperatures to maintain the buoyancy of the emitted plume. The overall effect of first cooling the effluent, in an attempt to minimize corrosion of control devices, may actually be to increase downwind pollutant concentrations experienced at ground level even when the device is properly functioning. This is because the less-buoyant plume stays closer to the ground as it drifts and disperses.

With these factors in mind, let us now turn to consider in more detail the main principles of commonly used devices.

Filters are an obvious means of cleaning particulate matter from an air stream. They consist of a porous structure—such as woven fabric, vegetable fibers, or a bed of aggregate such as sand or coarse gravel—which retains particles on the leading surface or within. To keep the interstices free, water may continually be sprayed onto the surface, or the filter may be shaken to dislodge the trapped particles.

In large installations, a fabric filter in the form of a long cylindrical bag, closed at the top and supported over a wire frame, may be mounted vertically so that air introduced through the open bottom is forced to pass outward through the side. Several of these may be arrayed side by side to decrease the resistance to gas flow, as illustrated in Fig. 9.1. If the assembly of bags is shaken by a mechanical device, the dislodged particles can be collected in a hopper at the bottom permitting the filter to be operated continuously. Such a device and the protecting outer structure is known as a "baghouse."

The size of the particles collected by a filter depends primarily upon the size of the interstices; however, these cannot be made arbitrarily small or the resistance to air flow will become unacceptably high. In operation, the collection efficiency of a filter increases as the pores become filled with trapped particles, so the actual efficiency depends upon the operating procedures as well as the design of the filter material. For small-scale use at near-ambient temperatures, paper filters are available which have more than 99% efficiency for removing particles as small as 0.3 micron. These filters permanently trap the particles

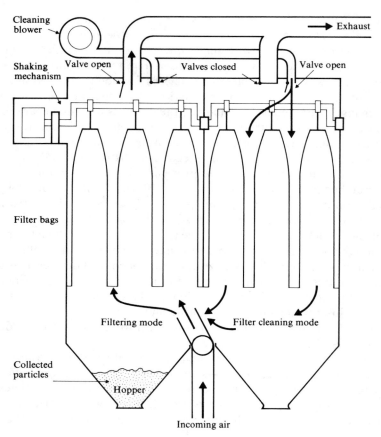

Fig. 9.1 A bag filter employing mechanical shaking and a periodic reverse flow of air on alternate sides for cleaning.

and therefore have a limited useful lifetime. This level of efficiency could not be achieved by a filter which would be appropriate for processing the high-temperature effluent from a power plant.

A *settling chamber* is frequently used as a primary collector to remove large particles before an effluent is passed on to secondary processing in other control devices. Its use, for example, can help to prevent clogging of filters designed to remove primarily small particles. As its name suggests, it is merely a large chamber where the gas spends sufficient time to allow gravitational sedimentation, and the fallout is collected in a hopper at the bottom. In practice, only particles whose radius exceeds roughly 50 microns are separated from the effluent.

Inertial collectors include both ducts which are baffled or louvered and a device more popularly known as a *cyclone*. The baffled collectors are designed so the sudden changes of direction of gas flow do not permit the more massive

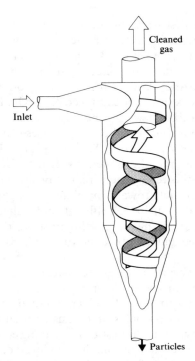

Cleaned gas

Inlet

Particles

Fig. 9.2   Outer and inner vortex flow in a cyclone, showing the helical pattern of air flow.

particles to follow the air stream, and these particles impact on the obstructions. In this way, particles with a size exceeding 10 microns can be collected on the baffles; they are commonly removed by a continuous water wash.

A *cyclone* is a device with no moving parts. Dirty air is introduced tangentially at the outer edge near the top of a cylindrical chamber, as shown in Fig. 9.2, and continues downward in a spiraling trajectory, thereby causing particles to move outward toward the wall as a result of their inertia (or the centrifugal force they experience). When they impact, they fall to the bottom and are collected. The design of the lower end can be arranged to send the air stream back upward in another vortex motion at the center of the chamber, thus increasing the path length that the air travels in its cyclonic motion. Then particles which work their way to the region between the two spirals find it possible to settle downward. The cyclone is a very efficient collector of large particles, but its efficiency drops off rapidly for particles of radius less than about 2 microns. As happens in the other devices mentioned above, the cyclone removes both solid and liquid matter. It can collect more than 99% of the droplets exceeding 50 microns radius in an air stream, a size commonly entrained in air passing over a boiling liquid.

Electrostatic precipitators are often employed to remove solid and liquid particles from the exhaust of power plants and other sources emitting large volumes of effluent. The first workable device was built in 1907 by Frederick G. Cottrell to collect sulfuric acid mist. In large contemporary models, the waste-gases, before entering the exhaust stack, pass between an array of alternating wires and plates, the wires being maintained at a high positive or negative electrical potential with respect to the plates, which are at ground potential. In smaller precipitators a single wire is surrounded by a coaxial cylindrical conductor, which also serves as the duct through which the gases pass. The strong electric field in which the field lines converge near the wires ionizes some of the gas molecules, forming what is called a *corona discharge*. This region near the wires is bluish-white or possibly red and is a source of mobile ions. If the wires are at a high positive potential, electrons are collected by the wire, but positive ions can migrate under the influence of the electric field toward the plates. As they move, some of the ions strike particles such as fly ash in the effluent and adhere to their surface, or remove electrons. Because of this continuing process, the particles themselves become positively charged and are attranted to the plates (as shown in Fig. 9–3), where build up a deposit.

More commonly, however, the wires are instead at a large negative potential, and the corona discharge provides electrons which migrate toward the plates. Most become attached first to neutral molecules (such as O_2), and subsequently these molecules collide with particles and transfer their charge. The negative-discharge electrode is not used in domestic air-cleaning equipment because it also forms ozone. When the plates are mechanically vibrated, the ash can be broken off, and the large agglomerations fall out of the air stream to a location below the precipitator, where they can be removed. Smaller particles may, however, become entrained in the effluent again and escape, producing puffs of smoke in the exhaust.

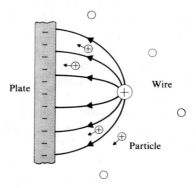

Fig. 9.3   Principle of the electrostatic precipitator. The curved lines denote electric field lines between the wire and plate. Particles, which are positively charged as ions produced near the wires attract electrons from them during collisions, are accelerated toward the plate by the electric field.

The collection efficiency depends on many factors, including particle size and resistivity, humidity and temperature of the effluent, and the flow rate. Under ideal conditions the efficiency can exceed 99% by weight. Even more efficient cleaning may be possible with two units operated in tandem. On the face of it, this is an impressive figure. But large sources such as power plants using pulverized coal emit particles at such a great rate, as indicated by the emission factors in Table 8–6, that the 1 percent or so which escape still represents an appreciable amount. The efficiency of electrostatic precipitators is least for particles of about 0.1 micron radius, so it is these which are primarily released to the environment.

Scrubbers operate on the principle that contact with a liquid can remove certain gaseous and particulate matter from an air stream. In fact, it may be possible to employ a liquid that reacts chemically with the contaminant and renders it harmless. One very simple scrubber is merely a device that bubbles the waste gas through an appropriate liquid, and this may be adequate for processing the effluent from small sources. However, to increase the efficiency as is generally required for large sources, devices are used which increase the contact area between the liquid and gas by introducing the liquid into the gas stream in droplet form. The droplets with their burden of pollutants are later removed by another control device such as a cyclone.

Figure 9.4 shows a popular device known as a venturi scrubber. The appropriate liquid is introduced in the region of high gas speed at the constriction of the tube and is broken up into fine droplets. They are then well mixed with the gas as it slows and collects where the tube expands. During this time a contaminant such as SO_2 will be absorbed in the droplets, perhaps a slightly alkaline water solution, and will continue to do so until the droplets are removed by a filter or inertial collector. The simplest scrubber consists of a chamber through which the effluent is passed while water is sprayed from the top and collected at the bottom.

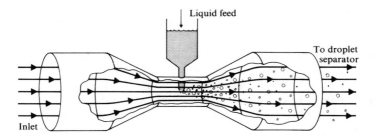

Fig. 9.4   Venturi scrubber. Introduction of a liquid at the constricted throat where the gas flow speed is greatest causes the formation of fine droplets which then mix with the gas.

Adsorbers consist of porous material such as charcoal or a zeolite with a large surface area to which contaminant gases adhere. The adsorbed contaminants can later be driven off by heat or steam, and the material reused. Such regenerative systems are necessary from an economic standpoint for large sources from which great quantities of a pollutant must be removed. Some adsorptive materials such as those composed of complex oxides are more selective than charcoal in their adsorptive capabilities and can be used to separate out only one or a small class of pollutants. The nonuniform distribution of charge over the molecules, and the accompanying nonuniform distribution over the surface gives them an affinity for electrically polarized gas molecules. This offers the possibility of tailoring an adsorber to collect specific pollutants, and if the amount of pollutant in the effluent is not great, for example, a malodorant, even a costly nonregenerate adsorber may be economically attractive.

Chemical reactors include both simple combustion chambers and catalytic reactors. The principle of these devices is to convert a pollutant into a more acceptable compound. This is especially useful for organic compounds, which can be burned at 850°C (about 1500°F) in a furnace; some malodorant mercaptans can be beneficially altered in this way. But if the concentration of organic molecules is insufficient to support combustion, the oxidation process must be aided by a catalyst, such as a metal of the platinum series (including iron, cobalt, nickel, platinum, and some alloys) or rare-earth compounds. A catalytic reactor is designed to bring the effluent into contact with the active surface so that adsorption occurs. Conditions must be appropriate for the desired chemical reaction to go to completion and the reaction products to be desorbed. In many cases, the reactions can be carried out at relatively low temperatures of 300°C or less. Some catalysts may be deactivated by reactions with certain components of an effluent; lead is notorious for its ability to "poison" many catalysts.

9.2 FOSSIL FUEL POWER PLANTS

The greatest anthropogenic sources of SO_2 in most industrialized nations are fossil-fuel-burning power plants, and more than half of the SO_2 emissions in the United States have this origin. Thus we are motivated to begin our survey of major sources by considering them. As indicated by the emission factors in Section 8.1, contaminants emitted in greatest amounts from oil and coal-burning power plants are the sulfur oxides, nitrogen oxides, and particulate matter.[2] Carbon dioxide can also be included in the list, insofar as it may have undesirable effects on the earth's energy balance, but we shall here be concerned only with the other pollutants which have a demonstrable adverse effect near the source. For gas-burning plants, SO_x and particulate emissions are usually quite small, and the emission of NO_x is considered the major problem. We shall discuss methods for reducing NO_x formation in Chapter 10 when we take up the subject of photochemical smog. Here we are limiting our consideration to the control of particulates and SO_x.

To a degree, particulate matter in the effluent can be controlled by burning oil or coal with a low fly ash content. However, the current trend toward greater use of pulverized coal goes counter to this policy. Thus more effective collection devices are required to clean the effluent, a task usually accomplished by one or more electrostatic precipitators.

Removal of SO_x from the exhaust is another matter, because there is no proven method to accomplish this for very large sources. Use of low-sulfur oil or coal has been the way usually chosen to minimize the amount of SO_x formed during combustion, but supplies of such fuels are limited. Methods are available for desulfurizing oil by catalyzed hydrogenation processes, in which the sulfur is removed as hydrogen sulfide for subsequent recovery as elemental sulfur or sulfuric acid; and we can anticipate that if public concern over environmental pollution remains active, the added costs for such processing may be found acceptable. However, removal of sulfur from coal is less straightforward, since the organic sulfur—which constitutes about half of the total content—cannot be removed by known processes except by gasifying the coal at high temperatures.[2] Much effort in research and development is now being invested in seeking practical methods to accomplish this, for the present techniques are capable of gasifying only small quantities of coal.

A second line of attack on the SO_x problem has been to seek ways to clean the gases from the exhaust. One of the numerous possibilities being explored is the dry limestone process in which the effluent is filtered through a bed of pulverized limestone or dolomite (calcium-magnesium carbonate), and the SO_x is removed as it reacts to replace the carbonate by sulfate.[3] It is hoped that the $CaSO_4$ can then be processed to yield sulfuric acid for commercial sale.

There are two consequences that might ensue from the widespread adoption of such methods; one is environmental, and the other economic. We shall mention them here to remind the reader that our attention to air pollution should not cause us to lose sight of the total picture. The environmental consequence would result if a way cannot be found to utilize the sulfate content of the spent limestone, for the accumulation of waste sulfates near a power plant might very well lead to water pollution through the runoff. Sulfates are not highly soluble, so their removal from the site by the alternative method of direct discharge into a river or lake would require great quantities of water for the necessary dilution.

On the other hand, if the sulfur from the effluent of most power plants were collected and commercially sold as elemental sulfur or sulfuric acid, there would be a profound economic impact on the sulfur market. On a global basis, about 63 million Tonnes of sulfur is released each year into the atmosphere by anthropogenic sources and most of this is in the form of SO_2.[4] About 34 million Tonnes is released in the United States alone. By comparison, the annual world production of sulfur for commercial sale amounts to only 27 million Tonnes.[5] Thus widespread adoption of sulfur-collection techniques at power plants and other large sources could be expected to influence the market price of this byproduct and consequently affect the net cost of SO_x emission controls.

Some techniques for SO_2 removal have not always proven successful. At the London power plants at Battersea and Bankside a scrubber technique was introduced whereby much of the SO_2 was washed out of the effluent by a spray of water. Unfortunately, as a consequence of the resulting cooling of the flue gases, the emitted plume had little buoyancy and even on occasion sank to the ground only a short distance from the stack. Because insufficient time had elapsed for the remaining SO_2 in the plume to become well diluted by mixing with ambient air, high SO_2 concentrations were still experienced at ground level. In fact, scrubbing is ceased during prolonged inversion conditions, so the plume can penetrate the inversion layer and disperse in the overlying unstable air. Another disadvantage of this particular design is the water pollution caused by the liquid effluent from the scrubber.

Because there is no proven way economically to clean sulfur from coal or to remove SO_x from the exhaust of large plants, power plants currently under construction use tall exhaust stacks 300 m or more high to minimize ground-level concentrations.[6] Furthermore, there is a trend toward construction of giant-economy-size plants, with a large heat emission rate in the plume. The resulting plume buoyancy further enhances the effective release height of the pollutants. In some instances, such as at Chavalon, Switzerland, power plants have been constructed on mountain sides to gain a higher release height for the effluent and avoid having it trapped in deep radiational inversions that frequent the valleys.

Closely associated in principle with the pollution problem of a power plant is the contribution of contaminants from the home space heater or from any furnace fired by coal or oil. For homes, the economically advantageous solution appears to be use of cleaner fuels, for the introduction of scrubbers or other controls necessitates a fairly high capital investment and an elaborate system to operate the device; then too, there is the problem of what is to be done with the collected pollutants. For light industries, pollution control can often be accomplished with a combination of the control devices enumerated in the previous section.

9.3 COPPER, LEAD, AND ZINC SMELTERS

Ores of copper, lead, and zinc commonly have the desirable element occurring as sulfides. The metal can be won from the ore by heat treatments which oxidize the sulfur and cause its liberation as SO_2. During the early 1800's, means were sought in England to recover the SO_2 from the exhaust gases for conversion into marketable sulfuric acid. Then, as now, the main problem was the fact that often the SO_2 concentration in the flue gas was *too low* to permit profitable recovery, and the principal reason for lack of emission control has been economic, not technical. The reason is that the cost of removing SO_2 increases with the total amount of gaseous effluent that must be processed, whereas the offsetting revenues from sale of the end product (usually sulfuric acid) are determined by the amount of SO_2 alone.[7] Thus the percentage of SO_2 in the exhaust is a key parameter determining the economic viability of sulfur recovery.

In the United States, most high-purity sulfuric acid is manufactured by the *contact process* in which dry air containing SO_2 is brought into contact with a solid catalyst such as vanadium pentoxide (V_2O_5). When heated to a temperature of about 500°C, the SO_2 is rapidly oxidized by oxygen in the gas stream according to the reaction

$$2\,SO_2 + O_2 \rightarrow 2\,SO_3,$$

with the release of 189 kJ (45 kcal) of energy per mole of molecular oxygen consumed. In practice, no heat need be provided to warm the incoming gas to a temperature where the catalyzed oxidation occurs at an acceptable rate provided that the SO_2 concentration exceeds 3.5%. At higher concentrations, the energy liberated by oxidation itself is sufficient for this purpose. This criterion has an important bearing on the economics of recovery.

The SO_3 produced by the catalyzed oxidation quickly unites with water vapor in the air stream to produce sulfuric acid mist. This is then cooled to reduce the water vapor content and leave concentrated H_2SO_4. A concentration of 93% H_2SO_4 can be produced without refrigeration if the SO_2 concentration of the initial gas mixture exceeds 4%, and a higher concentration can be obtained with the necessary investment in refrigeration. The usual efficiency for removal of SO_2 from the effluent is 98.5%, but this can be increased to 99.5% by recirculation of the effluent to process it at least twice.[7] Plants are often operated with a throughput exceeding design specifications, so this conversion efficiency may not be met, and the effluent of the acid plant could itself cause an air pollution problem.

Various methods of smelting copper, lead, and zinc ores can be employed and the gaseous products differ. Even for a given metal, several processes and their variations will be found throughout the world. We will consider only the most prevalent. Copper smelters, of the three, emit the greatest quantity of SO_x, amounting to about 77% of the total.[8]

In smelting *copper*, as depicted in Fig. 9.5, the ore is first roasted in a furnace to eliminate some of the sulfur content as SO_2, without forming any soluble compounds of the iron present. Typically, the effluent contains SO_2 at a concentration of about 8%. The resulting molten material, known as *calcine*, is a mixture of sulfides and oxides of copper and iron, worthless rock or *gangue*, and

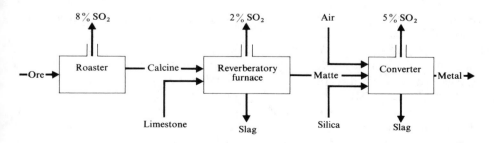

Fig. 9.5   Major steps in smelting copper.

other nonvolatile impurities. The calcine is fed into a reverberatory furnace together with fluxes such as limestone or silica, chosen so that they combine with metallic oxides and other impurities composing the gangue to form a slag which floats on top of the molten copper and iron sulfides; the latter, known as the *matte*, can be drawn off through a tap hole in the bottom of the furnace. During the heating process, the gaseous effluent consists of only 1–2% SO_2, a concentration inherent in the process.

The final stage is a batch process in a converter in which the matte is heated while introducing a stream of air. Iron and sulfur have a greater affinity for oxygen than does copper, so after a sufficient period the iron is converted to an oxide, much of the sulfur to SO_2, and the copper to copper sulfide. Then introduction of silica causes the iron oxide to form a slag which is skimmed off. With a continued supply of air, some of the remaining copper sulfide is oxidized, and the copper oxide reacts with the sulfide to produce metallic copper and SO_2. Because of the large quantities of air introduced, the SO_2 concentration in the effluent is only about 5%; ordinarily even this would be susceptible to economic recovery, but not in the case of the converter because it involves a batch operation which is difficult to match with the contact process. A more regular output could be realized by use of several converters operating on a staggered schedule.

Many *lead* minerals are known, but galena (PbS) is the source of virtually all lead obtained for commercial sale. It is first roasted, as an air current is directed onto the bed of ore, and the SO_2 so liberated may be found in the exhaust at a "low" concentration of only 1–3% owing to the presence of excess air. Recirculation of the exhaust back through the roasting ovens can raise this to 4% or more, which is suitable as feed for a sulfuric acid plant. Some of the lead oxidized to PbO combines with galena to yield metallic lead and more SO_2. Further treatment is carried out in a blast furnace to reduce the oxides of desired metals and produce liquid lead in which is dissolved copper, gold, silver, zinc, antimony, and other metals. However, comparatively little SO_2 is released in this last portion of the smelting process.

The two main sources of SO_2 emissions in *zinc* smelting are the ore roasters and the sintering machines. On a global basis, at least 90% of metallic zinc is obtained from zinc blende (ZnS). When recirculated, the air from the roasters can have an SO_2 concentration of 6–10%. The fine calcine which remains is sintered into larger pieces, so that the oxides of the metals which remain can more conveniently be moved to further heat treatments to be reduced to the metals.

Waste gases from the smelting operations we have just described contain large quantities of dust and fumes, the latter being formed by vaporization and subsequent condensation of volatile compounds such as hydrogen chloride (HCl), hydrogen fluoride (HF), and arsenic trioxide (AsO_3). These must be removed from the exhaust so that sulfuric acid of acceptable purity can be subsequently produced. Gas leaving the roasters has a temperature in excess of 600°C and generally is cooled before further treatment. Up to 90% of the mass of particulate matter can be removed by a cyclone and, if followed by an electrostatic precipitator, an additional factor of 98% or so can be eliminated. A wet scrubber can then be employed to convert SO_3 to an acid mist and remove this as well as the other fumes. In this regard, it is desirable to minimize SO_3 formation in the process preceding the scrubber, so that most of the SO_x will pass through the scrubber as SO_2 gas. It can then be removed from the relatively pure air for conversion into usable H_2SO_4.

In instances where the SO_2 content of the effluent is only 1%, such as the exhaust from some sintering ovens or from the H_2SO_4 plant, it may be desirable to remove the pollutant directly by a cyclic absorption system. For example, ammonium sulfite-bisulfite can be used as the absorbant. If the SO_2 cannot be used locally, it may be reduced to elemental sulfur for economical transport.

Other scrubbing processes are also used. For example, the smelter at Trail, British Columbia, now combines SO_2 with NH_3 to produce ammonium sulfate as fertilizer.

It should be clear from the preceding discussion that stiffening requirements on air pollution control can be expected to influence the development of metallurgical technology in the next decade.[8] One important design goal will be to decrease the volume of exhaust gas, so as to maximize the SO_2 concentration for more economical recovery.

9.4 IRON AND STEEL MILLS

Metallurgical processes have traditionally been associated with the emission of particulate matter and waste gases which are released during high-temperature treatment of the raw materials. We shall here consider another common example, the production of pig iron and steel.

A necessary ingredient for producing usable iron is coke or carbonated coal. This is formed from crushed coal by driving off in an oven the volatile constituents such as tar, benzene, and naphtha; if the effluent is controlled, these can later be condensed and recovered for sale. The coke is then removed from the oven and the still-incandescent material is quenched in a water spray and dried.

Iron is commonly obtained from the ore in a blast furnace, a steel chamber lined with a refractory material, into which iron ore, coke, and limestone are fed and reacted at high temperature with a high-velocity stream of hot air. In addition to the desired molten iron, the end products include a slag, waste gases, and particulates entrained in the air stream. The particulate matter is mainly Fe_2O_3, accompanied by silicates, Al_2O_3, and CaO. Other pollutants are emitted by the furnace heaters, which generally operate on coal or oil. A representative system of air pollution control devices for a blast furnace is shown in Fig. 9.6. In this illustration, large particles are trapped in a combination settling chamber and inertial collector, and the remainder are further treated in a scrubber and two-stage electrostatic precipitator. The dust burden can be lowered by a factor of more than 1000 by such an arrangement. Molten iron is removed from the furnace and is either directly conveyed to a steel mill or is cast into pigs for later working in a foundry.

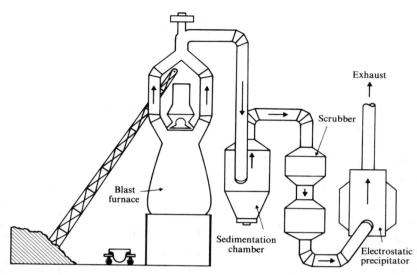

Fig. 9.6 Blast furnace and conventional system for controlling gaseous and particulate emissions.

   In the United States, steel is commonly made in open hearth furnaces, although the basic oxygen furnace is fast gaining popularity. In the open hearth process, iron, scrap metal, and limestone are melted in a hearth by a burning mixture of natural gas and air at the surface. Sometimes iron ore or another source of iron oxide is also added. Carbon in the molten mixture combines with oxygen and is carried out as CO or CO_2. The calcium-rich limestone combines with other impurities to form a slag on top, and the molten steel is allowed to drop through a tap hole into a ceramic-lined vessel below, where other materials can be added to obtain a specific composition. The waste-gas effluent with its particulate burden is generally at a high temperature of over 800°C and therefore must be cooled before it can be sent to an air pollution control device. Over three-quarters of the particulate matter is Fe_2O_3, and is accompanied by much smaller amounts of aluminum, manganese, calcium, and magnesium oxides, as well as silicates.

   The basic oxygen furnace is a variation of the above process in that a stream of oxygen is blown onto or into the molten material. Unfortunately, considerable quantities of particulate matter are released in the process; these and waste gases are gathered by a hood overhead and are led to cleansing devices where CO and other combustibles are burned and the dust is collected.

   Other steel-making equipment such as the Bessemer converters and electric arc furnaces are commonly found in older mills. But the main characteristics of the effluent are similar to those we have described above for the open hearth and basic oxygen processes.

Generally, the presence of sulfur is detrimental to the quality of steel, so low-sulfur coal is sometimes used as the basis for coke. Nevertheless, some sulfates remain and may be converted to SO_2 during the purification of iron and formation of steel. Heaters burning fossil fuels used to produce the melt are also a source of sulfurous and particulate pollutants.

9.5 PETROLEUM REFINERIES

Whereas in many communities power stations and space heaters are the major contributors to sulfurous smog, other sources may also be important. Emissions from petroleum refineries are often a cause of discontent among neighboring residents, primarily because of malodors. Pollutants emitted in the greatest amounts are CO, hydrocarbons, NO_x, and SO_x.

There are several stages of petroleum refining, and the amount of each emitted pollutant depends upon the peculiar design features of a given installation. The major operations of refining are *separation*, *conversion*, *treating*, and *blending*. Petroleum is first separated into broad fractions according to boiling temperatures as listed in Table 8.2. Depending upon the conditions of the market, it may be economically advantageous to convert a proportion of some of the fractions into other types of molecules. This is commonly achieved by catalytic cracking under heat and pressure, in which some molecules are broken up into smaller ones or combined to form larger ones. Several types of catalytic-forming techniques may be used. During separation and conversion, the products can be treated to remove sulfur compounds and other undesirable materials. Finally, the resulting compounds are blended with each other and with additives to produce the marketable product. As several of the processes are performed at high temperatures, the emissions from a refinery include not only the organic molecules released from pressure relief valves, drains, seals, and cleaning operations, but the emissions from numerous furnace heaters. There is an economic advantage for a refinery to fuel these heaters with its own cheapest product, residual oil, which often has a high sulfur content. Substantial hydrocarbon emissions come from the evaporation of petroleum products through leaks in storage tanks, and during transferring operations.

Since refineries remove sulfur-bearing compounds from many of their products, there is a temptation for the management to release the unwanted mercaptans and SO_2 into the air as the simplest means of disposal. Numerous anecdotal accounts are told of how this often seems to be done in the dead of night when most of the neighbors and the air pollution control authorities are asleep. Unfortunately, this is often the time of greatest atmospheric stability.

Because many of the organic sulfur compounds have extremely low odor recognition thresholds, their effective control requires that great care must be exercised to prevent leaks, and that unwanted gases be treated by chemical reactors.

9.6 PULP AND PAPER MILLS

Another source of sulfur-bearing contaminants are mills for processing pulp and paper; here odorants and particulate matter are usually the chief concern. To render wood into pulp, the desired fibers of cellulose must be separated from a surrounding amorphous matrix of polymeric material. One polymer known as hemicellulose contains some glucose but is mainly composed of other natural sugars, and their composition differs according to the genus of the tree. Another polymer is lignin, the main constituent in the matrix between the cellulose cell walls. This molecule is a three-dimensional network based on building blocks of the phenylpropane groups: benzene rings with three-carbon side chains. The most important method from an economic standpoint for removing lignin and hemicellulose in chemical pulping is the kraft process. In this, wood chips are cooked in an aqueous solution of sodium hydroxide and sodium sulfide at high pressure, and the dissolved lignin and hemicellulose in the spent cooking liquor is drawn off. The pulp continues on to several stages of bleaching and washing and is made into paper.

A significant feature of the kraft process is that economical operation requires the liquor to be treated to recover the original reagents. In fact this reprocessing is a major part of the mill's chemical operations. First the liquor is concentrated by evaporation at high temperature, and then the organic components are burned off. The remaining inorganics, mainly sodium sulfide and sodium carbonate, are then reacted with calcium hydroxide in a causterizer to produce sodium hydroxide and calcium carbonate; the former is recycled and the latter further treated to yield calcium hydroxide.

Unfortunately, when the hot liquors are first released from the wood digester, and again when the liquor is evaporated and the organics are burned, small quantities of hydrogen sulfide and mercaptans—formed by the demethylation of cellulose and reaction with the sulfide during digestion—are released into the air, and they give the odors that are commonly found around kraft mills. Some control of these can be achieved by collecting the gases and incinerating them in a lime kiln, which converts H_2S and mercaptans into SO_2.

Fallout of large particles released from bark burners and the recovery furnaces may be unfavorably received by neighbors, but most of this would be controlled by use of inertial collectors or electrostatic precipitators. The effluent of the recovery furnace also contains sodium sulfate ash which after collection can be sent back for reuse in the cooking liquor.

As is the case for petroleum refineries, the control of malodors remains a technical challenge due to the fact that organic sulfur compounds have such a low recognition threshold. Not much material needs to escape to be noticed.

9.7 INCINERATORS

Disposal of solid waste has become a major problem in many communities.[9] A city of 1 million people in the United States must find a way for disposing of

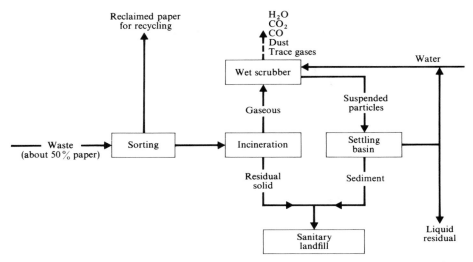

Fig. 9.7   Example of a proposal for disposal of solid residues in the New York City region.

about 20,000 Tonnes of refuse each week. The cost to the City of New York, for example, is surpassed only by outlays for education and welfare. Landfill and incineration are the traditional means of disposal, and even the latter is not a final solution, because not all refuse is combustible. But the solid refuse that remains when used for landfill occupies one-twelfth the space that would be required for the raw refuse. For this reason, incineration has been widely practiced in highly congested areas. An elementary proposal for waste disposal in the New York region is illustrated in Fig. 9.7. An additional advantage is gained by incineration, because much of the sensible heat can be used to produce steam for generating electricity. At present, an incinerator which processes 6000 Tonnes a day should be capable of generating 150 MW of electrical power.

The solid waste and air pollution problems are not separable. Incineration produces numerous toxic gases and large quantities of particulates. In New York City, approximately 17,000 private incinerators in apartment houses, hospitals, schools, etc., and seven municipal incinerators are believed to account for about 30% of the total particulate emissions. These uncontrolled sources are responsible for much of the soot whose fallout is responsible for a marked degradation of the environment and exacts a considerable economic toll. Particulate emissions of all sizes when uncontrolled commonly amount to 12–20 kg per Tonne of refuse.

Constituents of the refuse delivered to municipal incinerators depend, of course, upon a number of factors determined by the makeup of the community. Not surprising perhaps is the fact that the major variable is the moisture content, which is related to the weather. A representative breakdown of the physical composition and chemical constituents of refuse processed at a suburban municipal incinerator are given in Tables 9.1 and 9.2. The calorific value of the refuse characterized in these tables is about 11×10^6 J/kg (4600 BTU/lb) for organic

TABLE 9.1

Physical composition of typical municipal refuse

	%
Corrugated boxboard	6
Newspaper	13
Magazines, books	2
All other paper	26
Plastic shapes	2
Plastic film	1
Rubber, leather	1
Textiles	3
Wood	3
Food waste	10
Grass, leaves	5
Dirt, under 1 inch	6
Glass, ceramics, stones	12
Metal	10
	100

*E. R. Kaiser and A. A. Carotti, "Municipal Incineration of Refuse with 2% and 4% Additions of Four Plastics: Polyethylene, Polystyrene, Polyurethane, and Polyvinyl Chloride," A Report to the Society of the Plastics Industry, Department of Chemical Engineering, School of Engineering and Science, New York University, June, 1971.

TABLE 9.2

Chemical analysis of typical municipal refuse

Component	%
Moisture	25.00
Carbon	25.54
Hydrogen	3.34
Oxygen	22.00
Nitrogen	0.52
Sulfur	0.10
Chlorine	0.50
"Inerts" (glass, metal, ash)	23.00
	100.00

*E. R. Kaiser and A. A. Carotti, "Municipal Incineration of Refuse with 2% and 4% Additions of Four Plastics: Polyethylene, Polystyrene, Polyurethane, and Polyvinyl Chloride," A Report to the Society of the Plastics Industry, Department of Chemical Engineering, School of Engineering and Science, New York University, June, 1971.

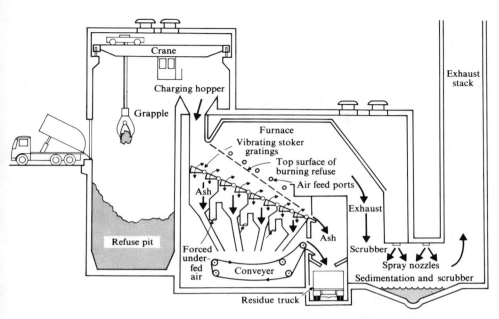

Fig. 9.8 Sectional view of a contemporary suburban municipal incinerator.

matter alone, with a possible additional 3.5×10^5 J/kg (150 BTU/lb) from the partial oxidation of metals.[10]

Figure 9.8 shows a representative municipal incinerator with elementary—and not particularly efficient—emission controls. Refuse dropped into the charging hopper is conveyed through the furnace by a stoker consisting of individual gratings which can be rotated upward in a sequence that pushes the burning material along.

The major gaseous pollutants emitted from an uncontrolled effluent are hydrogen chloride (*ca* 500 ppm), NO_x (50–100 ppm), SO_x (50–130 ppm), and organic acids (10–150 ppm when calculated as acetic acid, CH_3COOH). The NO_x and SO_x emissions total about 2 kg per Tonne of refuse. Fly ash contains mainly oxides of aluminum, silicon, and titanium, but with high concentrations of oxides of calcium, iron, magnesium, sodium and up to 1% of lead.[10]

Hydrogen chloride makes the effluent distinctly acidic, and about 60–65% of the chlorine constituent of the refuse is liberated as this gas. Most organic fractions of rubbish are found to contain chlorine, probably much of it as NaCl. Less than half of the chlorine is bound in chlorinated plastics such as polyvinyl chloride. Most plastics consist mainly of carbon and hydrogen, although some contain nitrogen and oxygen in sizable amounts. The chlorine-bearing vinyls, however, contribute to the evolution of HCl. Chlorine and phosgene gases are not known to be emitted. If the plastic component of the refuse is nearly doubled by addition of 2% polyvinyl chloride, the HCl concentration in the effluent rises to about 2000 ppm.[10]

A wet scrubber can reduce the HCl concentration by 85%, and more efficient removal should be obtainable by addition of soda ash (sodium carbonate) or sodium hydroxide to the scrubber water. Electrostatic precipitators and inertial collectors such as a cyclone might be acceptable means for controlling emissions in municipal incinerators, but they are likely to be far too costly for individual private incinerators. In several countries, municipal incinerators now operate with pollution controls and produce small amounts of electricity.

In the fall of 1970 enforcement of pollution control regulations was initiated in New York City to upgrade the performance of private incinerators. Three options were available to owners of incinerators:

1. Improve combustion in existing incinerators and add pollution control devices, including a wet scrubber.

2. Abandon the incinerator and install a mechanical refuse compactor, reducing the refuse volume to one-quarter of the original, with refuse to be hauled away by the city collectors.

3. For apartments with fewer than 40 living units, abandon the incinerator and utilize regular refuse collection.

The latter two options are in effect subsidized by the city, since such refuse collection is provided without direct cost to the apartment house management.

Over 30% of the city's incinerators have been upgraded or eliminated. Poor reliability of the first installed scrubbers, arising from corrosion problems and the difficulty of finding competent maintenance personnel, have encouraged apartment managers to favor the compactor by a margin of $2\frac{1}{2}$ to 1. If that continues, the air pollution problem will be greatly reduced, but the solids disposal problem will be aggravated.

SUMMARY

Devices for controlling particulate emissions exploit the physical properties of the pollutants—size, inertia, gravitational sedimentation, and the possibility of retaining an electric charge—in order to separate them from a gas stream for collection. Devices for controlling a gaseous pollutant may depend upon its solubility in a liquid, the electrical affinity for adsorption to a surface, or its chemical properties. No device can control all sizes and types of particulates, or all types of toxic gases, so a series of devices is often utilized. A continuously operating system is usually preferred from the standpoint of economy. Disposal of the collected pollutant may itself be a problem unless it is converted into a saleable material. If the known large coal reserves are exploited, a means must be found either to remove sulfur before combustion or to scrub SO_2 from the effluent. Practical devices exist for removing SO_x when present at high concentrations in an effluent with a low rate of flow, but the economics appear not to justify adaptations of these techniques to the much lower concentrations found in the exhaust of fossil fuel power plants. Removal of SO_2 from power plant exhaust remains a challenge to industry.

NOTES

1. H. J. White, *Industrial Electrostatic Precipitation*, Reading, Mass: Addison-Wesley, 1963.

2. A Levy, E. L. Merryman, and W. T. Reid, "Mechanisms of Formation of Sulfur Oxides in Combustion," *Environ. Sci. and Tech.*, **4**, 653 (1970).

3. A. M. Squires, "Clean Power from Coal," *Science*, **169**, 821 (1970).

4. E. Robinson and R. C. Robbins, "Gaseous Sulfur Pollutants from Urban and Natural Sources," *J. Air Poll. Control Assoc.*, **20**, 233 (1970).

5. *Industry Week*, August 17 (1970), p. 3.

6. A. V. Slack, G. G. McGlamery, and H. L. Falkenberry, "Economic Factors in Recovery of Sulfur Dioxide from Power Plant Stack Gas," *J. Air Poll. Control Assoc.*, **21**, 9 (1971).

7. K. T. Semrau, "Control of Sulfur Oxide Emissions from Primary Copper, Lead, and Zinc Smelters—A Critical Review," *J. Air Poll. Control Assoc.*, **21**, 185 (1971); "Two New Processes for Recovery of Sulfur Oxides from Smelter Gases," *Eng. Mining J.*, **172**, 115 (1971).

8. K. T. Semrau, "Sulfur Oxides Control and Metallurgical Technology," *J. Metals*, March (1971).

9. For recent reviews see the *Proceedings of the National Incinerator Conference*, American Society of Mechanical Engineers, 1972.

10. E. R. Kaiser and A. A. Carotti, "Municipal Incineration of Refuse with 2% and 4% Additions of Four Plastics: Polyethylene, Polystyrene, Polyurethane, and Polyvinyl Chloride," Report to the Society of the Plastics Industry, New York, 1971.

FOR FURTHER READING

W. STRAUSS, Ed.   *Air Pollution Control, Part I* and *Part II*, Environmental Science and Technology Series, R. L. Metcalf and J. Pitts, Jr., Eds., New York: Wiley-Interscience, 1970.

H. J. WHITE   *Industrial Electrostatic Precipitation*, Reading, Mass: Addison-Wesley, 1963.

K. T. SEMRAU   "Sulfur Oxides Control and Metallurgical Technology," *J. Metals*, March, 1971: "Control of Sulfur Oxide Emissions from Primary Copper, Lead, and Zinc Smelters—A Critical Review," *J. Air Poll. Control Assoc.*, **21**, 185 (1971).

J. M. CONNOR   "Economics of Sulfuric Acid Manufacture," *Chem. Eng. Prog.*, **64**, 59 (1968).

For reviews covering incinerator technology, see recent issues of the *Proceedings of the National Incinerator Conference*, American Society of Mechanical Engineers; these conferences are usually held in even numbered years.

QUESTIONS

1. What physical principles have been used as the basis for air pollution control devices?

2. In what respects must a control device be designed to match processing equipment it follows?

3. If you had responsibility for defining operating specifications for a device to collect particles from a fertilizer processing plant, what parameters would you specify?

4. Why is the SO_2 concentration in an effluent a useful measure of the economic feasibility of recovery of that pollutant?

5. Indicate some approaches which should be explored to minimize SO_2 emissions from fossil fuel power plants. What are the advantages and disadvantages of the approaches you suggest?

6. Would a settling chamber usually be preceded or followed by filters in a particle collection system in which both are to be employed? Explain your answer.

7. How does a cyclone work? For what size particles is it useful?

8. How can a scrubber make use of the venturi tube?

9. What is the predominant particulate constituent from a steel mill? What control devices might be appropriate for removing it from the exhaust? Can you think of a method which has not been mentioned in the text?

10. Estimate the mass of combustible solid waste that you discard each day, directly and indirectly. If, for simplicity, it is assumed that this is all carbon, how many liters of CO_2 (at 1000 mb and 0°C) would be liberated if it were all incinerated?

11. Collection hoods of a foundry which forms aluminum castings entrap air with hydrochloric acid fume; what type of control device would be most appropriate to cleanse the effluent? Explain.

12. Suppose that an emission inventory for a county shows that SO_2 emissions from a power plant constitute 65% of the total emissions of that gas from all sources. In what respects may this percentage be deceptive, indicating either a greater or smaller than actual relative contribution to the pollution level experienced by a person in the neighborhood of the plant?

PROBLEMS

1. A smelter emits SO_2 at a concentration of 5% by volume and in a day emits 300 Tonnes of the gas. How much more (or less) SO_2 is released by a coal-burning power plant emitting 10^6 liters/sec of gases having an SO_2 concentration of 400 ppm?

2. If all the SO_2 from the plants mentioned in the preceding problem is cleaned from the exhausts, and if the processing cost is assumed to be proportional to the

volume of gas which is treated, what for the power plant is the ratio of the cost of gas cleaning to the value of the sulfur collected? Assume that the cost of treating a cubic meter of effluent is the same for the two plants, and that the smelter breaks even.

3. Suppose that a lead smelter produces 200 T/day of lead and does not control its pollutant emissions. Estimate the amount of SO_2 released.

4. If a municipal incinerator processing 6000 T/day of refuse is capable of generating 150 MW of electrical power, what is the efficiency of the process? Use appropriate data given in the text.

5. Is the incinerator in the previous problem more or less efficient than a fossil fuel power plant producing electrical power at the same rate? Give a reason why this is so.

CHAPTER 10

PHOTOCHEMICAL SMOG

Whereas the effects of sulfurous smog have been known for over 600 years, photochemical smog has only recently appeared in the urban environment. The characteristic symptoms of a brownish coloration of the atmosphere, reduced visibility, plant damage, eye irritation, and respiratory distress first became prominent during the middle 1940's in Los Angeles, California. But it was not until the early 1950's that research conducted primarily by A. J. Haagen-Smit and colleagues succeeded in identifying the phenomenon as arising from photochemical reactions in the lower atmosphere. The prime ingredients for these reactions were found to be hydrocarbons and nitrogen oxides exhausted by automobiles, and some stationary sources.

Local features of heavy automobile traffic, frequent and intense sunlight, geography, and meteorology reinforce to plague the Los Angeles area with some of the most concentrated and chronic photochemical smog. More recently the symptoms have appeared in many other metropolitan regions in both the Northern and Southern Hemispheres. Its presence may even have come to represent a status symbol for a society in an advanced state of technology! For the formation of substantial quantities of nitrogen oxides depends upon combustion at higher temperatures than can be achieved in open fires. The causes and possible methods of curtailing photochemical smog have been the focal point of considerable public interest. Photochemical smog has spread much more rapidly in urban regions than the growth of the population would suggest. This has occurred for two reasons:

1. Emissions of the primary pollutants have increased more rapidly than the population owing to the greater individual use of motor vehicles and electrical power.

2. The downwind area covered by pollution increases more rapidly than the emission rate of an area source (as we have discussed in Section 7.8).

Although for more than a score of years research efforts have been directed toward elucidation of the chemical process of photochemical smog, the phenomenon is understood only in broad terms and in a few details. The overall effect is suggested in Fig. 10.1. The atmosphere initially contains dilute concentrations of gaseous organic molecules (predominantly hydrocarbons) and nitrogen oxides, which consist of both nitric oxide NO and nitrogen dioxide NO_2, and are known

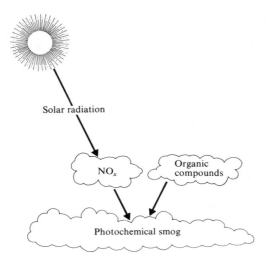

Fig. 10.1 Ingredients of photochemical smog.

collectively as NO_x. The energy of solar radiation triggers an initial reaction in NO_2, releasing products which can combine with the organic molecules. A series of complex reactions follow between NO_x, organic compounds, and ambient oxygen. One well-known secondary pollutant produced by these reactions is ozone, O_3. Another is an aerosol which restricts visibility. The totality of gaseous and particulate primary and secondary pollutants is known as photochemical smog.

One characteristic of this smog is that it is an *oxidizing atmosphere*. Molecules such as ozone readily give up an oxygen atom in chemical reactions with their surroundings. The resulting oxidation is responsible for many detrimental effects, such as the cracking of stressed rubber. In contradistinction, the sulfurous smog discussed in the previous chapter is a *reducing atmosphere*. The SO_2 and SO_3 take up oxygen from the surroundings to be converted into products such as sulfuric acid. (A chemist will recognize that the terms "oxidizing" agent and "reducing" agent are relative terms. Although sulfurous smog is a reducing atmosphere in the sense just mentioned, it can also be an oxidizing agent. For example, SO_3 and H_2SO_4 oxidize metals, forming sulfates.)

A perspective

Important photochemical processes are not found solely in the lower atmosphere. In Section 4.3 we discussed the formation of ozone in the stratosphere by photochemical reactions and the role played by this process in the absorption of ultraviolet solar radiation. It was pointed out that circulation in the upper atmosphere carries some ozone downward into the troposphere, with the result that the natural ambient concentration at lower levels is about 0.02 ppm. Within the stratosphere, above the altitude where the maximum concentration of ozone is found, other

photochemical reactions take place. These result in formation of low concentrations of N_2O, NO, and NO_2, whose presence is revealed by the absorption spectrum of the atmosphere (Fig. 4.5).

Reactions between the nitrogen oxides and oxygen within the upper atmosphere have received considerable attention because of their relation to airglow
phenomena. But within the lower atmosphere the ambient concentration from
downward mixing is negligibly small. Only anthropogenic sources emit nitrogen
oxides in sufficient quantity to increase ambient concentration to the point where
photochemical smog is possible.

10.1 FORMATION OF NITROGEN OXIDES

Nitrogen oxides are produced by the oxidation of atmospheric nitrogen during
combustion at high temperatures. Common sources thus include automobile
engines and furnaces burning fossil fuels. This "fixation" of nitrogen at high
temperatures occurs by the overall conversion of molecular nitrogen and oxygen
to nitric oxide:

$$N_2 + O_2 = 2NO. \tag{10.1}$$

However, this equation alone does not describe the chemical reactions, for its rate
is too slow to account for the observed production of nitric oxide. The process
actually depends upon the dissociation of nitrogen and oxygen molecules during
combustion; and the formation of NO occurs primarily through two simultaneous
reactions known as the Zeldovich mechanism:[1]

$$N + O_2 = NO + O$$
$$O + N_2 = NO + N. \tag{10.2}$$

The production of NO depends upon the concentration of atomic nitrogen and
oxygen and therefore ultimately upon the flame temperature in combustion; it is
insensitive to the composition of the fuel, as long as the temperature is sufficiently
high, above about 1000°C. Below this temperature, reactions involving the formation of N_2O become important.

If the combustion effluent is slowly cooled as it is emitted, NO converts back
into O_2 and N_2. However, in the usual combustion apparatus, the thermal energy
of the gaseous products is quickly converted into useful work, thereby causing
their rapid cooling. In this case, the NO does not have time to decompose and
remains "fixed" as NO at lower temperatures.

NO_x control

The most important contributers of NO_x to atmospheric pollution are the
sources which utilize high combustion temperatures, for example, motor vehicles
and fossil fuel power plants. Methods for decreasing nitrogen oxide production
have concentrated mainly upon changing combustion processes so as to lower the
flame temperature. Some automobile emission control systems achieve this by
recirculating a portion of the exhaust gas back into the intake. This dilutes the

mixture of fuel and air in the combustion chamber and provides more of a heat sink to reduce the temperature of the burning gases. Another technique concentrates on restricting the amount of oxygen available for NO formation by using a rich mixture of fuel to air. However, a disadvantage with this is that the hydrocarbons are incompletely burned.

Both techniques can be employed in modern power plants which burn gas or oil. Fuel is sprayed into the hot gases in the combustion chamber, and through turbulence it is mixed with air introduced through neighboring ports. Combustion occurs in regions in which the mixture is near stoichiometry, with a peak flame temperature nearly as high as would be expected for an adiabatic process (2000°C); the combustion products, losing energy to the surrounding gas, within 0.02–0.05 sec, cool below 1500°C, where NO formation is slow. Continued mixing within the chamber sees any remaining fuel burn at lower temperatures as turbulence produces small regions of approximate stoichiometry. Reductions of 50–80% in NO production can be achieved by a combination of the two techniques mentioned previously: (1) the temperature of primary combustion can be reduced by adding recirculated combustion products as inert components, which absorb a portion of the energy released, and (2) insufficient oxygen is provided for the primary combustion, so there is a reduction in the proportion of fuel which is burned at the peak flame temperature, decreasing the supply of atomic nitrogen and oxygen available for NO formation.[2] Excess air is then introduced, so that the remaining fuel is oxidized a short time later at lower temperatures. Some systems incorporate two separate combustion chambers for this process, a technique known as *two-stage combustion*.

Atmospheric oxidation of NO

When the effluent is cooled rapidly after combustion, with little oxygen present, the NO concentration accounts for practically all of the NO_x emissions and is identical to the concentration formed in the combustion chamber. Very little NO_2 is emitted. Perhaps 10% of the NO_x from power plants and only 2–3% of the NO_x from automobiles is NO_2. For automobiles and power plants without exhaust controls, the concentration of NO_x in the effluent is roughly 1000 ppm, but this concentration is highly variable, depending upon design features of the apparatus used and its method of operation. Some well-controlled power plants with two-stage combustion may have an emission concentration of only 100 ppm or less.

Once the effluent begins to mix with oxygen, a second process commences to play a role—the oxidation of NO. This reaction can be summarized by the overall equation:[1]

$$2NO + O_2 = 2NO_2. \tag{10.3}$$

Several reactions contribute to this overall process, and if equilibrium could be established, most of the NO would be converted to NO_2. As is suggested by Eq. (10.3), whose initiation necessitates a simultaneous collision between three molecules, the reaction speed increases rapidly with both NO and O_2 concentration. At normal ambient temperatures and O_2 concentrations, the time required for oxidation of half of the NO at an initial concentration of 500 ppm is about 10 minutes. In practice, the low oxygen concentration in the plume immediately following combustion greatly retards oxidation. Then as the plume mixes with ambient air, the oxygen concentration builds up, but by that time the NO concentration has decreased, so the net result is that oxidation proceeds only very slowly. According to Eq. (10.3), as the NO concentration decreases from, say, 500 ppm to 50 ppm, the rate of oxidation for a given oxygen concentration decreases by a factor of 100.

Once NO has mixed with the atmosphere, it continues to oxidize slowly. Poor ventilation owing to meteorological conditions may discourage dispersal of the NO sufficiently to allow oxidation to proceed at a measurable rate. In polluted urban regions, the resulting concentration of NO_2 may approach 0.1 ppm, or even higher in exceptional circumstances. The presence of NO_2 in the atmosphere establishes conditions which can lead to photochemical smog.

10.2 GENESIS OF SMOG

Nitrogen dioxide strongly absorbs radiation with wavelengths between 0.60 and 0.38 micron; as a result of its preferential absorption of short wavelengths it has a brownish color. By contrast, NO is colorless. The component of solar radiation with wavelengths less than about 0.4 micron causes molecular dissociation of NO_2, which can be described by the overall reaction:[1]

$$NO_2 + E = NO + O \qquad (10.4)$$

The resulting atomic oxygen quickly combines with molecular oxygen in a three-body reaction identical to the one that is important for the production of ozone in the stratosphere:

$$O + O_2 + M = O_3 + M. \qquad (10.5)$$

Here M represents any third molecule which is available to carry off excess energy. Both of these reactions are fast. In full sunlight the first reaction Eq. (10.4) acting alone would deplete half of the NO_2 within 2 min. Owing to the high ambient con-

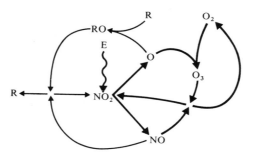

Fig. 10.2 Fast reactions (heavy arrows) and slow reactions (light arrows) in photochemical smog. The symbol E represents solar radiation.

centration of O_2, the second reaction is also rapid, yielding a half-life of about 13 μsec for atomic oxygen. But this is not all that happens, because ozone from the second reaction rapidly oxidizes NO:

$$NO + O_3 = NO_2 + O_2. \tag{10.6}$$

The NO which participates in Eq. (10.6) of course includes not only what is produced by reaction (10.4) but also the unoxidized NO remaining in the air after its emission from combustion sources. With ambient concentrations of O_3 and NO that are each 0.1 ppm, the half-life of NO from this reaction alone is 0.35 min.

However, none of the above reactions operates alone, for the three complete a closed sequence of reactions as indicated by the heavy arrows in Fig. 10.2. So they would proceed until an equilibrium is established, fixing the relative concentrations of NO_2, NO, and O_3.

But calculations for the expected concentrations of O_3 fail to account for the large concentrations measured in the photochemical smog of Los Angeles. For example, with a concentration of NO_2 amounting to 0.1 ppm, the predicted ozone concentration should be 0.03 ppm; whereas the observed ambient concentration is more nearly a factor of 10 higher.[3] Therefore the nitrogen oxide–ozone cycle is not by itself sufficient to explain the formation of photochemical smog.

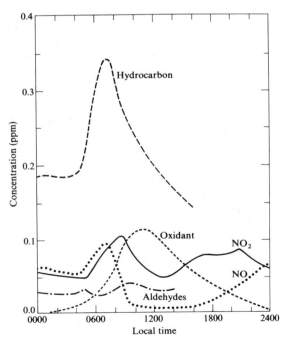

Fig. 10.3 Evolution of photochemical smog as indicated by average levels of pollution in downtown Los Angeles. Data for hydrocarbons, aldehydes, and oxidant (essentially all ozone) are from 1953–1954; nitrogen oxides from 1958. (Philip A. Leighton, *Photochemistry of Air Pollution*, New York: Academic Press, 1961.)

In fact, the fast reactions are supplemented by a second process involving the participation of organic molecules, the predominant types being hydrocarbons. In most cities, the main source of these hydrocarbons is the exhaust from automobiles. Hydrocarbons, indicated in Fig. 10.2 by the symbol R, compete for the free oxygen released by photodecomposition of NO_2. If the resulting oxidized hydrocarbon RO competes with O_3 for the NO, and the reaction yields NO_2 as an end product, both O_3 and NO_2 concentrations will build up at the expense of NO. The reactions envisioned for typical hydrocarbons are much slower than the nitrogen dioxide-ozone reactions, so only a slow increase in NO_2 and O_3 concentrations should be observed.

Figure 10.3 shows that this is indeed the case for the observed development of smog in Los Angeles. This figure illustrates the importance of the dynamical aspect of photochemical smog. Early in the morning as commuter traffic commences, the ambient concentrations of NO and hydrocarbons begin to rise. Shortly after, upon sunrise, the smaller amount of NO_2 begins to suffer photodissociation. This provides atomic oxygen for formation of ozone and oxygenated hydrocarbons such as aldehydes; and the O_3 concentration (indicated approximately by the curve labeled "oxidant") initially keeps pace with the increasing NO concentration. However at about 0700 hours the effect of the slow increase of RO becomes apparent as it begins to combine appreciably with NO to accelerate the NO_2 build-up, while reducing the NO concentration. The higher NO_2 concentration and more intense sunlight accelerate production of atomic oxygen by dissociation, and the higher concentration of atomic oxygen leads to a very rapid increase in O_3 and other oxidants at about 0800 hours. When NO is substantially depleted, O_3 builds up quickly. The strongly oxidizing atmosphere is meanwhile converting hydrocarbons into other types of organic molecules, so the hydrocarbon concentration decreases.

As the smog matures, hydrocarbons and NO_2 are finally removed by side reactions which yield other secondary pollutants such as aerosols.[4] Not much is known about these aerosols other than that they have a high nitrate content and an organic constituent. About 95% of the particles have a radius less than 0.5 micron and so are effective in scattering light. They are a major cause of the reduction of visibility. Nitrates can also be formed by reactions between NO_2 and sea salt aerosols of maritime origin.[5] The reaction with NaCl releases gaseous HCl as an end product.

Ozone does not persist at high concentrations in the air. It reacts with chemicals in the atmosphere, the ground surface, and especially plant surfaces; thus toward the end of the day, as the intensity of solar radiation diminishes, the ozone concentration also decreases.

It is often said that the NO_2 is the "catalyst" for the photochemical reactions and the hydrocarbons are the fuel. This is true to an extent, for it is the atomic oxygen from the dissociation of NO_2 that appears to trigger the sequence of reactions with the hydrocarbons. Once started, the reactions are sustained until the charge of contaminants—including much of the NO_x—is used up.

Hydrocarbons therefore perform a key function in the evolution of photochemical smog. It has also come to be realized that the *oxidized hydrocarbons* (represented by RO in Fig. 10.2) are responsible for many of the adverse effects of smog, such as eye irritation and damage to some vegetation. We shall take up these aspects in the next section. To clarify the relationship between the structure of various hydrocarbons and their relative importance in smog, it may be helpful for the reader to review first some terminology and features of organic chemistry summarized in Appendix G. The nomenclature of the following section can then be readily understood.

Recently it has been found that the pollutant carbon monoxide also plays a role in the evolution of photochemical smog. It does this through a series of reactions whose net effect is to convert CO, NO, and O_2 into CO_2 and NO_2; thus oxidation of NO is accelerated. The importance of this mechanism is debated; we shall defer a detailed discussion until Section 10–7, when we discuss carbon monoxide and its adverse effects.

The role of hydrocarbons

Some hydrocarbons are more reactive than others, and any rational scheme for smog control must recognize this. To convey an appreciation of why this is so, we will outline below the reactions which are believed to be most important in smog formation. The relevance of some of these is speculative, even though laboratory experiments indicate that the reactions occur and that they have some characteristics which are similar to those found in community air.

Let us consider in more detail how gaseous hydrocarbons in the air commence their active role. The initial step, once free oxygen is present, appears to be the formation of very reactive, oxygen-bearing free radicals:

$$R + O \rightarrow R'O \cdot + \text{products such as } R'' \cdot, \tag{10.7}$$

where R' and R'' represent hydrocarbon groups which may differ from R. Numerous reactions of this type are possible, depending upon the structure of the hydrocarbon R.

If R is an *olefin*, a hydrocarbon with one or more carbon-carbon double bonds (such as ethylene, $CH_2{=}CH_2$), we could also represent it by the formula $R^1{=}R^2$, where R^1 and R^2 are each hydrocarbon groups, though perhaps different ones. Oxygen may easily be added by breaking one of the double bonds and pairing one of the carbon electrons with one oxygen electron. The new molecule, indicated in Eq. (10.7) as R'O would in this case have the formula $R^1{-}R^2{-}O\cdot$, or possibly $O{-}R^1{-}R^2\cdot$ if R^1 and R^2 differ. This new molecule has an unshared valence electron, so it is a free radical, as denoted by the dot. For the specific example of ethylene, we would have

$$\begin{array}{c} \text{H} \\ \vert \\ \overset{\text{H}}{\underset{\text{H}}{\diagdown}} \text{C} {=} \text{C} \overset{\diagup\text{H}}{\underset{\diagdown\text{H}}{}} + \text{O} \longrightarrow \text{H} - \overset{\cdot}{\underset{\vert}{\text{C}}} - \overset{\vert}{\underset{\vert}{\text{C}}} - \text{O}. \\ \text{H} \quad \text{H} \end{array}$$

The ease with which the double bond can be oxidized makes olefins a very reactive component of smog.

A second reaction, not illustrated in Fig. 10.3, is the oxidation of hydrocarbons by ozone:

$$R + O_3 \rightarrow R'O\cdot + \text{products such as } R''\cdot. \tag{10.8}$$

Both Eqs. (10.7) and (10.8) produce free radicals in the atmosphere, the presence
of which is characteristic of photochemical smog. They also yield oxygenated
hydrocarbons such as the aldehydes, formaldehyde ($H-\overset{\overset{\displaystyle O}{\|}}{C}-H$) and acrolein
($CH_2=CH-\overset{\overset{\displaystyle O}{\|}}{C}-H$).

Still another possible type of reaction is the photochemical formation of free
radicals. This may be accomplished by photodissociation of an aldehyde, for
example acetaldehyde $CH_3-\overset{\overset{\displaystyle O}{\|}}{C}-H$ or propionaldehyde $CH_3-CH_2-\overset{\overset{\displaystyle O}{\|}}{C}-H$.
The result is

$$R-\overset{\overset{\displaystyle O}{\|}}{C}-H + E \longrightarrow R\cdot + H-\overset{\overset{\displaystyle O}{\|}}{C}\cdot \tag{10.9}$$

with the production of both an alkyl radical (for example $CH_3\cdot$ or $C_2H_5\cdot$, respectively) and a formyl radical. This reaction is perhaps most effective in a maturing
smog, when the concentration of oxygenated hydrocarbons has been built up. The
comparatively great dilution of radicals in the atmosphere, at concentrations on
the order of 0.1 ppm, results in half-lives of minutes or even hours for these reactive constituents.[3]

Which of the many possible reactions dominate smog evolution is not yet
known. But one important consequence of the presence of free radicals comes
from their oxidation by molecular oxygen; for example,

$$R\cdot + O_2 \longrightarrow ROO\cdot. \tag{10.10}$$

This is a particularly significant step if $R\cdot$ is not a hydrocarbon but an acyl unit
which can be represented by $R'-\overset{\overset{\displaystyle O}{\|}}{C}\cdot$ where R' is an alkyl. These radicals can be
formed by reactions such as Eq. (10.9) where R' appears in place of the hydrogen.
They are especially reactive because the carbon is already oxidized and therefore
is susceptible to further oxidation. The end product in Eq. (10.10) is a peroxy
radical.

The presence of peroxy radicals based on the acyl unit may account for much
of the competing mechanism which converts NO into NO_2 in Fig. 10.2.[1] This might
be accomplished by the following reaction:

$$R'-C\overset{\displaystyle O}{\underset{\displaystyle O-O\cdot}{\Big\backslash}} + NO \longrightarrow R'-C\overset{\displaystyle O}{\underset{\displaystyle O\cdot}{\Big\backslash}} + NO_2 \tag{10.11}$$

When reactions such as this have increased the NO_2 concentration sufficiently, another reaction would begin to compete for the peroxy radical:

$$R'-C\overset{\displaystyle O}{\underset{\displaystyle O-O\cdot}{<}} + NO_2 \longrightarrow R'-C\overset{\displaystyle O}{\underset{\displaystyle O-O-NO_2.}{<}} \qquad (10.12)$$

The end product is called by the generic name peroxyacyl nitrate (PAN). Numerous PAN's could be formed, corresponding to the different possible R' groups. Whether Eqs. (10.11) and (10.12) are in fact the way PAN's are formed in smog has not been established. These molecules are the center of much interest on account of their vigorous oxidizing activity which is known to be responsible for many adverse effects of photochemical smog.

Not all of the important types of reactions in smog have been included in the preceding discussion. Nevertheless, we have considered what we believe to be the key ones. Our ignorance of the relative importance of other reactions indicates that little value would be gained by going into more detail. Recent calculations by S. K. Friedlander and J. H. Seinfeld[6] have shown that the time variation of the ambient concentration of a hydrocarbon, NO, NO_2, and O_3 can be predicted by a dynamic model based on only reactions of the type in Eqs. (10.4) through (10.8), (10.11), and (10.12). They found that the temporal variation of the reactive components was similar to what is observed in urban atmospheres; thus reactions of this type do indeed explain qualitatively the evolution of smog.

Experimental studies of simulated atmospheric pollution within large reaction chambers have also been enlightening.[7] Studies such as these demonstrate the essential elements of smog chemistry, since reactions can be followed under controlled conditions. One important goal is to learn what is responsible for the high rate at which hydrocarbons disappear in smog. Calculations to date can explain only half of the observed rate. It may be that the formation of organic aerosols can explain the rest. Very little is known about these aerosols and whether they may be important in photochemical reactions.

10.3 SOME EFFECTS OF SMOG

What is responsible for the undesirable aspects of photochemical smog? Plant damage, material damage, coloration of the atmosphere, eye irritation, and toxic effects are all principally caused by *oxidizing agents*—ozone, nitrogen dioxide, and the PAN's. These are known as the *oxidants* of photochemical smog. Visibility reduction results from the aerosol which is formed as smog matures. In some instances, adverse effects are caused by a single agent, in others by several acting synergistically. It is important to be specific about causes and effects, for photochemical smog is a complex mixture of chemicals and particles. We will take up a few in this chapter. Detailed reviews of toxic effects have been given by J. R. Goldsmith,[8] H. E. Stokinger and D. L. Coffin,[9] and others.[10]

But before we discuss some of the adverse effects of smog, it is worthwhile to clarify the significance of a parameter by which smog intensity is commonly measured—the *total oxidant*. This term refers to the group of substances that can oxidize the iodide ion of potassium iodide or some other chemical reagents in solution when a sample of air is bubbled through. In fact most measurements of "total oxidant" give in reality the "net oxidant." The distinction is important in localities which have both photochemical smog and high SO_2 concentrations. In many types of measurements the reducing properties of SO_2 counterbalance to an extent the effects of the oxidants, resulting in a biasing of the data known as *interference*. This gives a misleading indication of the amount of ozone, NO_2, and PAN actually present. And the concentration of oxidant measured by different methods may not agree because of differing response of various agents. Hence a complete specification of the "total oxidant" concentration should include a description of the technique of measurement.

Eye irritation

The mechanism by which components of smog irritate the eyes is unknown, but in heavily polluted regions, such as Los Angeles, 75% of the populace may experience eye irritation. As a defense, the eye may produce copious tears in an attempt to wash away the irritating agent. Many of the oxidants of photochemical smog cause eye irritation, particularly the PAN's, but contrary to popular belief ozone appears to have no effect at normal levels of pollution. Two aldehydes, formalde-

hyde $\overset{\text{O}}{\overset{\|}{H-C-H}}$ and acrolin $\overset{\text{O}}{\overset{\|}{CH_2=CH-C-H}}$ are known to be irritants; the

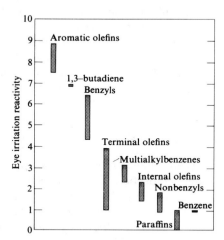

Fig. 10.4 Relative eye irritation reactivity for different classes of hydrocarbons. Reactivity is defined by the formula $10(1-t/240)$, where t is the time in seconds which elapses between initiation of a photochemical reaction between NO_x and a fixed initial concentration of the indicated hydrocarbon and the onset of irritation. "Terminal" and "internal" refer to whether or not the double bond of an olefin is shared by a carbon on the end of a chain. (J. M. Heuss and W. A. Glasson, *Environmental Science and Technology*, **2**, 1109, 1968; copyright by the American Chemical Society; by permission)

former in particular is an important constituent of automobile exhaust.

A strong correlation has been found between molecular structure and eye irritation. There are three principal methods by which the irritation potency of a chemical can be judged:

1. The intensity of sensation after exposure for a given interval of time.

2. The threshold concentration of the irritant.

3. The response delay, or the time which elapses between exposure and onset of sensation.[11]

Still another method, which may be more significant in a practical sense, is the time required for a photochemical reaction with a particular initial ingredient to proceed to the point where eye irritation is first noticed. This so-called "eye irritation reactivity" is a combined measure of the rate at which the reactions proceed and the potency of the products. By this measure it has been found that the initial hydrocarbon ingredients which are the most potent precursors of irritation are aromatic hydrocarbons having paraffinic (alkane) side chains or olefinic (alkene) side chains. Examples include toluene and styrene. Figure 10.4 shows a comparison of several classes of hydrocarbons according to their relative eye irritation reactivity.

TABLE 10.1

Simplest members of the PAN series

Name	Symbol	Structure
Peroxyformyl nitrate	PFN	$H-C\overset{O}{\underset{O-O-NO_2}{}}$
Peroxyacetyl nitrate	PAN	$CH_3-C\overset{O}{\underset{O-O-NO_2}{}}$
Peroxypropionyl nitrate	PPN	$CH_3CH_2-C\overset{O}{\underset{O-O-NO_2}{}}$
Peroxybenzoyl nitrate	PBzN	(benzene ring)$-C\overset{O}{\underset{O-O-NO_2}{}}$

On investigating the effect of individual products of photochemical smog, it has been found that the most potent eye irritants are the PAN's. The one with the lowest threshold is a derivative of an aromatic acid—peroxybenzoyl nitrate (PBzN). The PAN's are so important in causing adverse effects that they merit a special word. The chemical and structural formulas of the simplest are listed in Table 10.1. These are unstable compounds, and the first has never been observed. The PAN's are extremely reactive oxidizing agents. PBzN can be readily produced under simulated smog conditions by photoreaction of styrene (or other aromatic olefins) with NO_x. However, it has not yet been detected in urban air.

The irritation thresholds for some constituents of smog and PBzN are indicated in Table 10.2. Eye irritation occurs in smog at a threshold when the total oxidant is about 0.2 ppm, but the high perceived intensity of eye irritation in smog is not explained by the known concentrations of the major constituents. The potent effect of PBzN suggests that only a small amount of some compounds might be sufficient to explain the discrepancy. Techniques for measuring the minute ambient concentration of PBzN which would account for the difference are not available and there is a significant possibility that minute quantities of other potent—and as yet undiscovered—irritants are being experienced in smog.

Experiments indicate that PAN formation may be suppressed if hydrocarbon concentrations are kept sufficiently low. If they are so low as to prohibit total conversion of NO into NO_2, reaction Eq. (10.12), for example, is not favored over Eq. (10.11) and PAN formation may be insignificant.[1] Thus there may be a threshold concentration of hydrocarbons for their production.

Odor

The important odorant of smog is ozone. It has been found to have a threshold of 0.02 ppm,[12] and so is easily noticed at ambient levels of 0.2 ppm or more, common in smog. However, odor is not a cause of frequent complaint among residents of smoggy cities such as Los Angeles. This partially results from odor adaptation in which the intensity of sensation decreases with exposure time; another important factor is the distracting effects of eye irritation, stimulation of the lachrymal glands, and irritation of the respiratory system. Nonetheless a person first experiencing a heavy smog or one moving into a region of higher ozone concentration finds the odor distinctly unpleasant.

TABLE 10.2

Approximate eye irritation thresholds for
five-minute exposures

Compound	Threshold (ppm)
Formaldehyde	1
PAN	0.7
Acrolein	0.5
PBzN	0.005

Toxicology

In discussing the effects of photochemical smog on physiological functions and health, we shall begin with toxicology studies which permit the association of adverse effects with specific pollutants. Little is known except for the toxic effects of relatively high concentrations of ozone and nitrogen dioxide, but we shall examine what significant results are available. Then we shall take up the epidemiologic aspects.

The respiratory system may respond to very low concentrations of ozone. One important and remarkable aspect of respiratory irritation from ozone and photochemical smog is the extreme variability in sensitivity among members of an exposed population. There is a definite increase in airway resistance in humans exposed to 1 ppm for an hour; but some subjects responded to concentrations as low as 0.1 ppm, a level of ozone very commonly experienced in major cities. Measurement of airway resistance is mainly sensitive to the decreased compliance of the central airways and is insensitive to the response of the alveoli deep in the lung. Thus it is not clear whether a similar variability exists in the susceptibility to injury of the lung. There is no evidence of irreversible change in the respiratory system from exposure to ozone concentrations commonly found in community air.

As yet there are no data which have established that long-term exposure to ozone at ambient concentrations can cause chronic respiratory disease in humans. But animal experiments suggest that such effects can occur for daily exposure to somewhat higher levels, about 1 ppm for at least a year.[13] Irreversible changes in pulmonary function in exposed animals reflect the effect of chemical changes at the cellular and molecular level. Studies of the lungs of rabbits exposed to 1 and 5 ppm of O_3 for only an hour have revealed some of these.[14] It was found that carbonyl compounds were formed in the structural proteins elastin and collagen. A possible hazardous consequence is that the carbonyls may crosslink between structural proteins, and since crosslinking is a characteristic of aging tissue, it has been *speculated* that O_3 accelerates the aging of lung tissue. Fibrotic changes in the lungs of some animals exposed to 1 ppm of O_3 for one year have definite attributes of accelerated aging.[13] However, there is no pathologic evidence yet that this occurs in the human lung when exposed intermittently to ozone levels in urban atmospheres.

Exposure of animals to various levels of O_3 has also produced a greater susceptibility to infection from such agents as aerosols of *Klebsiella pneumoniae*, a bacterial organism. Whether human populations similarly become more susceptible to infections has not been determined. An interesting effect from the exposure of animals to high concentrations of ozone is development of a measure of tolerance to subsequent exposures when mortality is used as an indicator. Ordinarily lethal concentrations of ozone can be tolerated by some animals if they are first exposed to somewhat lower concentrations. Such a tolerance mechanism may be accompanied by irreversible change which impairs lung function.[9]

Several *nitrogen oxides* coexist in photochemical smog. The most prevalent on a global basis is not a pollutant but is the naturally occurring N_2O, which is normally found at a concentration of 0.3 ppm. This is not toxic and has long been used in concentrated form as an anesthetic. Of the common pollutants, only NO_2 is toxic and for concentrations well in excess of 10 ppm can produce mortality. Both ozone and NO_2 are known to irritate the alveoli of the lung, and animal experiments show that chronic exposure to 0.5 ppm of NO_2 for several weeks produces loss of cilia, distention of the alveoli, and changes in the bronchiolar epithelium resembling an early condition of emphysema.[15] For the much lower urban levels of pollution, there is no evidence that exposure produces a chronic disorder.

Epidemiology

There have been no known episodes of photochemical smog which have produced increased mortality, unlike the examples of Donora and London for sulfurous smog. Epidemiologic studies seeking associations between chronic illness and levels of total oxidant have so far failed to uncover any definite effect.

Many epidemiologic studies have sought a relationship between the occurrence of pronounced smog and an increase in the symptoms associated with irritation of the respiratory system, and some have been found. For example, comparison of over 500 telephone company employees who worked outside demonstrated a significantly greater incidence of persistent cough, but no difference in pulmonary function, for those who worked in Los Angeles as compared with less-polluted San Francisco.[16] Again, several thousand residents of Los Angeles with sinus trouble, hay fever, bronchitis, asthma, and several other respiratory disorders have reported that their condition worsened as a result of air pollution more often than did over a thousand residents of San Francisco with similar disorders experiencing similar levels of total oxidant.[17] But how much of this was a psychological effect has not been determined.

Another symptom suggestive of a change in respiratory function from exposure to smog has been demonstrated by a study of high school athletes in competitive events, where performance was found to be reduced when oxidant concentrations had been higher earlier in the day.[18]

Irritation of the respiratory system during smoggy days is most noticeable to those who undergo vigorous exercise. There is no evidence that the temporary distress of a healthy person exposed to an oxidant level of 0.3 ppm or so produces long-term effects. Similarly, there is no firm evidence that aggravation of the symptoms of patients with chronic lung disease leads to an ultimate deterioration of their condition. Studies of hospital admissions for respiratory illness show little or no correlation with oxidant or NO_2 levels.[19] There is, as yet, no supporting data to link mortality with ambient levels of total oxidant, even for patients with chronic respiratory diseases.

We should realize that lack of correlation may mean only that the ambient concentrations were not properly measured. Concentrations vary greatly from one part of a city to another. Few metropolitan areas have had more than one or two air-monitoring devices in operation for the past decade, so the data is hardly representative of a large region. The lack of complete or reliable data has been a continuing difficulty with epidemiologic studies.

Plant damage

PAN has been accepted as the toxicant which is most responsible for "smog" type damage. Exposure may produce a collapse of internal cells in leafy plants. The initial attack is on cells just inside the stomata, causing a characteristic silvering or glazing of the underside of the leaf. Levels as low as 0.01 to 0.05 ppm will damage some plants after as little as an hour's exposure. Once it has passed through the stomata of a leaf, PAN can affect membrane structure and permeability, metabolic processes, and photosynthesis. Ozone produces many of the same effects, but is generally not as potent. When the guard cells of a leaf are affected, they respond by closing the stomatal opening; however, this protective action may be too late to be successful. Changes in the internal functioning of a leaf sometimes result in formation of a pigment, such as stipple on grape leaves. They may also result in loss of pigment, observed as fleck on tobacco. Some sensitive plants respond to 0.05 to 0.1 ppm of ozone within one or two hours. But the sensitivity of plants to O_3 and PAN varies greatly. Most leafy vegetables are very sensitive, a fact brought home through great economic loss to agriculturalists in many areas of California.

Not much is known about the toxic effects of low levels of NO_2, perhaps because the other components of smog are so much more toxic. Continuous exposure of citrus trees to 0.5 ppm for several months has been found to produce a 10–20% decrease in fruit yield. Toxic effects are apparent at much higher levels, but these higher concentrations are rarely experienced in smog. Exposure of navel orange trees to levels commonly experienced in the Los Angeles environment produced no appreciable change in yield.

One component of smog not popularly regarded as a toxic agent is the hydrocarbon ethylene ($CH_2{=}CH_2$). This olefin is a component of automobile exhaust and acts as a hormone which affects plant growth. Exposure to as little as 0.05 ppm for several weeks can significantly retard the growth of sensitive plants such as orchids. The demise of the cut flower industry in Los Angeles and San Francisco was due in part to the effects of ethylene.

10.4 REACTIVITY

The preceding two sections suggest that the dynamics and effects of smog depend upon more than just the initial amount of NO_x and organic compounds in the air. They are also influenced by the type of organic molecule, or its *reactivity*. It is thus important to examine more closely the chemical properties of these compounds. Several criteria for reactivity are possible, including the following:

TABLE 10.3

Relative reactivity of various classes of hydrocarbons. The most reactive are listed first,
with less reactive structures in descending order.

Internally bonded alkenes (olefins)
Terminally bonded alkenes (olefins)
Dialkyl and trialkyl benzenes (aromatics)
Diolefins
Ethylene
Toluene and other monoalkyl benzenes
$C_6{}^+$ paraffins
C_1–C_5 paraffins, acetylene, and benzene

1. The rate at which an organic molecule will convert NO into NO_2.

2. The rate at which hydrocarbons are consumed.

3. The yield of aerosols.

4. The eye irritation reactivity.

These factors have been investigated for the broad class of organic molecules most abundantly represented in community air—the hydrocarbons.

On practically all counts—except eye irritation—the relative order of reactivity of various types of hydrocarbons agree. A brief listing is given in Table 10.3 [19,20] The most reactive hydrocarbons are olefins, especially those whose double bond is not associated with an end carbon. Next in reactivity are aromatic compounds with two or more side groups. Those with only one side group are only slightly reactive; the least reactive (C_1–C_5 paraffins, acetylene, and benzene) produce essentially no effects.

A quantitative means of characterizing reactivity is provided by the rate at which a particular hydrocarbon oxidizes NO to NO_2 as measured in a smog chamber experiment. The comparisons are shown in Fig. 10.5 for several structural types of hydrocarbons. According to the reaction rates, the reactivity has been ranked somewhat arbitrarily into six classes, ranging from the least reactive in Class 1 to the most reactive in Class 6. The bars in this figure show the range of reaction rates observed for tested compounds of each structural type. Again, olefins and aromatics with alkly side groups are found to be the most reactive, whereas the lighter paraffins and benzene are essentially unreactive.[21] Not included in this study but of high reactivity are the oxygenated hydrocarbons such as formaldehyde.

Thus control of hydrocarbon emissions should logically be based on the overall reactivity of hydrocarbons rather than just the amount which is emitted. Automobiles are the major source of these pollutants in most communities, but it is ironical that a common optical technique for measuring their hydrocarbon emissions (which we shall describe in Chapter 12) is most sensitive to the unreactive paraffins and insensitive to many olefins and aromatics.

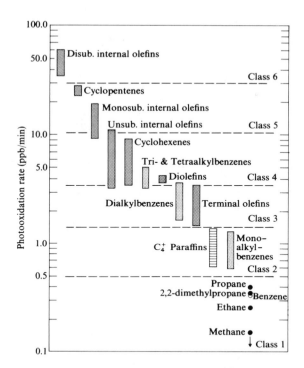

Fig. 10.5 Hydrocarbon reactivity as indicated by the rate at which NO is oxidized to NO_2 for an initial mixture of 1 ppm hydrocarbon, 0.38 ppm NO, and 0.02 ppm NO_2. (W. A. Glasson and C. S. Tuesday, *Environmental Science and Technology*, **4**, 916, 1970. Copyright © 1970 by the American Chemical Society. Reproduced by permission.)

10.5 MOTOR VEHICLES

The automobile is the main source of four atmospheric pollutants—carbon monoxide, organic compounds, nitrogen oxides, and lead compounds. Its importance as a contributor to photochemical smog, relative to stationary sources, varies according to the makeup of the particular urban area, but in most communities it is dominant. Automobile emissions come from more than just the tailpipe, as Fig. 10.6 illustrates. Hydrocarbons are also released through evaporative loss and escape through vents in the fuel tank (A), carburetor (B), and the crankcase (C). In the first two instances, the normal vapor pressure of gasoline ensures a continual evaporation unless the storage space is completely closed. Then evaporation will cease when sufficient molecules have evaporated to establish the saturated vapor pressure above the liquid. Loss from the crankcase is possible, because when the engine is running, the high pressure in the combustion chamber forces some hydrocarbons and other organic molecules through the small space between the piston ring and cylindrical walls of the combustion chamber, and they enter the crankcase below. This leakage is

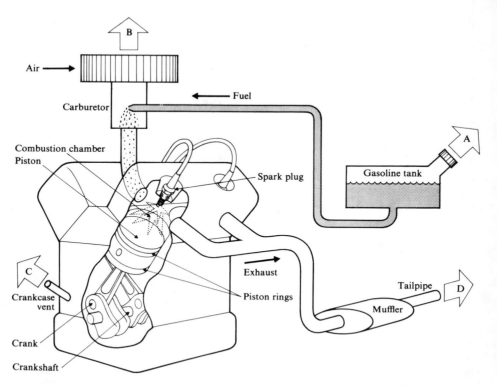

Fig. 10.6 Organic emissions from an internal combustion engine: (A) evaporative loss from gasoline tank, (B) evaporative loss from carburetor, (C) blowby into crankcase and out of vent, (D) tailpipe exhaust.

called "blowby." About 80% of the blowby is unburned gasoline, and the remainder is partially oxidized hydrocarbons.[22]

Typically about 55% of the hydrocarbon emissions of an automobile come from the exhaust, 20% from the gasoline tank and carburetor, and 25% from the crankcase, assuming that no emission control devices are installed. The relative importance of these sources depends mainly upon the condition of the engine, the ambient temperature, and typical driving cycle of the operator. Cars used in New York City during the winter, for example, generally have minimal evaporative loss when compared with those used in Los Angeles during summer, where the higher temperature produces a higher gasoline vapor pressure and evaporation rate. On the other hand, exhaust emission of some other pollutants in stop-and-go New York traffic will greatly exceed those from cars in steady Los Angeles traffic patterns. As we shall point out, current pollution control methods have enabled almost complete control of evaporative loss from automobiles. Therefore the main source of pollution remains the organic emissions in the exhaust.

The conventional internal combustion engine works on the principle of the

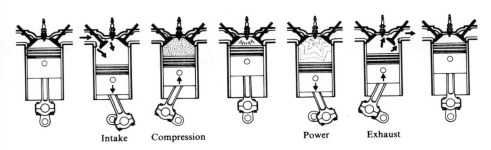

| Intake | Compression | | Power | Exhaust |

Fig. 10.7   Otto four-stroke cycle.

Otto four-stroke cycle, which converts thermal energy, produced by ignition of a gaseous mixture of fuel and air, into the mechanical kinetic energy of piston motion. The sequence of intake, compression, power, and exhaust repeats cyclically in each cylinder of the engine. In Fig. 10.7 we show this cycle for one cylinder. The piston is connected by tie rods to a crank on the crankshaft below. And the shaft is geared through the transmission to the driveshaft of the vehicle and thus to the wheels. In the first step of the cycle, the piston is pulled downward by the crank as the crankshaft rotates in response to the force supplied by the other pistons. Also an inlet valve is opened to permit an air-fuel mixture prepared by the carburetor to enter and fill the created vacuum. The normal ratio of air to fuel for approximately correct chemical stoichiometry is 14.5 to 1 by weight. Then as the piston reaches its point of fullest withdrawal, the inlet valve closes, and the piston is forced upward by the crank to compress the air-fuel mixture. An electrical discharge across the electrodes of the spark plug ignites the compressed mixture when the piston is near the top of its stroke. The ratio of the volume of the mixture at this point to the volume when the piston was at its lowest position is called the *compression ratio*. A ratio of 8.5:1 is commonly found in new American-made cars. The flame created in the mixture by the spark plug spreads at subsonic speeds, passing across the cylinder in about 5 msec as the piston comes to a halt and is accelerated downward.

The power stroke commences as the temperature of the burning gases rises. This temperature in the bulk of the gas is about 2500–3000°C. However, the flame does not penetrate into small recesses, such as the space between the side wall of the piston and the cylinder wall. Hydrocarbons in this region, as well as those near the relatively cool (about 200°C) surface of the cylinder, do not burn completely. As the high pressure of the gas forces the piston downward, the gas cools rapidly at a rate of about 80°C per msec by its expansion as it performs work on the piston. Nitric oxide formed in the high-temperature combustion has little time to revert back into N_2 and O_2. At the bottom of the power stroke, an exhaust valve is opened and the waste gases are forced out as the piston is once again thrust upward by the crank. The temperature of these gases at this point is typically 1000°C, but when they expand into the tailpipe they rapidly cool.

Because of the nonuniform conditions of combustion, a wide variety of partially oxidized hydrocarbons is exhausted with CO_2, N_2, and H_2O. Insuf-

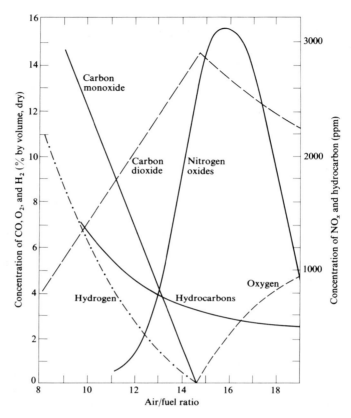

Fig. 10.8 Relationship of combustion products to air-fuel mixture. Hydrocarbon emissions can be 3 times or more the vales shown, depending upon the condition of the engine. (CO, CO_2, H_2, and O_2 from A. H. Rose, Jr., *Air Pollution*, A. C. Stern, ed., New York: Academic Press, 1962; and hydrocarbons and NO_x from W. Agnew, *Proc. Roy. Soc.*, **A307**, 153, 1968. Courtesy of the General Motors Corporation.)

ficient oxygen in some regions of the combustion chamber will also leave CO in the exhaust. Agglomerated hydrocarbons or "tars" are also emitted, but, although some are carcinogenic, their importance as a health hazard in community air has not yet been established. Fuel composition influences the type and amount. For an uncontrolled eight-cylinder automobile, the emissions were found to contain about 1 gram of tar for each kilogram of gasoline consumed.[22]

Many variables of engine design and operation can affect the type and quantity of pollutants which are finally emitted: the air/fuel ratio, the surface/volume ratio of the combustion chamber, the compression ratio, the time of ignition during the cycle (the spark timing), and the speed of crankshaft rotation. Some of these vary with the mode of driving, and hence so will the rate of emission of pollutants.

One example of this dependence is illustrated in Fig. 10.8 which shows the percentage of O_2, H_2, CO, NO_x, and hydrocarbons in the exhaust emissions and their dependence upon the air/fuel ratio. The precise concentrations of these products will be affected by driving speed and other factors, but the qualitative features of the curves remain much the same. It can be seen that correct stoichiometry minimizes the emissions of CO and nearly minimizes total hydrocarbons; but the more efficient combustion produces a higher flame temperature and nearly maximizes the production of NO_x. The NO_x curve follows roughly what is predicted from the rate for Eq. (10.1) on the basis of the maximum flame temperature.[23] From the data of Fig. 10.8 we see that there is clearly a trade-off between one pollutant and the other if only the air/fuel ratio is varied.

In normal operations, a vehicle with no emission controls emits about 1400 ppm of gaseous organic material, slightly over 3% of the amount of fuel supplied to the engine. Some 50–100 ppm of this consists of oxygenated hydrocarbons, formaldehyde and acetaldehyde being the most prominent. The bulk of the organic emissions are hydrocarbons. About 30% of these (by volume) are C_5 and heavier hydrocarbons which correspond closely to the composition of the fuel. (The aromatic content of the fuel strongly influences the amount of benzo(a)-pyrene in the exhaust.) The remainder, almost 70% of the emissions, are light molecules produced by cracking of the fuel during the compression and combustion portions of the engine cycle. These are also the more reactive constituents.[23] A high air/fuel ratio increases the aldehyde output. Thus the reactivity of automobile exhaust may be largely determined by molecules that are formed within the engine.

Emission controls

Initial efforts for reducing emissions from motor vehicles have concentrated on minimizing evaporative loss and improving combustion to minimize production of CO and hydrocarbons in the exhaust. To achieve the latter, automobile manufacturers have increased the air/fuel ratio to ensure complete combustion. Unfortunately, as Fig. 10.8 indicates, the NO_x production also increases.

Loss of hydrocarbons through blowby past the piston and subsequent venting from the crankcase has been eliminated by sealing the crankcase and extending the vent line to return the hydrocarbons to the carburetor for subsequent combustion. Evaporative emission can be controlled by use of a well-sealed gasoline tank and carburetor, and an adsorbant such as activated charcoal to avoid a pressure build-up when the engine is not running. Air is then passed through the adsorbant device when the engine is started, to desorb the hydrocarbons from the charcoal and convey them to the combustion chamber.

These improvements have been brought out in response to various state (notably, the State of California) and federal regulations. An example of these is given in Table 10.4. Initial controls were placed on only hydrocarbon and carbon monoxide emissions, since it was believed that this could be most directly accomplished and might provide the most immediate relief. However, in designing their product to meet these requirements, domestic automobile manufacturers allowed NO_x emissions to escalate by over 40%.

TABLE 10.4

Emissions standards for automobiles. Maximum emission rate for an average trip as specified by California and United States laws. The first pair of columns indicate the average emissions before controls were introduced in California, the last four, the requirements which new automobiles should meet as of the indicated year. Comparison of the third and fourth columns shows the substantial difference which can result when emissions are based on different driving cycles.

Pollutant	Prior to control (Calif. cycle)		1966 Calif. 1968 U.S. (Calif. cycle)		1972 Calif. (Calif. cycle)		1972 Calif. (U.S. cycle)		1976 U.S. (U.S. cycle)	
	g/km	g/mi	g/km	g/mi	g/km	g/mi	g/km	g/mi	g/km	g/mi
Hydrocarbons	6.8	11	2.1	3.4	0.9	1.5	1.9	3.2	0.26	0.41
Nitrogen oxides (as NO_2)	2.5	4	—	—	0.8	1.3	—	—	0.25	0.40
Carbon monoxide	50	80	21	34	14	23	28	46	2.1	3.4

The data in Table 10.4 represent the average emissions for a specified cycle of automobile operations, from a warm-up to the end of a "typical" trip. It is significant that the standards are expressed as the maximum allowable amount of pollutant per kilometer, rather than as a percentage by volume of the exhaust gases. Such a limitation on the mass emission rate is a democratic criterion, for it means that larger cars are permitted to emit no more pollutants than smaller ones.

The progressively stricter controls on hydrocarbon and carbon monoxide emissions indicated in Table 10.4 have had a measurable effect on the amounts of these pollutants emitted by late-model automobiles. Figure 10.9 shows the

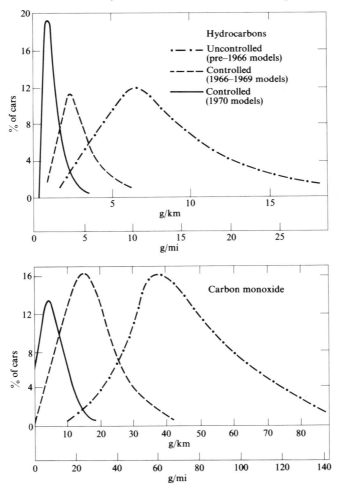

Fig. 10.9 Distribution of hydrocarbons and carbon monoxide emissions from controlled and uncontrolled cars. (*Air Pollution Control in California*, 1970 Annual Report, Air Resources Board, State of California.)

results of measurements made during 1970 on 10,000 California automobiles, each with an average mass of 1800 kg (a weight of 4000 lb), selected at random. The percentage of cars emitting a given amount of each pollutant is shown for uncontrolled models (pre-1966), controlled 1966–1969 models, and controlled 1970 models. The improvement for the later-model cars is remarkable.

Unfortunately, late-model motor vehicles comprise only a small fraction of the total number on the road, and no such major improvement has been found in the ambient levels of pollution. The large number of older cars in use means that many years must elapse between the introduction of controls on new ones and an alleviation of conditions, unless devices are also applied to the existing cars. In addition some tests indicate that emissions from new vehicles increase substantially after they have been driven for 20,000 km or so.

A limit of 2.5 g/km (4 g/mile) of NO_x for 1971 model automobiles placed on all cars sold in California was the first such restriction on emissions of nitrogen oxides. To achieve this, flame temperatures were reduced by returning some of the engine exhaust gases to the combustion chamber. This increases the air/fuel ratio, but the noncombustible gas acts as a heat sink.

Even stricter emission standards will necessitate the introduction of more elaborate control devices, or an alternative to the internal combustion engine. Some effort is now being placed on development of exhaust reactors or catalytic mufflers to complete combustion of organic compounds and CO and convert NO_x back into N_2 and O_2 once the exhaust has left the engine. Other efforts are directed toward finding methods to control the conditions of combustion better and to minimize the production of pollutants. One alternative is an engine which burns its fuel *continuously*, for then sensing devices can regulate the air/fuel mixture and combustion sequence to minimize CO and organic emissions. Continuous combustion also affords an opportunity to reduce the maximum flame temperature and reduce NO_x production. For these reasons, turbines and steam engines are attractive. However, it will be no mean engineering achievement to design such power sources that can beat the current performance standards of the internal combustion engine and can be manufactured in quantity at an acceptable price. It is not our purpose to discuss in detail the many alternatives to the internal combustion engine; their technical and economic aspects have such importance that reviews of the current status appear several times a year in the literature.

Diesel engines

The diesel engine differs from the spark-ignited gasoline engine in three important respects:

1. The flow rate of air into the engine is not throttled, with the result that it normally operates with excess air.
2. A carburetor is not required, because the fuel is injected directly into the combustion chamber after the air is compressed.
3. There is no spark plug, but ignition is achieved when the fuel meets the hot,

compressed air in the combustion chamber. The combustion temperature is about the same as for spark-ignited engines.

Fuels with a vapor pressure lower than that of standard gasoline can be successfully burned, and this is an economic advantage of the diesel. Theoretically the fuel oil can be completely burned, owing to the excess air in the chamber; but improper maintenance or an overloaded engine can result in emission of copious amounts of particulates and hydrocarbons. About 75–90% of the former is carbon. A properly maintained engine will not emit the dark plumes often seen coming from the exhaust pipes of large trucks, buses, and railroad engines. Emission of malodorants (most of which seem to be aldehydes) can be similarly minimized by proper engine care. But excess oxygen in the combustion chamber causes greater production of NO_x than occurs in a spark-ignited engine.

Jet engines

Emissions from jet aircraft, particularly on landings and takeoffs, are a source of bitter complaints from nearby inhabitants. In some cities such as Los Angeles with relatively low particulate emissions from other sources, the solid emissions from aircraft may constitute a substantial fraction of the total. In a few airports visibility has been dangerously restricted by particulate emissions and photo-chemical smog. Airlines have a considerable expense in cleaning the obnoxious odors of unburned fuel from aircraft air conditioning systems. But pilots favor exhaust plumes, because aircraft are thereby made more visible.

In a jet engine, air enters through the front and is compressed by rotating vanes as it is forced into combustion chambers arranged around the circumference as illustrated in Fig. 10.10. Fuel is steadily sprayed into the leading end of each chamber where it ignites in the hot compressed air, burns, and causes the air to expand. As the burning gases push on toward the rear, they strike turbine blades whose rotation drives the compressor to which they are connected by a shaft. At the exhaust nozzle, the burning gases are further compressed to provide a high-velocity exhaust, thus providing forward thrust to the aircraft.

Reliable data on engine exhausts are difficult to obtain, for engines operate under a variety of conditions. Proper engine design plays an important role in reducing pollution emissions, for engines of one manufacturer may emit sub-

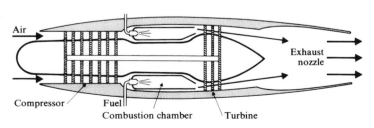

Fig. 10.10 Operation of a jet engine.

stantially more than a comparable engine of another company. The installation of "clean burner cans" on older aircraft has provided a substantial reduction in hydrocarbon, carbon monoxide, and particulate emissions because the fuel is more completely burned. However, nitrogen oxide omissions are barely affected.

10.6 SOURCES OF ORGANIC EMISSIONS

In many communities the prime source of gaseous organic molecules is the automobile; and the one next in importance is the usage of solvents. In some regions, vegetation may also release a significant amount. Each of these sources is responsible for the release of specific classes of molecules of differing reactivity. The organic component of the exhaust from a motor vehicle, as we have remarked, is partly characteristic of fuel cracking during combustion, but about 30% of the hydrocarbons are representative of the gasoline.

Gasoline

In Table 8.2 we indicated that the gasoline broad fraction of petroleum has over 100 different compounds. What constitutes the commercial gasoline used in automobiles very much depends upon the brand and quality. Refineries may alter the composition sold in a particular geographical region in accordance with seasonal climatic changes, so as to optimize motor vehicle performance. For this and other reasons, commercial gasolines in a given region may vary considerably; an example of this is illustrated by the compositions listed in Table 10.5 for three brands sold in the State of Michigan. For convenience, the hydrocarbons have been classified in this figure according to their structure, for each gasoline has over 50 different compounds. Of the predominant paraffins, isopentane occurred in the highest mole percentage for gasoline A and B, but still constituted only about 10% of the gasoline. All three brands had substantial amounts of n-pentane, n-butane, and neopentane. Among the aromatics, toluene at 6–7% was the most plentiful, and among the olefins, 2-methyl-2 butene comprises 2–3%.

TABLE 10.5

Composition of three brands of gasoline
(in mole percent)*

| | Brand | | |
	A	B	C
Paraffins	71	66	52
Aromatics	20	20	24
Olefins	9	14	24
Photo-oxidation rate (ppb/min)	1.7	2.4	3.2

*(From W. A. Glasson and C. S. Tuesday, *Journal of the Air Pollution Control Association*, **20**, 239, 1970.)

It is remarkable that the reactivities of the gasolines were significantly different. Laboratory experiments conducted by W. A. Glasson and C. S. Tuesday[24] compared the actual reactivity with the reactivity estimated from the known values for the constituents. Photochemical oxidation of NO, conducted at 1.0 ppm of total gasoline hydrocarbons, 0.38 ppm of NO, and 0.02 ppm NO_2, showed that the reaction rate for brand C was almost twice the rate for A. These results agreed closely with the predictions; the higher reactivity of C arose from the much higher content of olefins than in A. Thus the composition of gasoline may have an important bearing on the dynamics of photochemical smog.

Solvents

The second largest source of hydrocarbons in some metropolitan regions is the use of solvents. Chlorinated hydrocarbons, for example, are widely used as cleaning solvents, for they rapidly dissolve greases and oils, including substances which are not soluble in petroleum solvents. Perchloroethylene (tetrachloroethylene) is the most widely used chlorinated agent for dry cleaning, partly because it is nonflammable. Other nonchlorinated hydrocarbons are also used. On the average, in the San Francisco area, Los Angeles, and in Kent County, Michigan, the amount of chlorinated hydrocarbons released from dry cleaning plants is about 1 kg per person each year.[25]

Other uses of solvents also release hydrocarbons. Evaporation during the application and drying of paints and varnishes is the most common example. Industries that perform surface coating operations include automobile assembly plants, furniture manufacturers, plastic products manufacturers, tool assembly plants, and paint shops. Hydrocarbon emissions from solvents depend upon the types of prevalent operations and so vary from one location to another. In Los Angeles, daily emissions of organic solvents amount to about 0.1 kg per person and this constitutes about 20% of the total hydrocarbon emissions in the county.

Table 10.6 indicates the relative reactivity of common solvents.[26] It has been found that benzene, perchloroethylene, chlorinated saturated hydrocarbons, and acetone are virtually unreactive.

TABLE 10.6

Common solvents in descending order of reactivity, beginning with the left column.*

Xylenes and heavy aromatics	VMP solvent
Toluene	Stoddard solvents
Trichloroethylene	Isoparaffin mixtures
Naphthene	n-paraffin mixtures
Mineral spirits	

*From A. P. Altschuller and J. J. Bufalini, *Environmental Science and Technology*, **5**, 39, 1971.

Natural sources

On a global basis, the hydrocarbon found at the highest ambient concentration is methane CH_4, and it occurs at a level of about 1.5 ppm. Most of this is released as a product of the bacterial decomposition of vegetation, particularly in swamps, marshes, and paddies. The annual release is estimated to exceed 2×10^9 Tonnes, and is supplemented by much smaller amounts from natural leakage of natural gas from the ground.[27] By comparison, the total emissions of all types of hydrocarbons from antropogenic sources is less than one-tenth of this. However, methane is a paraffin and is thus relatively unreactive in photochemical smog; a comparison of tonnage outputs alone would be misleading insofar as its importance to photochemical smog is concerned.

The most significant natural source of reactive hydrocarbons is vegetation. Volatile organic molecules and particles may accumulate in such high concentrations over the jungles of northern Columbia and the Amazon basin that the air is colored by a blue haze. Organic molecules cause the pungent smell of the steppes and the aroma of pine forests. Also familiar are the odors of eucalyptus, cedar, sandalwood, and plants such as mint. A major group of volatile plant products are derivatives of isoprene, 2-methyl-1,3 butadiene (C_5H_8). The chemical formulas for these compounds have the generic form $(C_5H_8)_n$, so the carbon bonds evidently are not saturated with hydrogens. Each molecule usually has one or more carbon-carbon double bonds and therefore is reactive in photochemical smog. The C_{10} compounds are called *terpenes*, the C_{15} *sesquiterpenes*, the C_{20} *diterpenes*, and the C_{30} *triterpenes*, etc. They are generically known by the first-named, as the *terpenes*. A wide variety of molecular structures is possible; some are cyclic, and some open chain.

Other types of volatile organic molecules are common. One broad class is the oxygenated terpenes, including alcohols and aldehydes. The former includes rose and geranium odorants, the latter citrus odorants. Generally the concentration of naturally emitted hydrocarbons is negligible in urban areas, compared with the atmospheric burden from anthropogenic sources. However, this may not be the case elsewhere. There is reason for concern as fossil fuel power plants are now being located in rural areas, since their emissions of NO_x may set the stage for photochemical reactions with hydrocarbons from natural sources. Whether this is indeed a potential problem can only be determined for each individual locality.

Although the presence of nitrogen oxides in the atmosphere from advected urban air or bacterial action encourages the oxidation of terpenes, reactions between the organic molecules is possible without them. J. Tyndall has shown that many organic vapors agglomerate to form particles when exposed to strong sunlight.[28] Apparently photochemical reactions partially oxidize the organic compounds, after which they condense to form particles less than 0.1 micron. It is these particles which scatter light and thus produce the blue haze often seen over forests.

It is estimated by F. W. Went that as much as 2×10^8 Tonnes per year of terpenes and oxygenated terpenes are released globally into the atmosphere by vegetation and the decomposition of organic material.[29] However, this may be an underestimate, with the actual amount being a factor of 2 to 5 higher. This is indeed a considerable amount, and is only slightly less than the rate at which methane is released.

10.7 CARBON MONOXIDE

The gaseous substance emitted in greatest amounts from automobiles with the exception of carbon dioxide is *carbon monoxide*. Without emission controls, this odorless and colorless gas is released at a rate which is more than six times the rate of NO_x and organic gases (Table 10.4). We have seen that CO results when insufficient oxygen is present during combustion, perhaps due to improper mixing of air and fuel, so that the fuel is not completely oxidized. Combustion of fossil fuels generally produces CO, although well-regulated conditions can ensure that the amount is insignificant. Automobiles have a relatively high concentration of CO in the exhaust, because the ignition and combustion of the vaporized air-fuel mixture is rapid and nonuniform, and the combustion conditions are poorly controlled.

For many years, it was believed that CO did not participate actively in photochemical smog. However, recent experiments have demonstrated that its presence can in fact speed the oxidation of NO to NO_2, thus hastening the appearance of oxidants.[30,31] The important reaction involves the OH free radical and three parallel chain reactions:

$$OH + CO = CO_2 + H,$$
$$H + O_2 + M = HO_2 + M,$$
$$HO_2 + NO = OH + NO_2.$$

The overall reaction therefore is

$$CO + O_2 + NO = CO_2 + NO_2. \qquad (10.3)$$

Experiments in a photoreaction chamber with the hydrocarbon isobutene and 1.5 ppm initial concentration of NO showed that the presence of 100 ppm of CO advanced the time of ozone build-up from about 100 minutes to 80 minutes. What role CO plays in the more complex mixture of community photochemical smog remains to be assessed.

Public concern about CO ambient concentrations arises because this gas in high concentrations is itself toxic. Sunlight is not required for it to produce toxic effects, and consequently it occurs as a pollutant in many more cities than are affected by photochemical smog.

In communities which are heavily dependent upon motor vehicles for transportation, CO emissions may account for the major proportion of pollutants emitted each day (on the basis of weight). It is often said that 60% of the air pollution in the United States is caused by the 97 million automobiles. It is the massive release of CO by vehicles that justifies such a sweeping statement. How-

Fig. 10.11 Schematic representation of patterns of electron density in the hemoglobin molecule. Four interlocked protein chains each enfold a heme group, labeled *H*. (From "The Hemoglobin Molecule," M. F. Perutz. *Scientific American*, **211**, 64, 1964. Copyright © 1964 by Scientific American, Inc. All rights reserved. Reprinted by permission.)

ever, the assertion misses the mark when it comes to assessing the *effect* of air pollution. The reason is that the human body can tolerate much higher ambient concentrations of CO than of the other common pollutants, sulfur dioxide and nitrogen dioxide. Thus the statement is deceptive, for it takes no account of the different adverse thresholds for various pollutants.

The most marked pathologic effect of CO is its disabling of red blood cells from carrying out their function. Each blood cell contains about 280 million molecules of hemoglobin, a large almost spherical bundle of four intertwined chains of amino acid units. A representation of this molecule in terms of the density pattern of electrons is given in Fig. 10.11. Collectively the amino acid units constitute the protein part of the molecule which is called globin. In addition, each chain enfolds a group of atoms called heme centered around an iron atom. The iron of the heme serves as a binding site for oxygen, and the heme is the pigment for the entire hemoglobin molecule. When oxygen is exchanged for CO_2 in the extremities of the circulatory system, the color changes from bright red to a much darker color, which persists until the CO_2 is released and oxygen captured again in the pulmonary capillaries of the lung.

However, any CO that has diffused through the alveolar walls of the lung competes for capture by one of the iron sites in hemoglobin. In fact the affinity of the iron for CO is about 210 times greater than it is for O_2. Hemoglobin which has acquired a CO molecule is called *carboxyhemoglobin*, abbreviated COHb. Body cells cannot use CO in their process of oxidation, so the presence of this molecule in the red blood cells hinders proper body functioning. Not only does

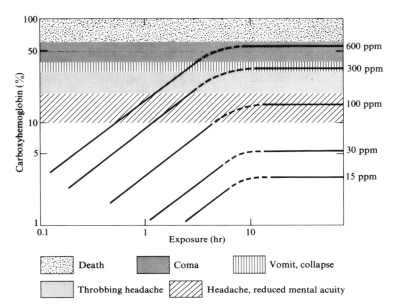

Fig. 10.12 Physiological response of normally healthy adults at light activity to various carbon monoxide exposures. (W. Agnew, *Proc. Roy. Soc.*, **A307**, 153, 1968; courtesy of General Motors Corporation.)

CO attached to hemoglobin reduce the concentration of oxygen carried in the blood stream, but its occupation of one of the binding sites makes the molecule hold more tightly to the O_2 it carries on other heme sites. This occurs despite the fact that the separation between the iron atoms is large, about 25 to 35 Å; and the explanation may be that occupation of a site by either CO or O_2 physically distorts the molecule in such a way as to increase its affinity for O_2 and possibly also CO. Since release of the tightly bound oxygen is more difficult, the efficiency of oxygen transfer from the lung to the vital organs is reduced even further.

Acute effects from the exposure to high concentrations are well known. Figure 10.12 summarizes in a schematic way the effects resulting from various levels of COHb. The family of curves illustrates the increase of COHb with exposure time for various concentrations of CO.[22] The leveling off of each curve after about an eight-hour exposure occurs when equilibrium is reached. Exposure for several minutes for 1000 ppm, a concentration not found in ambient urban air, may produce death. It is interesting that COHb has essentially the same bright red color as oxygenated hemoglobin; for this reason, the venous blood of victims of carbon monoxide poisoning has about the same coloration as normal arterial blood, and so the victim's face appears flushed.

It is not clear whether a threshold for toxic effects from CO exists. There has been some debate about the relationship of some adverse effects and dosage, but it seems that the circulatory system can quickly rid itself within 4–5 hours of small amounts of COHb once exposure to ambient concentrations are terminated. The body itself produces CO in amounts sufficient to maintain about 0.4% of the hemoglobin as COHb.

There is some statistical evidence that mortality in urban areas is directly related to fluctuations in ambient CO levels, and heart attack victims hospitalized where CO levels are relatively high are found to have a greater mortality rate than those at hospitals in areas with lower pollution levels.[32,33] But more studies will be needed before firm conclusions about a causal relationship can be made.

Our main concern focuses on the effects of CO at levels of about 10–30 ppm to which a large fraction of an urban population is exposed. For example, during eight-hour sampling periods in Chicago, from 1961 through 1967, the maximum average concentration at the sampling site was 44 ppm and the minimum, 0 ppm. The average concentration exceeded 12 ppm half of the time. Motorists especially are exposed to high CO levels; an average automobile trip in any of the 15 largest United States' cities would subject the occupants to an average CO concentration of 10–40 ppm, with short-term peaks possibly over 100 ppm.[34]

It is found that the increase in COHb would be about $1\frac{1}{2}\%$ for extended exposure (8 hours or more) to 10 ppm of CO; and for exposure to 20 ppm, the COHb would rise to about 3%. Thus in terms of the potential adverse effects, it is significant that as little as 2% COHb can affect a person's ability to estimate time intervals.[35] The functioning of the central nervous system thus appears to be impaired at these low levels of COHb. At 5% COHb visual sensitivity in the dark and performance on some psychological tests are adversely affected. (This level of COHb is found in the blood of many cigarette smokers who, by their habit, expose themselves to about 400 ppm of CO in the smoke.) For these reasons, there is much discussion about the effect of CO on driving abilities, since motorists are frequently subjected to high ambient levels. There has been speculation that air pollution may increase the incidence of traffic accidents, as the drivers' abilities deteriorate.

Despite increasing world-wide emissions of CO from anthropogenic sources, the mean atmospheric content has apparently been steady. Natural sources, especially the oxidation of CH_4 by OH in the troposphere, may produce as much as 4×10^9 Tonnes annually, more than ten times the production rate of anthropogenic sources. Concentrations over unpolluted land and oceanic sites during 1969–1970 were found to be 0.1–0.2 ppm. The ultimate fate of CO is not known with certainty, but calculations indicate that the concentration of hydroxyl radicals in ambient air, about 2×10^{12} molecules/m^3, is sufficient to oxidize CO to CO_2 and establish a lifetime of about 0.2 years for CO in the troposphere.[36] This lifetime is in good agreement with recent observations.[37]

10.8 LEAD

It has been found through experience that certain additives in a gasoline improve the smoothness of performance of high compression engines (reducing "knocking"). One class of additives commonly used is the tetraalkyl leads such as tetraethyl lead $Pb(C_2H_5)_4$. The performance of such additives is rated by a relative scale called the research octane number (RON). Generally, an RON of 105 is considered high, and 80 fairly low. As much as 0.8 grams of lead alkyl per liter of gasoline is currently incorporated in "premium" fuel with 100 RON, and about half of this amount is in "regular" grades of 90 RON. Halogen-bearing hydrocarbons (for example, ethylene dibromide and ethylene dichloride) are also added to scavenge the lead after combustion. They react with lead oxide produced during combustion to yield some volatile compounds such as lead bromide which are swept out with the exhaust. Lead-bearing particulate matter is a significant fraction of the total emitted particulate matter of 0.3 to 3 grams per liter of fuel, and consists of mixtures of complex lead halides. About 75% of the lead in the gasoline is eliminated through the exhaust, the balance being retained in the engine or scavenged by the engine oil.

There is much concern over the rising ambient concentrations of lead aerosols in urban environments. In Los Angeles monthly averages for 1968–1969 show an ambient lead concentration of about 4 $\mu g/m^3$, an increase of roughly 50% over the 1961–1962 values. And maximum concentrations in traffic may be as high as 50 $\mu g/m^3$, with an average concentration about half of this. Isotopic analysis of lead over San Diego clearly establishes its origin as the local gasoline.[38] There has been a world-wide dispersal of lead, as is evident from accumulations in the annual ice layers in the interior of northern Greenland. Analysis of lead concentrations by age shows a steep rise after 1940; the concentration in recent layers is well over 400 times above the natural background in 800 B.C.[40]

The public may be exposed to the hazards of lead-bearing air pollutants in two distinct ways: One is through respiratory intake, which is estimated at between 5 and 50 μg of lead per day, depending upon air quality. From 25 to 50% of this is absorbed, so that a daily intake of about 20 μg is typical for urban dwellers. The other intake is through ingestion of contaminated food, since lead accumulates in topsoils and surface waters, from which it can enter the food chain. The average dietary intake is estimated as 300 μg a day for American adults.[39] However, only about 10% of this is absorbed in the body, the balance being eliminated in the feces. Thus about 30 μg is retained, an amount comparable to that absorbed through the respiratory system.

When lead is absorbed by the body at a sufficiently high rate, as when the diet contains in excess of 1000 μg a day, the level of lead in the blood rises and adverse effects are experienced. In ordinary circumstances the concentration of lead in the blood of urban dwellers is typically 20 $\mu g/100$ g of blood and for rural inhabitants it is 15 $\mu g/100g$. That the difference is due to greater urban air pollution is suggested by the fact that automobile mechanics may have lead concentrations as high as 30 $\mu g/100g$. Other examples are given in Table 10.7. Mild symptoms of encephalopathy (brain damage) may be found at levels between 60 and 80 $\mu g/100g$.[39] The body's response to an increased absorption of lead is twofold: enhancement in the rate of elimination in urine and deposition of the excess in the body, primarily in bone

TABLE 10.7

Blood levels of lead and respiratory exposures*

Population	Exposure (μg/m^3 air)	Mean blood lead (μg/100 g blood)
Rural U.S.	0.5	16
Urban U.S.	1.0	21
Downtown Philadelphia	2.4	24
Cincinnati policemen	2.1	25
Cincinnati traffic policemen	3.8	30
Los Angeles traffic policemen	5.2	21
Boston automobile tunnel employees	6.3	30

*J. R. Goldsmith and A. C. Hexter, *Science*, **158**, 132, 1967. Copyright © 1967 by The American Association for the Advancement of Science.

marrow.[41] In this location, it inhibits enzymatic activity in the formation of heme for red blood cells, because the lead binds up free sulfhydryl ($-$SH) groups which are essential to the process. High concentrations of lead also affect the kidneys, brain, and peripheral nervous tissue, producing a diffuse inflamation with edema (swelling of the brain tissue), and eventually convulsions. Examples of such acute "plumbism" are usually associated with the ingestion and not inhalation of lead. Traditional sources have been lead-bearing ceramic glazes, solders, and storage batteries, and, more recently, household paint that is eaten by children. But thresholds for chronic exposure to low levels of lead have not been reliably established. Because of the storage of lead within the body, dosage may be the most important criterion determining adverse effects, and extreme prudence is warranted before the public is exposed for long periods. It is known for example that the concentration of lead in bones increases with age. An important unanswered question is whether a content of lead in the body insufficient to cause obvious symptoms can nevertheless give rise to slowly evolving irreversible adverse effects.

However, the chief motivation for recent trends toward use of unleaded or low-lead gasolines is not the threat to health but the knowledge that lead quickly deactivates most catalysts in catalytic reactors which are being developed to meet increasingly stringent emission standards for CO, NO$_x$, and hydrocarbons. General use of low-compression automobiles and low-lead fuels will permit more options for the types of catalysts which may be economically acceptable.

On the other hand, if substitute additives are introduced in gasoline to take the place of lead, there is a potential for aggravation of the photochemical smog problem. The current tendency is for refineries to blend more C$_6$–C$_9$ aromatics and fewer olefins into the gasoline for this purpose, since the former have a high value of RON. Toluene, for example, has a RON of about 120. Evidently the reactivity of the exhaust may be increased or decreased, depending upon the proportion of the olefinic and aromatic constituents. With a view toward minimizing the need for lead additives and aromatics, some states such as California will not permit the sale of new automobiles that require gasoline with a high octane number.

10.9 A CLASSIC EXAMPLE: THE LOS ANGELES BASIN

Meteorology, geography, and a multitude of automobiles conspire to make a classic example of the photochemical smog over Los Angeles. Not only was the phenomenon first observed in that city, but the conditions for its development have been so nurtured as to make the frequency and intensity of smog episodes among the most severe of any urban region.

The city of Los Angeles sprawls over a coastal plain and is encircled on the north and east, or inland sides, by a series of mountain ranges rising some 400 to 2000 m, forming a natural basin facing the sea. These topographical features are depicted in Fig. 10.13, which shows the Los Angeles basin and surrounding region that comprises what is known as the South Coast Air Basin; and from this we can see that the mountain ranges are broken by a few passes which cleave the ridge lines to join the Los Angeles basin with the urbanized San Fernando Valley on the northwest and the San Gabriel Valley and high inland deserts to the east. The mountains hinder advection eastward from the basin, although they do not pose an unsurpassable barrier.

Of dominant importance in fostering episodes of air pollution is the presence

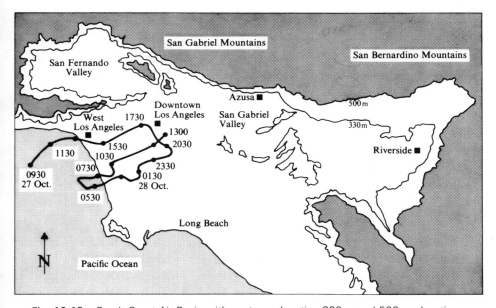

Fig. 10.13   South Coast Air Basin with contours denoting 330 m and 500 m elevations.

of a semipermanent inversion layer which sits over the basin, and restricts the mixing layer over a long stretch of the California coastline.[42] Situated at 32° N latitude, Los Angeles is approximately east of the Pacific subtropical anticyclone, and an inversion resulting from the subsiding air aloft is observed in the eastern end of this high-pressure region more than 75% of the time. The base of the inversion is found to descend sharply from an altitude of about 1500 m some 1000 km from the coast to less than 400 m over the coastal plane as a result of cooling of the mixing layer by the cold coastal currents and the attendant increase in the air density at low levels. Thus the overriding air at an altitude corresponding to a pressure of 700 mb is observed to descend most rapidly in the vicinity of the coast and the inversion is the strongest, with a temperature increase of 6°C or more from its base to its top, at 800–1000 m.[42]

The inversion rises and falls with diurnal regularity over the land, for the daytime thermals erode it from beneath. It generally slopes upward as one proceeds from the coast toward the elevated interior east of the city, where it is weakened by the thermals rising off the hot, semi-arid earth, and ceases to impose a continuous stable layer over the interior.[43] The eastern edge of the inversion does not extend much beyond the mountains.

The subsiding and warming air aloft lowers the relative humidity of that portion of the atmosphere, so the coast of southern California receives rain only infrequently, when the average pattern is disturbed. Predominantly clear skies ensure the transmission of intense solar radiation which is effective in triggering the photochemical reactions of smog.

Within the confines of the mountains and the inversion layer live over 7 million people. An inventory of air pollution emissions from anthropogenic sources in Los Angeles County is given in Fig. 10.14, and from this it is evident that motor vehicles are responsible for most of the ingredients of photochemical smog: 69% of the organic gases and 59% of the nitrogen oxides. They are also responsible for essentially all of the carbon monoxide which is emitted. One reason for this is that the automobile is practically the only means of transportation within the county; mass transportation by trains or subways is nonexistent. And the prevalence of single-family dwellings within the basin leaves the population distributed over a wide area, necessitating travel for a considerable distance for commuting to work or market. But automobiles are not the only source of pollution. It is noteworthy that the combustion of fossil fuels—principally by power plants and petroleum refineries—accounts for a significant fraction of the NO_x emissions, about a third of the total. Sulfur oxide pollution is not a major problem because power plants and light industry burn natural gas. In winter insufficient supplies necessitate the use of fuel oil, but pollution control regulations require that the oil have a sulfur content of less than 0.5%.

Emission inventories of this nature can be properly interpreted only from a viewpoint that also regards *how* the pollutants are emitted. For instance, the NO_x from power plants would not contribute to the basin's photochemical smog if the exhaust plumes were released at sufficient height and with adequate buoy-

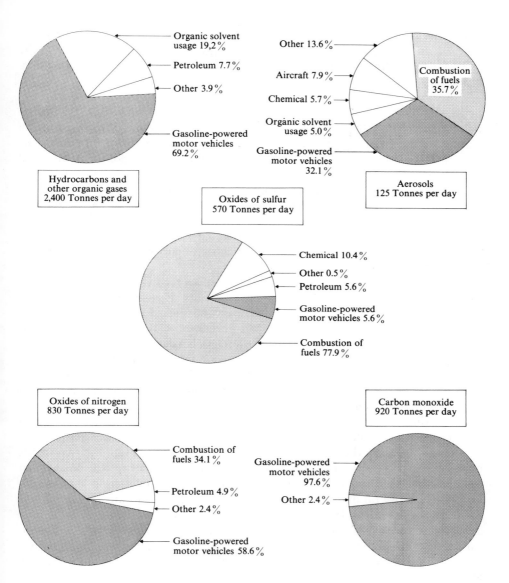

Fig. 10.14 Percentage contributions of air contaminants from major sources in Los Angeles County in 1969. (J. N. Pitts, Jr., *Journal of the Air Pollution Control Association*, **19**, 658, 1969.)

ancy to penetrate through the elevated inversion. This may in fact happen on occasion, but only when the inversion is abnormally low and weak. Most of the power plants have a relatively low stack (100 m or less) and only a moderate heat emission rate, so under usual circumstances these plumes would have difficulty even reaching upward 200 m–300 m to the base of the inversion. One reason stems from the fact that these stations are located on the coastline to take advantage of the economical supply of cooling water; but it is also at the coastline where the sea breeze is strongest, and the plume height is correspondingly lower.

In other ways as well, the sea breeze is a dominant influence on the dispersal of pollutants. Near the coast, the cool marine air maintains a strong temperature inversion by cooling the ground and discouraging the formation of strong thermals that would otherwise erode the inversion base. Of course, the marine layer adjacent to the ground is itself unstable, so pollutants are well dispersed *beneath* the inversion within the mixing layer, and it is there where exhausts from automobiles and factories are intermingled with the plumes from power plants. As the marine breeze penetrates inland, it is warmed by conduction and convection from the ground, and the erosion of the inversion from below becomes more pronounced.[43] Thus the cooling influence of the sea breeze is greatly diminished 30–40 km inland, although on many occasions the wind may still be strong.

The mountains surrounding the basin are often regarded as a barrier against the free movement of air. This is partially true; but they also help to ventilate the basin through the medium of the upslope wind during the day. Unstable air near the slope destroys the inversion and develops a chimney for the venting of some pollutants (see Fig. 7.3). This air and its burden of pollutants can thus escape from beneath the inversion and flow out over the sparsely settled deserts to the east.

Los Angeles generally experiences a prevailing west wind in late spring and early summer as inland continental regions warm and a local monsoon develops. An east wind is often found in winter, when the continent has cooled more than the ocean. An extreme example of the east wind is the Santa Ana described in Section 6.4. During the transition periods, especially in the fall, the two trends balance to a certain extent. This is even true on a diurnal or weekly basis. Then the nighttime land breeze is fairly pronounced and, in conjunction with the sea breeze, gives rise on many occasions to an almost cyclic motion of air back and forth across the basin. If this is by and large the same air, then pollutants may accumulate for several days. Meteorological balloons have been released in the Los Angeles basin to trace the motion of air parcels. The balloons, called *tetroons*, are constructed with a fixed volume and so maintain an altitude that has a constant air density as they follow the air currents.

A trajectory similar to those obtained by tetroon observations is illustrated in Fig. 10.13.[44] The figure shows the calculated trajectory of an air parcel during a period of $1\frac{1}{2}$ days, beginning at 0930 on October 27, 1965. After the sea breeze had blown pollutants inland and toward the east, the land breeze moved this polluted air back across the basin; the next morning many Angelenos did not arise to invigorating fresh air, but suffered under the dregs of the preceding day. By effects such as this, smog episodes may develop over a period of several days, becoming increasingly more severe until the meteorological pattern is broken.

During the two days illustrated in Fig. 10.13, the total oxidant in downtown Los Angeles reached 0.6 ppm, and the NO_2 concentration 0.38 ppm. Unusually low winds and a ground-based inversion during this period helped to confine pollutants. The cyclic land-sea breeze, although a dynamic factor, has the effect of allowing pollutants to accumulate as in the more common case of true air stagnation.

A common episode

Commuter traffic in Los Angeles begins shortly before 0600 hours. As the sun rises, the first stage in the photochemical processes is stimulated when the relatively low levels of NO_2, oxidized from the NO recently emitted or remaining from the previous day, are photodissociated with the release of atomic oxygen (Fig. 10.3). Solar warming of the interior of the basin causes the marine breeze to commence near the coast and progressively develop inland. It dilutes pollutants and conveys them at about 15 km/hr toward the inland valleys. Ambient concentrations of NO_x and hydrocarbons build up throughout the basin and inland regions as automobile traffic reaches its peak at about 0800 hours. Meanwhile, the atomic oxygen from photodissociated NO_2 has been oxidizing hydrocarbons and increasing the levels of aldehydes and free radicals. These in turn oxidize NO to NO_2, causing a marked reduction in the ambient concentration of the former despite the continuing emissions of this gas by motor vehicles.

When the NO is essentially gone, the oxidant level commences to rise, primarily because an increase in the ozone concentration is no longer impeded by an annihilating reaction with NO. On many occasions the oxidant reading at 1000 hours may be 0.3 ppm. Moderate to severe eye irritation affects many residents, and a commuter first entering the basin would easily notice a characteristic ozone odor for five or ten minutes before odor adaptation sets in. His visual range is reduced so that it is impossible to see the encircling mountains, and the atmosphere has taken on a yellow-brown hue, except overhead where the shallowness of the mixing layer permits a view of the blue sky. The color arises from the build-up of NO_2;[45] but in some cases the aerosol concentration is sufficiently great that the scattering of white light predominates and only a faint yellow hue is visible.

When oxidant levels reach 0.35 ppm, physical education classes for school children are suspended, following a regulation by the Los Angeles Board of Education passed on the recommendation of the Los Angeles City Medical Society. Respiratory distress is experienced by many people, particularly during vigorous exercise.

Later in the day the conversion of NO_x into the PAN's and aerosols begins to reduce the NO_2 concentration, so photodissociation yields a lower concentration of atomic oxygen. Simultaneously, the ozone production rate decreases, as indicated by a decline in measured total oxidant. This may be accelerated as the sun passes its zenith and the intensity of solar radiation diminishes. Thus in western Los Angeles when smog is locally produced, the total oxidant reaches

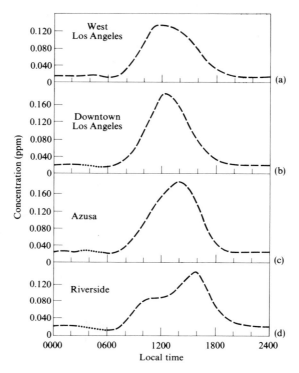

Fig. 10.15 Monthly mean hourly average concentrations of oxidant during October, 1965, at several locations in the South Coast Air Basin. (J. N. Pitts, Jr., *Journal of the Air Pollution Control Association*, **19**, 658, 1969.)

a peak at midday, as shown in Fig. 10.15(a). (See Fig. 10.13 for the approximate position of this monitoring station.) Pollutants accumulate as the smog drifts eastward toward the "downtown" of the city. Figure 10.15(b) shows that the cumulative burden of the air due to the area source upwind provides much more severe conditions downwind. Even further downwind the peak oxidant reading may occur later in the day as the highly polluted air from the basin reaches the monitoring station. Fig. 10.15(c) illustrates this for the city of Azusa, approximately 10 km from downtown Los Angeles. The smog continues to drift eastward, and at the city of Riverside, two distinct peaks in the oxidant curve of Fig. 10.15(d) become evident. The earliest peak represents smog which is home-brewed from the exhaust of the local traffic; but the second is apparently smog blown in from metropolitan Los Angeles. Sometimes this is seen to move into Riverside with a well-defined "smog front."

On severe days, several times a year, the oxidant reading in the San Gabriel Valley downwind from downtown will exceed 0.5 ppm. Between 1963 and 1965 the recorded maximum hourly average for oxidants in Los Angeles was 0.6 ppm; for NO_x, 3.7 ppm; and for NO_2, 1.3 ppm.

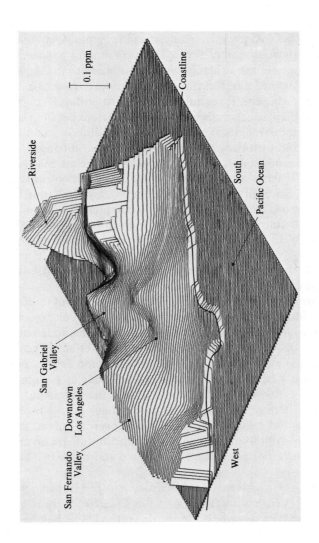

Fig. 10.16 Distribution of oxidant concentration in the South Coast Basin at 0800 hours on August 21, 1969. Oxidant concentrations for regions above the 500 m altitude boundary of the basin are arbitrarily set equal to zero, as they are for the region over the Pacific Ocean. (Environmental Systems Research Institute, Redlands, California.)

The patterns of smog evolution and motion are determined by two major factors: the distribution of sources and the micrometeorology. Figure 10.16 illustrates the complexities of the resulting pollutant levels in an eastward view of the basin; the height of the contours above the rectangular base give the oxidant concentration at 0800 hours on August 21, 1969. In this example, oxidant levels are high in the San Fernando Valley and back in the San Gabriel Valley and Riverside. The basin is experiencing a mild Santa Ana condition, with the drainage winds arriving mainly through two mountain passes, as evidenced by the two deep troughs in the oxidant contours where the wind blows the pollutants out toward sea (toward the foreground). It is clear from this illustration that there are wide variations in conditions between different parts of the basin. An adequate mathematical analysis would necessitate detailed consideration of the distribution and strength of various point and line sources, as well as varying wind speeds and turbulent conditions of pollutant mixing.

Smog episodes are common in Los Angeles. There are many more days each year when smog reaches a noticeable level in some part of the basin than there are days on which no adverse effects are noted. Eye irritation and respiratory distress are common, and the visual range is frequently reduced to 5 or 10 km. The aerosol mass concentration over the downtown reaches 200 to 300 $\mu g/m^3$, with perhaps only 10 to 50 $\mu g/m^3$ of this being sea salt of maritime origin. The markedly reduced visibility in the polluted air gives a sharp definition to the height at which the inversion base is located, as illustrated in Fig. 10.17 for a polluted inland valley.

Despite this aerosol burden, the upward movement of polluted air escaping from the basin along the mountain slopes may be difficult to see, for the depth of the upslope wind is usually not sufficient to contain enough aerosol to scatter an appreciable amount of light. Nevertheless, the effect is real, as evidenced by substantial ozone damage found on Ponderosa and Jeffrey pine trees on the slopes of the San Bernardino mountains at altitudes as high as 2000 m.[46] The upslope wind can also carry polluted air above the inversion, where, because of a wind shear, it may be caught and blown back across the basin *above* the inversion. Figure 10.17 illustrates such a case over an inland valley. The clear band of air in this photograph is the lower portion of the inversion layer, into which aerosols have not penetrated, but above and below which polluted air is found.

Angelenos have experienced the sensory and respiratory irritation of smog for over twenty years. Yet so far as statistics indicate, there is no firm evidence that the smog has increased the mortality rate. There has been no dramatic episode such as occurred in Donora or London, just a repeating cycle of episodes of varying intensities. Although damage to materials such as rubber is considerable, it goes largely unnoticed in the public mind. Neither has the elimination of many forms of agriculture, both in the Los Angeles County and downwind to the east, made an impact on public sentiments. But the outrage which develops when the eyes are streaming tears and the throat burning has had a more

Fig. 10.17 Inversion layer over an inland valley under which pollutants are trapped. Some aerosols have been carried around and above the inversion by upslope winds, and the pollutants have drifted back, leaving a clear strip that defines the lower portion of the inversion layer. (Photograph by John E. Taft. Courtesy of Environmental Films, Inc.)

constructive effect. The enactment of motor vehicle emission controls by the State of California was in large measure a result of the political pressure brought to bear by Southern Californians. And this in turn stimulated national and world efforts to control automotive emissions. Many concepts introduced by Los Angeles to control stationary sources have similarly been adopted by other pollution control agencies. Whether all of these efforts will be sufficient to reduce substantially the pollution episodes in Los Angeles remains for the future to determine. There is evidence that the situation is improving.

SUMMARY

For photochemical smog one needs intense sunlight and two chemical ingredients: nitrogen oxides and organic compounds. The photodissociation of NO_2 provides atomic O for oxidation of organic molecules and formation of free radicals. A series of atmospheric reactions creates oxidizing agents such as ozone, more NO_2, and the PAN's. These secondary pollutants are largely responsible for the adverse effects of smog. A maturing smog also evolves an aerosol consisting of organic material and nitrates; and certain particles as well as NO_2 are responsible for the characteristic color of a smog-supporting atmosphere. In most communities the major source of nitrogen oxides and organic molecules is the automobile. Control of the emission of nitrogen oxides has largely depended upon reducing the flame temperature of the combustion process. Control of organic emissions has come about through reduction of evaporative loss and more efficient combustion. Lead compounds and carbon monoxide are two primary pollutants which are also emitted by automobiles but are not known to contribute substantially to photochemical smog. Lead can be eliminated from the exhaust by removing it from gasoline. Carbon monoxide emissions can be reduced by more efficient combustion in the presence of sufficient oxygen.

NOTES

1. E. A. Schuck and E. R. Stephens, "Oxides of Nitrogen," in *Advances in Environmental Sciences*, J. N. Pitts, Jr., and R. L. Metcalf, Eds., Vol. I, New York: Wiley-Interscience, 1969, p. 73.

2. F. A. Bagwell, K. E. Rosenthal, D. P. Teixeira, B. P. Breen, N. Bayard de Volo, and S. Kerho, "Utility Boiler Operating Modes for Reduced Nitric Oxide Emissions," *J. Air Poll. Control Assoc.* **21**, 702 (1971).

3. A. J. Haagen-Smit, "Reactions in the Atmosphere," in *Air Pollution*, Vol. I, A. C. Stern Ed., New York: Academic Press, 1962.

4. N. A. Renzetti and D. J. Doyle, "Photochemical Aerosol Formation in Sulfur Dioxide-hydrocarbon Systems," *Intern. J. Air Poll.* (London) **2**, 327 (1960).

5. R. C. Robbins, R. D. Cadle, and D. L. Eckhardt. "The Conversion of Sodium Chloride to Hydrogen Chloride in the Atmosphere," *J. Meteorol,* **16**, 53 (1959).

6. S. K. Friedlander and J. H. Seinfeld, "A Dynamic Model of Photochemical Smog," *Environ. Sci. and Tech.,* **3**, 1175 (1969).

7. See, for example, J. N. Pitts, Jr., A. U. Khan, E. B. Smith, and R. P. Wayne, *Environ. Sci. and Tech.* **3**, 243 (1969).

8. J. R. Goldsmith, "Effects of Air Pollution on Human Health," in *Air Pollution*, Vol. I, A. C. Stern, Ed., New York: Academic Press, 1968, p. 547.

9. H. E. Stokinger and D. L. Coffin, "Biological Effects of Air Pollutants," in *Air Pollution*, Vol. I, A. C. Stern, Ed., New York, Academic Press, 1968, p. 446.

10. "Toxicologic and Epidemiologic Bases for Air Quality Criteria," *J. Air Poll. Control Assoc.*, Sept. 1969.

11. L. G. Wayne, *Atmos. Environ.*, **1**, 97 (1967).

12. D. Henschler, A. Stier, H. Beck, and W. Neuman, "Olfactory Thresholds of Several Important Irritant Gases and Manifestation in Man in Low Concentration," *Archiv. für Gewerbepathologie und Gewerbehygiene,* **17**, 574 (1960).

13. H. E. Stokinger, W. D. Wagner, and O. J. Dobrogorski, "Ozone toxicity Studies III. Chronic Injury to Lungs of Animals Following Exposure at a Low Level," *Arch. Ind. Health,* **16**, 514 (1957).

14. G. C. Buell, Y. Tokiwa, and P. K. Mueller, "Potential Crosslinking Agents in Lung Tissue. Formation and Isolation after *in vivo* Exposure to Ozone," *Arch. Environ. Health,* **10**, 213 (1965).

15. W. H. Blair, M. C. Henry, and R. Ehrlich, "Chronic Toxicity of Nitrogen Dioxide II. Effect on Histopathology of Lung Tissue," *Arch. Environ. Health*, **18**, 186 (1969).

16. M. Deane, J. R. Goldsmith, and D. Tuma, "Respiratory Conditions in Outside Workers: Report on Outside Plant Telephone Workers in San Francisco and Los Angeles," *Arch. Environ. Health*, **10**, 323 (1965).

17. R. Hauskneckt, "Experiences of a Respiratory Disease Panel Selected from a Representative Sample of the Adult Population," *Amer. Rev. Respirat. Dis.*, **86**, 858 (1962).

18. W. S. Wayne, P. F. Wehrle, and R. E. Carroll, "Oxidant Air Pollution and Athletic Performance," *J. Amer. Med. Assoc.*, **199**, 901 (1967).

19. G. W. Wright, "An Appraisal of Epidemiologic Data Concerning the Effect of Oxidants, Nitrogen Dioxide and Hydrocarbons upon Human Populations," *J. Air Poll. Control. Assoc.*, **19**, 679 (1969).

20. A. P. Altshuller, "An Evaluation of Techniques for Determination of the Photochemical Reactivity of Organic Emissions," *J. Air Pol. Control Assoc.*, **16**, 257 (1966).

21. W. A. Glasson and C. S. Tuesday, "Hydrocarbon Reactivities in the Atmospheric Photooxidation of Nitric Oxide," *Environ. Sci. and Tech.*, **4**, 916 (1970).

22. W. Agnew, "Automotive Air Pollution Research," *Proc. Roy Soc.*, **A307**, 153 (1968).

23. R. W. Hurn, "Mobile Combustion Sources," in *Air Pollution*, Vol. III, A. C. Stern, Ed.. New York: Academic Press, N. Y., 1968, p. 55.

24. W. A. Glasson and C. S. Tuesday, "Hydrocarbon Reactivity and the Kinetics of the Atmospheric Photooxidation of Nitric Oxide," *J. Air Poll. Control Assoc.*, **20**, 239 (1970).

25. R. L. Duprey, "Compilation of Air Pollutant Emission Factors," U.S. Dept. of Health, Education, and Welfare, Public Health Service, Bureau of Disease Prevention and Environmental Control, National Center for Air Pollution Control, Durham, N.C., 1968.

26. A. P. Altschuller and J. J. Bufalini, "Photochemical Aspects of Air Pollution: A Review," *Environ. Sci. and Tech.*, **5**, 39 (1971).

27. E. Robinson and R. C. Robbins, "Sources, Abundance, and Fate of Gaseous Atmospheric Pollutants," Final Report, Stanford Research Institute Project PR-6755 for American Petroleum Institute, New York (Feb. 1968).

28. J. Tyndall, *Phil. Mag.*, **37**, 384 (1869); also **38**, 156 (1869).

29. F. W. Went, "Organic Matter in the Atmosphere, and its Possible Relation to Petroleum Formation," *Proc. Natl. Acad, Sci.*, U.S., **46**, 212 (1960).

30. J. Heicklen, K. Westberg, and N. Cohen, in *Chemical Reactions in Urban Atmospheres*, C. S. Tuesday, Ed., Amsterdam: Elsevier. 1972.

31. K. Westberg, N. Cohen, and K. W. Wilson, "Carbon Monoxide: Its Role in Photochemical Smog Formation," *Science*, **171**, 1013 (1971).

32. J. R. Goldsmith and S. I. Cohen, "Epidemiological Bases for Possible Air Quality Criteria for Carbon Monoxide," *J. Air Poll. Control Assoc.*, **19**, 704 (1969).

33. A. C. Hexter and J. R. Goldsmith, "Carbon Monoxide: Association of Community Air Pollution with Mortality," *Science*, **172**, 265 (1971).

34. R. M. Brice and J. F. Roesler, "The Exposure to Carbon Monoxide of Occupants of Vehicles Moving in Heavy Traffic," *J. Air Poll. Control Assoc.*, **16**, 597 (1966).

35. R. R. Beard and G. A. Wertheim, "Behavioral Impairment Associated with Small Doses of Carbon Monoxide," *Am. J. Public Health*, **57**, 2012 (1967).

36. H. Levy II, "Normal Atmosphere: Large Radical and Formaldehyde Concentrations Predicted," *Science*, **173**, 141 (1971); B. Weinstock and H. Niki, "Carbon Monoxide Balance in Nature," *Science* **176**, 290 (1972).

37. B. Dimitriades and M. Whisman, "Carbon Monoxide in Lower Atmosphere Reactions," *Environ. Sci. and Tech.*, **5**, 219 (1971).

38. T. J. Chow and J. L. Earl, "Lead Aerosols in the Atmosphere: Increasing Concentrations," *Science*, **169**, 577 (1970).

39. J. J. Chisolm, Jr., "Lead Poisoning," *Scientific American*, **224**, 15 (1971).

40. M. Murozumi, T. J. Chow, and C. C. Patterson, *Geochim. Cosmochim, Acta*, **33**, 1247 (1969).

41. R. A. Kehoe, "Toxicological Appraisal of Lead in Relation to the Tolerable Concentration in the Ambient Air," *J. Air Poll. Control Assoc.*, **19**, 690 (1969).

42. M. Neiburger, "The Relation of Air Mass Structure to the Field of Motion over the Eastern North Pacific Ocean in Summer," *Tellus*, **12**, 31 (1960); M. Neiburger, D. S. Johnson, and Chen-Wu Chien, *The Inversion Over the Eastern North Pacific Ocean*, Los Angeles: University of California Press, 1961.

43. J. G. Edinger, "Changes in the Depth of the Marine Layer over the Los Angeles Basin," *J. Meteorol.*, **16**, 219 (1959).

44. W. J. Hamming, W. G. MacBeth, and R. L. Chass, "The Photochemical Air Pollution Syndrome," *Arch. Environ. Health*, **14**, 137 (1967).

45. H. Horvath, "On the Brown Color of Atmosphere Haze," *Atmos. Environ.*, **5**, 333 (1971); and **6**, 143 (1972).

46. J. G. Edinger, M. H. McCutchan, P. R. Miller, B. C. Ryan, M. J. Schroeder, and J. V. Behar, "The Relationship of Meteorological Variables to the Penetration and Duration of Oxidant Air Pollution in the Eastern South Coast Basin," Project Clean Air Research Reports, Vol. 4, Univ. of California, Sept. 1, 1970.

FOR FURTHER READING

P. A. LEIGHTON   *Photochemistry of Air Pollution*, New York: Academic Press, 1961.

J. G. CALVERT and   *Photochemistry*, New York: Wiley, 1966.
J. N. PITTS, Jr.

J. N. PITTS, Jr. and   *Advances in Environmental Sciences*, New York: Wiley-Interscience,
R. L. METCALF, Ed.   1969, Vol. 1.

A REVIEW: Toxicologic and Epidemiologic Bases for Air Quality Criteria, *J. Air Poll. Control Assoc.*, **19**, 629–732 (1969).

M. TRESHOW — *Environment and Plant Response*, New York: McGraw-Hill, 1970.

J. J. CHISOLM, Jr. — "Lead Poisoning," *Scientific American*, **224**, 15 (1971).

A. P. ALTSCHULLER and J. J. BUFALINI — "Photochemical Aspects of Air Pollution: A Review," *Environmental Sci. and Tech.*, **5**, 39 (1971).

E. S. STARKMAN, Ed. — *Combustion Generated Air Pollution*, New York: Plenum Press, 1971.

R. G. TEMPLE — "Control of the Internal Combustion Engine," in *Air Pollution Control*, Part I, W. Strauss, Ed., New York: Wiley-Interscience, 1971.

C. S. TUESDAY, Ed. — *Chemical Reactions in the Urban Atmosphere*, New York: American Elsevier, 1971.

QUESTIONS

1. Explain the meaning of NO_x, total oxidant, peroxyacyl nitrate, olefin, paraffin, and aromatic.

2. During the development of smog, why does the concentration of ozone begin to increase appreciably only after the ambient NO is essentially eliminated?

3. What affects the appearance of an atmosphere which supports photochemical smog?

4. Why is photochemical smog said to be an oxidizing atmosphere? What are the important oxidizing agents?

5. Why is the ambient concentration of aldehydes found to increase as the NO_2 concentration initially increases during the formative states of photochemical smog?

6. How does the chemical composition of gasoline affect the reactivity of organic molecules in the exhaust?

7. What are the known eye irritants of photochemical smog and their threshold concentrations for producing irritation?

8. Explain from the standpoint of molecular structure why some organic molecules are more reactive than others in participating in smog. Why is benzene not appreciably reactive despite that fact that its chemical formula C_6H_6 suggests that it is an unsaturated hydrocarbon?

9. Can the emission of CO, NO_x, and hydrocarbons simultaneously be minimized in automobile exhaust by adjusting the air to fuel ratio? Explain your answer.

10. What role does CO play in photochemical smog? How does it affect humans?

11. If the constituents of gasoline are altered so as to reduce the reaction rate for NO oxidation by a factor of 2, what might be the consequences for the development of smog in various regions of a city across which a wind blows?

PROBLEMS

1. To compare the importance of NO_x emissions from uncontrolled power generating stations and automobiles in a community, assume that a population of 1 million inhabitants is supplied by a gas-burning power station rated for 1 GW electrical capacity, operating at 40% efficiency with no NO_x controls. If 250,000 pre-1966

automobiles operate each day, how far must each go on the average to have the NO_x emitted by all automobiles equal the NO_x emitted by the power plant during capacity operation for 12 hours?

2. At the equinox, with the sun in the equatorial plane of the earth, what is the maximum rate at which solar radiant energy is incident on each square meter of surface area of the city of Los Angeles (32°N latitude) compared with London (51°N)? On a winter day in the Northern Hemisphere when the sun is directly overhead at 20°S, what is the ratio of the incident solar energy for the two cities when the sun passes over their respective longitudes?

3. Early in the day the NO_x concentration at the monitoring station is found to be 0.30 ppm. If this is believed to come mainly from 1976 automobiles, what would you expect the CO concentration at the same point to be?

4. What is the reactivity of 1,3,5-trimethylbenzene, *o*-xylene, and toluene? What do the differences between rates suggest about the nature of the reaction?

5. Suppose that air were polluted with 1 ppm of 2-methyl-hexene-2 and 0.4 ppm of NO_x (about 5% of this being NO_2). How long would be required under typical conditions of sunlight to oxidize half the NO to NO_2? How long would it take for 2-methyl-hexene-1 under similar conditions? How long for 2,3-dimethyl-hexene-2? What is a chemical reason for these differences?

CHAPTER 11

AEROSOLS

In its most general sense, the word "aerosol" refers to dispersed solid or liquid matter in a gaseous medium. We have previously noted that the aerosol burden of the atmosphere consists of particles of various sizes and composition. Some are primary pollutants and some secondary. Some have anthropogenic origins and some are naturally produced. It is estimated that on a global basis, about 90% of the particulate matter in the air is naturally occurring.[1] The proportion, however, can be tipped in the other direction near highly industrialized cities.

The composition and size distribution among the particles varies from place to place and may differ markedly between air which has been over the oceans for many days and air which has been conditioned in traveling a long distance over a continental land mass. The evidence indicates that aerosols are continually being emitted and removed from the air. Some estimates conclude that the mean time for which a particle is a resident in the lower troposphere is about five days. This period is sufficiently long for emitted particulate matter to travel an appreciable distance before it is scavenged from the air, and thus the effect of aerosol release from a given source may be widespread. Yet the average residence time is sufficiently short that the emission or formation of aerosols in the lower troposphere may be regarded as producing a transient effect; in other words, natural processes of aerosol scavenging from the troposphere should cleanse the air within a few days once the emission of the aerosol is terminated. We recall that this is not true of aerosols injected into the stratosphere, where the residence time in these upper regions is two or three years, or the upper troposphere, where the residence time is about one month.

Aerosols play an important role in air pollution for several reasons. First, chronic exposure to high concentrations of particulate matter may be injurious to the lung. Second, aerosols are known to play a synergistic role in potentiating the toxic effects of certain other pollutants such as SO_2. Some ashes will also catalyze the oxidation of SO_2 into sulfuric acid. Third, aerosols are of interest because they increase atmospheric turbidity and reduce visibility. Possibly there is also an effect on the energy balance of the earth, through a change in the earth's albedo or through absorption of radiation. Fourth, many gaseous pollutants end their lives by conversion into aerosols. The formation of sulfuric acid droplets by oxidation of SO_2 is one example. Another occurs in photochemical smog where gaseous SO_2 is oxidized into very small particles. Still another—not in-

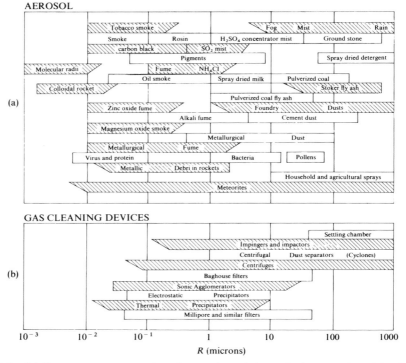

Fig. 11.1 (a) Some common examples of aerosols and the typical range of particle size for each. (b) Devices which may be used to collect aerosols from a gas stream for the indicated range of particle size. (From G. M. Hidy and J. R. Brock, *The Dynamics of Aerocolloidal Systems*, Oxford: Pergamon Press, 1970.)

volving SO_2—is the organic aerosol formed when a photochemical smog matures. Thus aerosols are a common secondary pollutant of smogs. If we ask what the ultimate fate of atmospheric pollutants is, we must in many cases turn to examine the dynamics of aerosols.

There are two general ways that aerosols are formed, the *breakup of material* and the *agglomeration of molecules*. Figure 11.1 lists many common examples of aerosols and the typical sizes of the particles. Some of these are formed by breakup and some by agglomerations. We also show in this figure control devices which are used for cleaning particles of various sizes from gases. The principles of most of these devices were discussed in Section 9.1.

Biological materials such as fungi and pollen do not fit uniquely into either category. Pollens are responsible for many allergic reactions. They have a typical size of 10–50 microns and many are hygroscopic.[2] But they are only a minor component of aerosols on a global basis, although their distribution is extensive. They may travel for several thousand kilometers dispersed in the air. Some spores of a number of molds have been found at altitudes above 12 km! In this chapter, however, we shall be concerned with the dominant components of atmospheric aerosols and how they evolve, so we shall not deal with the biological component. We shall commence by considering a mode of aerosol formation with which everyone is familiar.

11.1 BREAKUP

Man's utilization of mineral resources is dependent upon changing the physical form of these resources. Pulverization of coal, rock crushing, and cement manufacturing are everyday examples. If not properly controlled, the resulting emission of particles can cause a severe local nuisance and complaints. However important such emissions may be within a neighborhood, they are nevertheless unmatched on a global basis in comparison with the breakup and emission of particles by natural means. Attrition through rock weathering by physical and chemical processes has left fine grain top soils and sands over large areas of the earth. Under the proper circumstances, the finer particles may be entrained in the wind and carried to great heights. Debris from frequent sand storms in the Sahara desert has been carried by the prevailing trade winds, so that over the years large deposits have developed in the Atlantic Ocean west of north Africa. Sahara sand has been carried as far as Florida and on occasion to England.[3]

The removal of ground cover by careless farming practices has in some instances left the land prey to violent winds. Dust storms in midwestern United States after the 1920's were one result (perhaps such events ought really to be classified as an anthropogenic source). Even now, the increased activity in the arid regions of western China is reportedly causing an increase in atmospheric turbidity over the eastern portions of that country.

A second example of breakup of material is evident in the ejecta of volcanic eruptions. During eruption, molten basalt is subjected to such turbulent motion that the liquid is separated into many fine particles before it solidifies. The larger ones fall quickly to earth, but as we noted in Section 4.8 the smaller ones may remain in the air for a considerable period.

Sea salt

Sea salt is found nearly everywhere in the troposphere. Its formation as an aerosol is another example of the breakup process. Fine droplets of water come from rupture of the liquid film which forms the upper surface of bubbles on sea water as illustrated in Fig. 11.2.[4] The rupture is so violent that about 200 particles may be formed for each bursting bubble, with the size of a typical particle being 0.1 micron.[5] These particles are caught and drawn upward in turbulent air and may be conveyed for many thousands of kilometers. Bursting bubbles also release particles of a much larger size. As the film disintegrates and the water subsides into the air cavity, a jet develops in the center as shown in Fig. 11.2. When the jet breaks up, it may form as many as 10 larger droplets, each having a radius of about 5 microns. Their upward momentum carries them as high as 0.1–0.2 m

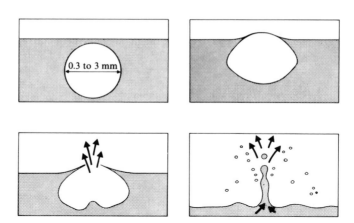

Fig. 11.2   Several stages in the formation of sea salt aerosol.

where they can be entrained in the wind. The extensive surface of the oceans ensures that the formation of aerosols by bursting bubbles is a major contribution to the atmospheric burden.

What of the particles formed by breaking waves? Visitors to an ocean shore can readily appreciate that locally the concentration of droplets released from the surf is considerable, but most of these droplets are quite large. They quickly settle to earth, usually within a few kilometers of the shoreline. Only the burst of a small liquid film of a bubble seems able to supply sufficient energy in a small enough volume to break up water into fine particles. The mighty crash of the surf, although much more impressive because of its scale, does not do this as effectively.

The composition of the sea salt aerosol includes not only water but the dissolved and suspended constituents of the ocean. The primary constituent is of course dissolved NaCl which constitutes 3.3–3.6% of sea water by weight. As listed in Table 11.1 other material in minor amounts includes sulfates, carbonates,

TABLE 11.1

Average composition of the main constituents of sea water [*]

Constituent	Concentration (g/kg)	Constituent	Concentration (mg/kg)
Cl	18.980	K	380
Na	10.561	C[a]	28
S	0.890	CO_3	140
SO_4	2.650	Br	65
Mg	1.272	Sr	13
Ca	0.400	B	4

[a]Inorganic carbon.

[*]C. E. Junge, *Air Chemistry and Radioactivity*, New York: Academic Press, 1963. Copyright © 1963 by Academic Press and reprinted by permission.

and small amounts of potassium, calcium, magnesium, and organic matter. The liquid particles lose considerable bulk as they diffuse through the air, for the atmosphere generally is not saturated with water vapor. Evaporation may shrink the droplets by more than a factor of 2 or 3 before they come into equilibrium with the water vapor content of the air. In regions where the relative humidity is less than 60%, the water component evaporates completely, leaving solid particles.

11.2 AGGLOMERATION

The physical or chemical breakup of solid and liquid matter is perhaps the most familiar method by which aerosols are formed. But a second method is just as important for their creation. This is the *agglomeration* of gaseous molecules to form solid or liquid aerosols. One example of agglomeration is the formation of water droplets in humid air by condensation, resulting in the development of clouds or fog. Water vapor condenses preferentially on hygroscopic particles such as sea salt or on hygroscopic molecules such as SO_3. As agglomeration proceeds, the water droplets grow.

Growth may be enhanced by chemical reactions, such as occur when SO_2 is absorbed and oxidized to sulfate in a droplet, thereby permitting further SO_2 absorption. The addition of atmospheric NH_3 by agglomeration and its entry into solution provides a buffer to increase the pH and encourage still more SO_2 absorption and droplet growth. Analyses of the composition of the particulate matter over many European cities[3] indicate that 70–80% by weight is material which is insoluble in water—about two-thirds of this material is ash and the remainder is tar and other combustible organic matter. The water-soluble compounds include sulfate ion $SO_4^=$, sodium Na^+, chlorine Cl^-, and nitrate NO_3^-, sulfate often accounts for about half of the total soluble material. Ammonium NH_4^+ is also observed, especially in aerosols whose particle size lies between 0.1 and 1.0 micron. The concentration of NH_4^+ compared with the concentration of $SO_4^=$ in these particles was approximately the same as occurs in $(NH_4)_2SO_4$, suggesting that trace amounts of ammonia NH_3 in the atmosphere have combined with sulfuric acid droplets. Atmospheric NH_3 will also unite with gaseous HCl to form NH_4Cl aerosols. The polar nature of the electric charge distribution in this molecule encourages their agglomeration when unlike charges come into proximity during collisions between molecules. This may hold true for agglomeration with other polar molecules such as water vapor.

Another example of chemical agglomeration occurs when fossil fuels are burned, and volatile compounds escape combustion. Carbon and complex organic molecules then combine in the air within the combustion chamber to form small tarry particles and soot, well-known constituents of the atmosphere wherever bituminous is used.

From the foregoing, it is correct to conclude that the aerosol composition over continental regions is heterogeneous. It is a mixture of emissions from both anthropogenic and natural sources. Included also is a small quantity of sea salt introduced by air blown in from maritime regions. This is illustrated by an analysis of cloud droplets over northern Japan; the rich composition can be divided into three broad classifications of particulate matter: sea salt, combustion products, and soil material.[6] The most abundant in this region were combustion products, followed by sea salt and then dust. Despite the heterogeneous nature of this and continental aerosols, their evolution and demise have several features in common which depend primarily upon the size of the particulate matter. For this reason, aerosols are usually classified according to their size or radius.

11.3 PARTICLE SIZE

Atmospheric aerosols are divided into three groups according to the following system:

Particle radius	Name
$R < 0.1$ micron	Aitken particles
$0.1 < R < 1.0$ micron	Large particles
$1.0 < R$ micron	Giant particles

The divisions between the different classes of aerosols are somewhat arbitrary; the numerical criteria for the radii are defined mainly for ease in recalling them. The classification in a general sense corresponds to the different physical behavior of particles of different size. *Large particles* are effective in scattering visible light, because their dimensions are comparable to the wavelengths of visible light. *Giant particles* are distinguished by the fact that the air around them can be treated as a fluid which exerts a uniform pressure on all sides. This means that they execute practically no Brownian motion. However, both *large particles* and *Aitken particles* are sufficiently small that Brownian motion is important in influencing their behavior, so they migrate through air in a random fashion. As we shall see, this mobility and the resulting frequent collisions between the moving particles is an important feature in the dynamics and evolution of aerosols.

The class of smallest particles is named after J. Aitken who early in this century studied them extensively.[7] He was able to observe their presence by using an expansion type cloud chamber, in which humid air containing a sample of the aerosol of interest is first introduced into the chamber whereupon the chamber is sealed off. The volume of the chamber is then rapidly increased by use of a bellows arrangement, and during this expansion the contents cool adiabatically. As the temperature passes below the dew point, condensation of the water vapor occurs preferentially on the aerosols; and the water droplets thus indicate where particles are located. In most aerosols found over continental

land masses, the concentration of particles less than 0.1 micron in size is overwhelmingly greater than the concentration of large and giant particles. Therefore the number of droplets in the expansion chamber yields a measure of the
number of only these smallest particles, known as *Aitken particles*, or *Aitken
nuclei*.

Despite the preponderance of Aitken particles in continental aerosols, they
normally constitute only 10–20% of the aerosol mass. Most of the mass is due to
the relatively few large and giant particles, shared about equally between particles in the two classifications.

11.4 COAGULATION

Through Brownian motion, Aitken particles and large particles continually
migrate through the air and occasionally collide with other particles. A certain
fraction of collisions will result in *coagulation* of the particles, leaving them united
as a larger one. By this process, there is a continual removal of particles from
the Aitken and large particle classifications and an enhancement of the number
of giant particles. The rate at which this occurs depends upon the composition
of the particles, the rate at which they migrate, and the cross-sectional area of
each particle which defines the size of the target it presents. But the factor of
overwhelming importance is the concentration of particles.

We can easily see why the concentration is the most important parameter.
Let us first consider an aerosol consisting of identical particles at a number
density n. The rate at which each particle strikes others is proportional to the
number density of targets or n; and since there are n projectiles in each unit
volume, the total rate at which collisions occur must be proportional to n^2.
Thus the rate increases rapidly with increase in aerosol concentration. Hence an
aerosol with particles covering a range of sizes, in which the representation of
small particles is much greater than large ones, does indeed find small particles
colliding and coagulating with other small particles more frequently.

Coagulation is most important in affecting the size distribution of Aitken
particles and appears to be at least partly responsible for defining the lower end
of the aerosol size distribution. Particles less than 0.01 micron coagulate so
rapidly that they are quickly removed in the formation of larger ones. Brownian
motion is most important in bringing the particles into contact with each other,
although atmospheric turbulence also enhances the diffusive motion of particles
and plays a relatively more important role for large particles whose Brownian
motion is less pronounced. Agglomeration of gas molecules onto particles also
works to remove particles from smaller size categories.

As particles increase in size, perhaps first through their formation by the
agglomeration of molecules in condensing vapors and then by the coagulation
of pairs of particles, the effect of gravity becomes more important than the Brownian motion. Sedimentation then is an important factor which removes these
particles from the air.

11.5 SEDIMENTATION

In a dry atmosphere, the upper limit for aerosol sizes is determined by sedimentation. This is such an important process that we shall give it a closer examination than we have in previous chapters. It will be sufficient to restrict our attention to the behavior of spherical particles. The gravitational force F_g experienced by such a particle of radius R and density ρ_p is given by

$$F_g = \tfrac{4}{3}\pi R^3 \rho_p g, \tag{11.1}$$

where the acceleration due to gravity is $g = 9.80$ m/sec^2. And if only this force were acting on the particle, it would indeed accelerate downward at the rate g. However once the particle has acquired a downward velocity, it will also experience a retarding force due to friction with the air. This so-called drag force F_d is directed upward, opposite to the velocity, as illustrated in Fig. 11.3. (There is also an upward buoyant force, but since the density of the particle greatly exceeds the density of air we can neglect it.) It is reasonable to expect that the drag force will increase in magnitude as the downward velocity of the particle increases; thus as the particle continues to gain speed, the drag force will become equal to the constant gravitational force at some point, and the particle will no longer accelerate. Then the particle will continue to fall with a constant velocity called the *terminal velocity*. The terminal velocity is quickly

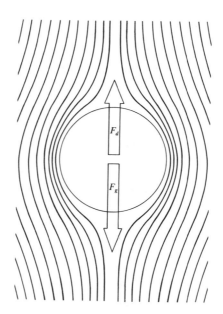

Fig. 11.3   Forces on a falling particle. Lines indicate how air is displaced as the particle falls at low velocity.

attained by a falling particle and for this reason it is an important parameter because it determines how fast a particle of a given size will settle. Therefore we shall derive an expression for this size dependence.

We first recall that the drag force represents the rate at which the particle loses momentum to the air and thus must be dependent upon the viscosity of air. It would not be appropriate to carry out a detailed calculation of this viscous effect, but we can give a simple dimensional argument to show how F_d might depend upon the viscosity coefficient η, the particle's downward speed V, and the particle's radius R. Then we will state the exact theoretical result. If we grant that F_d should be proportional to η, we can express the drag force by the following formula

$$F_d = A\,\eta.$$

The dimensions of the proportionality constant A must satisfy the following dimensional equation associated with this formula:

$$\left[\frac{(\text{mass})\,(\text{length})}{(\text{time})^2}\right] = (\text{dimensions of } A) \left[\frac{(\text{mass})}{(\text{length})(\text{time})}\right].$$

The left-hand side of this dimensional equation has the dimensions of a force, F_d. The square brackets on the right-hand side enclose the dimensions of the viscosity coefficient η. Thus from this equation we see that A must have the dimensions of $(\text{length})^2(\text{time})^{-1}$, which we expect should correspond to an appropriate combination of R and V. Just such a combination is provided by the product RV, and therefore we anticipate that F_d can be expressed in the following form:

$$F_d = (\text{dimensionless constant})\,\eta\,RV. \qquad (11.2a)$$

This is as far as we can go with dimensional arguments, for they cannot provide a means of evaluating the dimensionless constant. Nevertheless, the above result is a little surprising by itself, because it indicates that F_d is *proportional to the radius* of the sphere and not the cross-sectional area [which would be the case if Eq. (11.2a) contained a factor R^2 instead]. We might have naively and incorrectly guessed that the area would be the important parameter, on the basis of the amount of air the particle displaces as it falls. But the validity of Eq. (11.2a) is borne out by exact calculations describing the effects of the patterns of air flow as the particle moves through the air. Such calculations have been made for particles whose speed V is relatively small, so that the inertial effect of air flowing around the falling particle can be neglected. (Thus we assume that the Reynolds number, introduced in Section 7.6, is small, so there is no turbulence in the wake of the particle.) The calculations were first carried out by G. C. Stokes in 1851, and he found that the drag force is given by

$$F_d = 6\pi\eta\,RV. \qquad (11.2b)$$

Thus the dimensionless constant appearing in Eq. (11.2a) is the factor 6π. As expected, this result shows that the drag force increases as the particle's downward velocity increases. When the total force on the particle is zero ($F_d = F_g$), it ceases to accelerate and continues downward with its terminal velocity, which we shall denote by the symbol V_t. The value of the terminal velocity is obtained by equating Eqs. (11.1) and (11. 2b), and then we find

$$V_t = \frac{2R^2\,\rho g}{9\eta}. \tag{11.3}$$

From this we can see that the terminal velocity, which determines the rate of sedimentation, increases rapidly with particle radius R. If a particle with a 1.0 micron radius has 0.01 cm/sec for the value of V_t, a particle of similar composition, but with a radius of 10 microns, has V_t increased to 1 cm/sec. The terminal velocity is also related to the particle's density, but only by a less sensitive linear dependence. The density of particles in an aerosol may vary by a factor of two or three at most, averaging somewhere between 2 and 2.5 g/cm^3. As a consequence of the much broader spectrum of sizes in an aerosol, particle radius is by far the more important parameter determining sedimentation rates.

For particles larger than about 50 microns, the terminal velocity is sufficiently great that turbulence in the wake of the particle cannot be neglected, and F_d is found to be larger than is given in Eq. (11.2b). As a result, V_t is somewhat lower than is indicated by Eq. (11.3). For example, the actual terminal velocity of a particle with 200 micron radius is about one-third of the value given by Eq. (11.3).

From Eq. (11.3), including the retarding effects of turbulence, the time for a particle to fall from a given altitude can be calculated, and representative curves relating the sedimentation time with a particle's initial altitude are shown in Fig. 11.4 for several particle sizes. The curves in this figure show that particles larger than $R = 20$ microns rapidly settle out. The settling rate of course depends

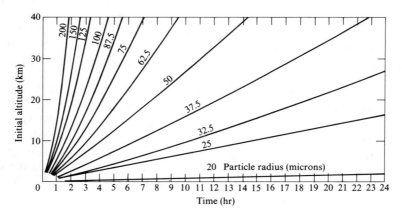

Fig. 11.4 Time for a particle of a given radius to fall from various altitudes. The particle density is assumed to be 2.5 g/cm^3. (From S. Glasstone, *The Effects of Nuclear Weapons*, revised edition, U.S. Atomic Energy Commission, 1962.)

upon the actual shape of the particle if it is not spherical, but it should not vary appreciably from the curves in the figure if R is taken to be an average radius. Sedimentation therefore determines the maximum size of giant particles that may remain for long periods in the air.

The drag force in Eq. (11.3) describes the effects of collisions between air molecules and a particle assuming that air behaves as a fluid. Measurements for this viscous effect on particles with radii ranging from 10–30 microns show that it is accurate to at least 0.5%. However, the concept that air behaves as a fluid is not valid for very small particles, when the mean free path between molecular collisions is comparable to the radius R of the particle, and in such circumstances where Brownian motion is important, Stokes' result in Eq. (11.3) is invalid.

11.6 SIZE DISTRIBUTION

With the process of coagulation determining the lower end of the size spectrum of aerosols, and the process of sedimentation the upper end, one might ask whether these processes alone can explain the size distribution of particles in atmospheric aerosols. Will a unique distribution be established and remain constant in time? There has been considerable interest in determining whether such a "self-preserving" distribution exists.[8] Under certain simplifying assumptions, theoretical descriptions do predict that the size distribution will evolve toward a stable form. However, measurements often fail to confirm that urban aerosols follow a self-preserving distribution.[9,10]

Studies of continental aerosols have revealed several characteristic features of the size distribution which are often found in community air. One feature is that there is usually only one peak in the size spectrum. As an aerosol evolves with time, the peak may shift toward larger or smaller radii, but the shape is qualitatively unchanged. When particles are continually added from anthropogenic sources, the distribution at nearby locations may exhibit several peaks, corresponding to the predominant sizes of the particles which are emitted. Within a few hours, as the aerosol drifts downwind, mixes, and evolves, such bumps are smoothed out.[10] Frequently the size distribution is then found to obey a log-normal distribution.

A striking feature of many continental aerosols is a regular decrease in the concentration of particles above 0.1 micron size. This is illustrated in Fig. 11.5 by the upper solid curve which gives the distribution observed by C. E. Junge for representative European aerosols some distance from their source. This particular curve summarizes data obtained from a monitoring station on the Zugspitze in Germany.[3] Junge has pointed out that the distribution of large and giant particles often can be described by the empirical formula

$$\Delta n = CR^{-(\beta+1)} \Delta R, \qquad (11.4)$$

where the symbol Δn is the concentration of particles whose radius lies between R and $R + \Delta R$, and C is a constant. For Eq. (11.4) to be valid, it is understood

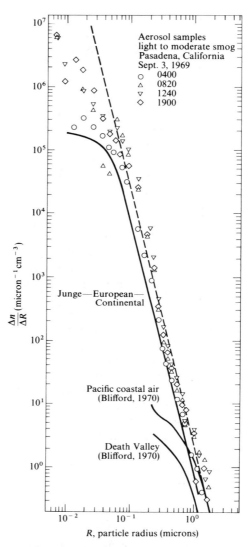

Fig. 11.5 Aerosol size distributions. The number density or concentration of particles per unit range of particle radius $\Delta n/\Delta R$ is given as a function of particle radius. The dashed line is an application of Eq. (11.4) to the Pasadena data, extended beyond its region of applicability. (From G. M. Hidy and S. K. Friedlander, "The Nature of the Los Angeles Aerosol," Second IUAPPA Clean Air Congress, Washington, 1970; Pasadena data: K. T. Whitby, Particle Laboratory Publication 141, Dept. of Mech. Eng., U. of Minnesota, 1970.)

that ΔR is small compared with R. The equation, which applies only in the regime $0.1 < R < 10$ microns, says that the size distribution can be approximated by a power law, and Junge found that the exponent β often has the value 3. The exact value may vary from city to city and with the time of day, ranging from 2.5 to

4 depending upon the type of pollution; for example, during a photochemical smog in the Los Angeles basin, it has been found on occasion to have the value 3.5.

Figure 11.5 shows data taken at several times during a day when photochemical smog was developing in the Los Angeles basin.[11] The size distribution for large and giant particles is in fair agreement with the Junge distribution. Early in the day the concentration of Aitken particles, for which Eq. (11.4) does not apply, is also similar to what was observed in the European aerosol. Later, as the smog matured, the concentration of Aitken particles rose markedly, the increase being more than a factor of 10 for particles of 0.01 micron size. Thus the size distribution of small particles very much depends upon the circumstances of local pollution.

Junge's distribution given by Eq. (11.4) is often fairly accurate for particles as large as about 10 microns; however, the observed concentration of larger particles generally falls off more rapidly than given by the formula in Eq. (11.4) as a result of a pronounced effect of sedimentation.

Regions not substantially affected by pollution may have aerosol distributions that differ markedly from Eq. (11.4). Two examples reported by Blifford[12] are shown in Fig. 11.5, one for the desert region of Death Valley, California, and the other for coastal maritime air. Unfortunately, the data do not extend for particles below 0.25 micron in radius, but it is clear that in both cases the concentration of large (and presumably Aitken) particles is substantially below that found in polluted community air.

Mass distribution

Equation (11.4) suggests that there is an interesting feature about the *mass distribution* of particles in an urban aerosol. This can be seen by rewriting the equation in the following form:

$$\Delta n = C_0 R^{-\beta} \Delta(\log_{10} R), \qquad (11.5)$$

where Δn now represents the number density of particles for which the logarithm of their radius lies within the interval $\Delta(\log_{10} R)$; and $C_0 = C \ln(10)$. If we make the approximation $\beta = 3$, we see that the right-hand side of the equation is inversely proportional to the mass of a particle of radius R. Thus we are motivated to multiply both sides of the equation by $4\pi R^3 \rho_p/3$, since the left-hand side would then be the mass concentration constituted by particles lying in the interval $\Delta(\log_{10} R)$. Thus for this aerosol, we would have a simple expression for the mass distribution given by

$$\Delta M = \frac{4\pi \rho_p C_0}{3} \Delta(\log_{10} R). \qquad (11.6)$$

In this expression ΔM is the mass concentration of particles for which the logarithm of their radius lies within the interval $\Delta(\log_{10} R)$. If we assume that

ρ_P is independent of particle size R, this equation means that the mass of the particles within a given size range is proportional to the difference in the logarithm of the maximum and minimum radius which define the endpoints of the range.

This is significant for the following reason. Where the Junge distribution is valid and the particles have nearly the same composition independent of size, the mass of all of the large particles, in the decade range 0.1–1.0 micron radius, must be equal to the mass of all the giant particles, in the decade range 1.0–10 microns. And the large and giant particles constitute nearly the entire aerosol mass, for the observed size distribution often falls below an extrapolated Junge distribution for radii above 10 microns and below 0.1 micron.

We are now in a position to understand why the prevailing visibility introduced in Section 2.3 is often found to depend only upon the total mass concentration of an aerosol. Visible light is scattered most effectively by the large particles, in the size range of 0.1 to 1.0 micron; and the amount of scattered light depends upon both the size distribution and concentration of these particles. But because the size distribution is often governed by Eq. (11.4) with $\beta = 3$, only the mass concentration of large particles varies from place to place, and this we have seen is a fixed fraction (about half) of the total aerosol mass. Thus we would expect the amount of scattered light to depend upon the total mass concentration of a typical aerosol, as has been observed experimentally. Our preceding discussion has carefully outlined the basic assumptions upon which this conclusion is reached, and whenever any of these is violated, we must anticipate that Eq. (2.14) may be found invalid.

It is appropriate at this point to consider a few quantitative aspects of typical aerosol concentrations in urban and nonurban regions. Many American cities are cluttered with suspended particulates whose mass concentration is typically 100–150 $\mu g/m^3$ on the average. This is considerably lower than the conditions over many European cities, according to the data of Green and Lane.[13] However, it is far above the average values of 40 $\mu g/m^3$ over agricultural areas and 20–30 $\mu g/m^3$ in areas remote from human activity.[14] The lowest observed aerosol concentrations are 5–10 $\mu g/m^3$ carried, on occasion, by ocean breezes.[15] This is also about the minimum value obtained on exceptionally clear days for some regions of the United States, and therefore appears to be the ever-present background level. It corresponds to a number density of about 200 particles per cubic centimeter.

11.7 COMPOSITION

The composition of an aerosol depends upon the sources where the supporting air mass was most recently conditioned. Thus the composition of continental aerosols differs from that of maritime aerosols. This is illustrated in Figs. 11.6 and 11.7 for large particles and giant particles, respectively, where the mass concentrations of their various important constituents are shown for a sequence

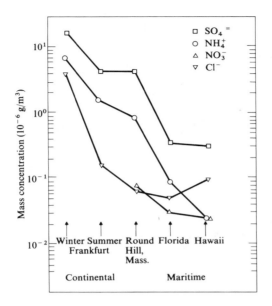

Fig. 11.6 Average chemical composition of large particles. (C. E. Junge, *Air Chemistry and Radioactivity*, New York: Academic Press, 1963.)

of monitoring locations, ranging from a continental urban region (Frankfurt, Germany) to a maritime environment (Hawaii).[3] Locations more affected by maritime air have a relatively greater concentration of chloride and lower concentrations of sulfate, nitrate, and ammonium ions; however, the sulfate concentration decreases only to the point where it represents the expected contribution from sea salt, as deduced from the measured chloride concentration and the known ratio of sulfate to chloride ions in sea water. As the vertical scales on these figures suggest, when compared with total aerosol mass concentrations, the amount of identified material is a small fraction of the total aerosol; what molecules constitute the balance is unknown.

Figures 11.6 and 11.7 show that the composition of an aerosol also varies with the size of the particles. For example, large particles over Frankfurt contained a greater percentage of ammonium sulfate than did giant particles; on the other hand, for maritime air over Hawaii, more chloride was found in the giant particles. It is possible that the large particles in continental air were created mainly by coagulation of agglomerated ammonium sulfate complex or by oxidation of sulfur dioxide in droplets containing dissolved ammonia, whereas the giant particles originated as such in maritime air in the form of sea salt from bursting bubbles.

The composition of continental aerosols also varies from urban to rural areas, according to the types of natural and anthropogenic sources and the

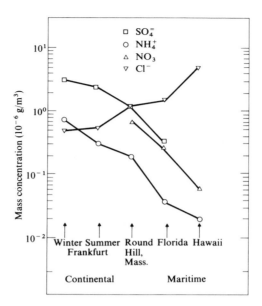

Fig. 11.7 Average chemical composition of giant particles. (From C. E. Junge, *Air Chemistry and Radioactivity*, New York: Academic Press, 1963.)

distance from these sources. An average trend is illustrated in Table 11.2 for data obtained from a number of monitoring stations in the United States. The pollution in the "urban" areas is predominantly sulfurous, so for comparison we also show data for the photochemical type from downtown Los Angeles in 1958 and 1966. The table gives the absolute and percentage concentration for sulfates, benzene-soluble organic matter, nitrates, iron, and lead. (The classification "benzene-soluble" includes polycyclic hydrocarbons, but not protein-based organic matter such as spores and pollens.) There is a marked contrast in the aerosol composition over cities dominated by sulfurous smog and those with photochemical smog. In the latter, there is a considerably higher concentration of benzene-soluble material and nitrates and very often a lower concentration of sulfates (not apparent in this table). Only a small fraction of the material in urban aerosols has been identified, as is evident from the percentages included in the table. In Los Angeles, perhaps 20–40% of the aerosols are primary pollutants and 30% are secondary; the remainder appears to be representative of the natural background concentration, mainly of marine origin, but this is highly variable.[16] In cities with sulfurous smog, much of the aerosol matter is ash or waste from chemical processes such as we discussed in Chapter 9. The table shows a marked drop in the percentage of the lead component as one goes from urban to remote areas.

TABLE 11.2

Comparison of compositions of urban and nonurban aerosols*

	Total mass ($\mu g/m^3$)	Sulfate		Benzene soluble		Nitrate		Iron		Lead	
		$\mu g/m^3$	%	$\mu g/m^3$	%	$\mu g/m^3$	%	$\mu g/m^3$	%	$\mu g/m^3$	%
Remote (1966–1967) 10 stations	21.0	2.51	11.8	1.1	5.1	0.46	2.2	0.15	0.71	0.022	0.10
Nonurban (1966–1967) 15 stations	40.0	5.29	13.1	2.2	5.4	0.85	2.1	0.27	0.67	0.096	0.24
Urban (1966–1967) 25 stations	102.0	10.1	9.9	6.7	6.6	2.4	2.4	1.43	1.38	1.11	1.07
Downtown Los Angeles (1958)	213.0	16.0	7.5	30.4	14.2	9.4	4.4	—	—	—	0.1–0.9
Downtown Los Angeles (1966)	119.0	14.0	11.8	15.2	12.8	13.0	10.9	1.4	1.2	1.6 (at Burbank)	1.3

*U.S. National Air-Sampling Network.

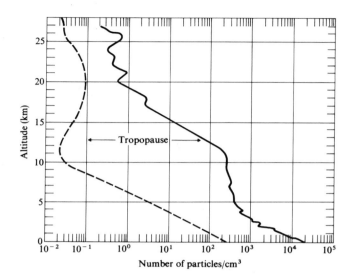

Fig. 11.8 Solid curve: a schematic representation of a concentration profile for particles over the central United States. (Adapted from C. E. Junge, *Air Chemistry and Radioactivity*, New York: Academic Press, 1963.) Broken line: total aerosol concentration for clean air, when the meteorological range exceeds 40 km. (L. Elterman, *Applied Optics*, **3**, 745, 1964.)

11.8 CONCENTRATION PROFILE

Most primary pollutants are emitted into the air at ground level. Even the secondary pollutants evolve for the most part in the mixing layer of the atmosphere within a kilometer or two of the ground. We might therefore expect that the concentration of aerosols decreases with increasing altitude. In fact, this has been observed, as is illustrated by the solid curve in Fig. 11.8. The concentration of each size of particle is usually found to decrease roughly exponentially with altitude within the troposphere, and the decrease is much more rapid than the decrease in the density of air (compare this figure with Fig. 3.8). Seasonal variations have been found in the concentration profile; in winter, when the lower troposphere is generally more stable, the decrease in concentration with altitude is more pronounced.

The importance of atmospheric turbulence in moving particles away from their point of origin and in establishing the concentration profile is perhaps most evident from observation of the conditions existing in an inversion layer where mixing does not occur. Figure 10.17 clearly indicates the nature of the barrier formed by inversion layers, where the maximum altitude attained by the light-scattering large particles is marked by a transition to a region of good visibility. Also the sedimentation of particles into the inversion from above is virtually nil, as shown by the sharp upper boundary.

The solid curve in Fig. 11.8 shows that above about 4 km the aerosol concentration ceases to decrease with altitude as rapidly as it had at lower levels, and the ratio of particle concentration to air density changes very little with altitude. Thus there is a marked difference between the aerosol dynamics within this layer of the upper troposphere and the mixing layer beneath. The boundary between the two regions varies with the instability of the underlying air; it rises when turbulent behavior prevails, as in the summer, and lowers during more stable periods, as in winter.

There is some evidence that the concentration of Aitken and giant particles decreases more rapidly with altitude in the lower troposphere than the concentration of large particles, a trend that is especially apparent above about 3 km.[17] This trend, prevailing at these higher altitudes in the presence of presumably older aerosols, is consistent with the notion that coagulation tends to reduce the population of small particles while sedimentation removes the larger ones. Thus the aerosol concentration at the tropopause of 200–300 per cm^3 represents mainly large particles between 0.1 and 1.0 micron size.

Within the lower stratosphere, the solid curve of Fig. 11.8 shows that the concentration of particles again rapidly decreases with increase in altitude. This decrease indicates that the lower stratospheric aerosol has a tropospheric origin. A steady-state upward and downward movement of particles is established when the upward diffusion of particles is balanced by sedimentation.

A surprising feature of the observed concentration profile is the existence of a relative maximum in the concentration of *large* particles at about 20 km. The broken curve of Fig. 11.8 for clean air illustrates this particularly well. The mean radius of these aerosols is about 0.15 micron, and their concentration is fairly uniform at 0.1 per cubic centimeter for latitudes between 60°S and 70°N, displaying little seasonal variation. Sulfate $SO_4^=$ is the dominant chemical constituent, accounting for about 90% of the aerosol mass, or $9 \times 10^{-3}\ \mu g/m^3$. Ammonium ion NH_4^+, and possibly sodium as well, is present in appreciable quantity. C. E. Junge has suggested that this stratospheric aerosol is formed *in situ* by the oxidation of SO_2 or H_2S.[3] As these gases mix upward from the troposphere, they would come into contact with the stratospheric ozone and experience intense ultraviolet solar radiation. Both ozone and radiation would favor oxidation. Through agglomeration and coagulation as they undergo reaction, the size of the particles increases until sedimentation eventually causes their elimination from the region. The downward flow of sulfur in sulfur compounds from the stratosphere is estimated as about 3×10^4 Tonnes per year, less than 10^{-3} of the total annual emission of sulfur into the troposphere from other sources. Thus, while apparently important as a source of stratospheric aerosol, the oxidation of sulfur in the stratosphere could not be expected to be a major source of aerosol in the troposphere.

11.9 WET PRECIPITATION

In addition to sedimentation, there are other important mechanisms for scavenging aerosols from the troposphere. One is *impaction*, in which particles carried with the wind strike an obstacle and are deposited. Another is *diffusion*, either eddy diffusion or molecular diffusion, which can also see particles migrate to a surface such as the ground, vegetation, or a structure where they impinge and remain.

But the most important mechanisms, in a global sense, are two processes of wet precipitation: *rainout* and *washout*. The first involves various processes taking place inside clouds, where the particles may become condensation or ice nuclei upon which water or ice has condensed. The second refers to the removal of materials below cloud level by falling rain or ice. In this section we shall first deal with the important aspects of droplet nucleation and then turn to consider the effectiveness of washout.

Large and giant particles serve preferentially as condensation nuclei for water droplets. Condensation on Aitken particles is not favored because the liquid would initially have a sharply curved surface, creating great surface tension and requiring more energy to form the droplet. Much less energy is required to form the more gently curved surface over large and giant particles.[18] As a consequence, the formation of clouds and subsequent precipitation should preferentially eliminate the large and giant particles from the air by rainout.

Once formed, the mean droplet radius appears to depend upon the composition and concentration of the nuclei; in dense cumulus clouds, the radius may range from slightly less than 10 microns in the arid southwestern United States to 15 microns in the central United States and 20 microns in maritime Caribbean air.[19] Together with the decrease in mean size of the droplets formed by continental aerosols occurs an increase in the concentration of droplets: Several hundred per cubic centimeter may be formed, whereas clouds over the oceans generally have only 30–40 per cm^3. Equation 11.3 indicates that the terminal velocity of such small droplets is very low, and consequently clouds can remain airborne for long periods.

The characteristics of condensation nuclei also influence the development and persistence of fog. It is generally agreed that urban regions experience more frequent fog and, furthermore, that the fog usually persists longer than in the surrounding countryside. Pollution provides a high concentration of condensation nuclei (perhaps 10^5 per cm^3) so when nucleation occurs, many fine droplets are formed. These settle much more slowly than would be the case if fewer nuclei were available and the droplets were larger.

Droplets in clouds can grow by a number of mechanisms, one of which is *coalescence*. Turbulence ensures that droplets continually collide with each other and, when they coalesce, the small ones disappear to form larger ones. Precipitation does not occur until the droplets have grown to a size of 500 to 1000 microns radius, so there must indeed be a remarkable increase in size. The details of how this comes about are not completely understood.

An interesting feature of some clouds is the fact that droplets may be sustained at temperatures as low as $-20°C$ and even $-40°C$ without freezing. This "supercooled" condition provides an opportunity for an important mechanism by

Fig. 11.9   Greater saturation vapor pressure over a supercooled water surface than over ice at the same temperature causes droplets to evaporate and ice crystals to grow.

which precipitation develops, a mechanism named after Tor Bergeron, a Swedish meteorologist who first developed a theory for it. The Bergeron process depends upon the fact that water droplets and ice may coexist in the same cloud. Some aerosols are effective nuclei for ice formation at low temperatures. The reason for this has not been explained completely, although it may have to do with the crystal form of the constituents of the particles which is compatible with the crystal lattice of an ice crystal. Many soils and sands are effective freezing nuclei at temperatures between −20° and −10°C. Another freezing nucleus, which is used in artificial seeding of clouds to form rain, is silver iodide aerosol. Thus ice and water droplets nucleate on aerosols and may coexist within the same cloud, although the ice may initially be found in the cooler upper regions. But through the turbulence that exists in clouds, ice crystals and supercooled droplets become mixed. At a given temperature, the saturation vapor pressure over water is greater than it is over ice. The result is illustrated in Fig. 11.9: The ice crystals in the Bergeron process will then grow at the expense of the droplets because of the transfer of water molecules in the gas phase. Ice crystals also grow as they are struck by droplets which then freeze onto the surface. If conditions are proper, the crystals grow to sufficient size to precipitate, and as they enter warmer air below, they melt to provide rain. Not all precipitation originates by the Bergeron process; rain clouds are frequently found to be too warm to support ice and thus must apparently develop solely by coalescence.

In addition to cleansing the liquid condensation and freezing nuclei from air by rainout, precipitation also removes particles by *washout* as they are struck by falling drops. This mechanism is depicted in Fig. 11.10. Large and giant aerosols in the path of the falling raindrop may be too massive to respond to the air as it parts in front of the drop; such particles will be struck and during the collision may be captured and then carried to the ground. There is a critical

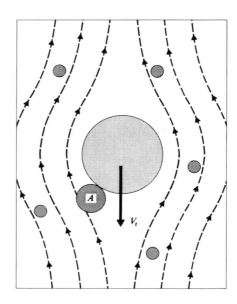

Fig. 11.10 A falling drop impacts against a particle (*A*) whose mass is so great it does not respond with the air which diverges in front.

lower size for particles which can be struck by a drop, for below a certain size a particle is sufficiently small to be carried out of the way with the parting air. This can be examined theoretically, and it is found that under certain circumstances a particle will escape from impact if its radius is less than about 0.2 of the radius of the drop, assuming that both are spherical. This fraction is only a rough guide; the exact value depends upon the size of the falling drop.[20] In addition, falling drops are not spherical but are considerably flattened on the leading surface. In any event, experiments do show that small particles escape from falling drops; for example, particles smaller than about 2 microns are not collected by drops of 40 micron size.[21] Calculations indicate that droplet aerosols, such as those in sulfurous smog, are not washed out by rain if the particle size is inferior to 0.5 micron.[22] We expect therefore that washout may not scavenge all particles from the air, especially the smaller ones.

However, the situation is not so clear-cut as we have indicated. The reason is that the Brownian motion of the very small Aitken particles, which are brushed to the side by the diverging air ahead of a drop, can enable them to migrate to the drop during the short time it passes by and become entrapped. This effect is particularly important for soluble gas molecules such as SO_2, so washout can very effectively cleanse the air of these gaseous pollutants and small Aitken particles below the region where clouds are formed.

A further complication is introduced by electrical effects, since many—if not all—cloud particles are charged. This is clearly evident during thunderstorms. Unfortunately, the magnitudes of the charges on raindrops and aerosols are not known, nor are the mechanisms by which the charges are initially separated. Experiments have shown that charged drops are dramatically more efficient in scavenging aerosols as a result of attractive electric forces; but quantitative appraisals of the inportance of such effects in washout have not yet been developed.

Much remains to be determined about how the atmosphere is scavenged of pollutants. The pattern of aerosol formation, evolution, and elimination is a major feature in the end stages of sulfurous and photochemical smogs. Yet very little is known about the relative importance of the various mechanisms of sedimentation, impaction, rainout, and washout.

SUMMARY

Aerosols occur in a variety of compositions, shapes, and sizes. However, mature aerosols share several common features. The size distribution of the particles usually has one peak, corresponding to a maximum concentration of aerosols with a size of 0.02–0.1 micron. The concentration of smaller particles is sharply down, because coagulation quickly eliminates them by transformation into larger particles. On the other hand, large particles whose radius exceeds 10–20 microns are eliminated by sedimentation. Thus the size distribution is influenced by the effects of Brownian motion, turbulent mixing, and gravity. The composition of maritime and continental aerosols differs as a result of the distinct sources of contaminants in the two regions of the globe and the relatively short aerosol residence time of perhaps five days in the lower troposphere. Polluted continental air has a much higher concentration of Aitken particles than maritime air. Some aerosols are secondary pollutants; for example, those from photochemical smog, those from sulfurous smog, and the stratospheric sulfate aerosol. Particles may be scavenged from the air by sedimentation, direct impaction upon surfaces, or by rainout and washout. We do not have a detailed quantitative assessment of the relative importance of these various scavenging mechanisms.

NOTES

1. G. M. Hidy and J. R. Brock, "An Assessment of the Global Sources of Tropospheric Aerosols," 2nd Clean Air Congress, IUAPPA, Washington, D.C., Dec. 1970.

2. J. P. Détrie, with collaboration of P. Jarrault, *La Pollution Atmosphérique*, Paris: Dunod, 1969.

3. C. E. Junge, *Air Chemistry and Radioactivity*, New York: Academic Press, 1963.

4. F. Knelman, N. Dombrowski, and D. M. Newitt, "Mechanism of the Bursting of Bubbles," *Nature*, **173**, 261 (1954).

5. B. J. Mason, *Nature*, **174**, 470 (1954).

6. D. Kuroiwa, "The Composition of Sea-fog Nuclei as Determined by Electron Microscope," *J. Meteorol.*, **13**, 408 (1956).

7. J. Aitken, *Collection of Scientific Papers*, London: Cambridge Univ. Press, 1923.

8. See, for example, the work by S. K. Friendlander *et al.*, "The Self-Preserving Particle Size Distribution for Coagulation by Brownian Motion-III Smoluchlovski Coagulation and Simultaneous Maxwellian Condensation," *J. Aerosol Sci.*, **1**, 115 (1970) and references therein.

9. W. E. Clark and K. T. Whitby," Concentration and Size Distribution Measurements of Atmospheric Aerosols and a Test of the Theory of Self-Preserving Size Distributions," *J. Atmos. Sci.* **24**, 677 (1967); C. Junge; *J. Atmos. Sci.*, **26**, 603 (1969).

10. G. M. Hidy and J. R. Brock, *The Dynamics of Aerocolloidal Systems*, Oxford: Pergamon Press, 1970.

11. K. T. Whitby, *et al.*, "Aerosol Measurements in Los Angeles Smog," Progress Report Particle Lab. Publ. No. 141, Dept. of Mech. Eng., Univ. of Minnesota (1970).

12. I. H. Blifford, Jr., "Tropospheric Aerosols," *J. Geophys. Res.*, **75**, 3099 (1970).

13. H. L. Green and W. R. Lane, *Particulate Clouds: Dusts, Smokes, and Mists*, 2nd ed., Princeton, N. J.: D. Van Nonstrand, 1964.

14. J. H. Ludwig, G. B. Morgan, and T. B. McMullen, "Trends in Urban Air Quality," *Trans. Amer. Geophys. Union*, **51**, 468 (1970).

15. W. M. Porch, R. J. Charlson, and L. F. Radke, "Atmospheric Aerosol: Does a Background Level Exist?," *Science*, **170**, 315 (1970).

16. G. M. Hidy and S. K. Friedlander, "The Nature of the Los Angeles Aerosol," 2nd Clean Air Congress, IUAPPA, Washington, D.C., December 1970.

17. I. H. Blifford, Jr. and L. D. Ringer, "The Size and Number Distribution of Aerosols in the Continental Troposphere," *J. Atmos. Sci.*, **26**, 716 (1969); see also J. M. Rosen, "Stratospheric Dust and Its Relationship to the Meteoric Influx," *Space Sci. Rev.* **9**, 58 (1969).

18. R. D. Cadle, *Particles in the Atmosphere and Space*, New York: Reinhold, 1966.

19. R. R. Braham, Jr., *Bull. Am. Meteorol. Soc.*, **49**, 343 (1968).

20. L. M. Hocking, "The Collision Efficiency of Small Drops," *Quart. J. Roy. Meteorol. Soc.*, **85**, 44 (1959).

21. R. G. Pickett, in *Aerodynamic Capture of Particles*, New York: Pergamon Press, 1960.

FOR FURTHER READING

C. E. JUNGE	*Air Chemistry and Radioactivity*, New York: Academic Press, 1963.
R. D. CADLE	*Particles in the Atmosphere and Space*, New York: Reinhold, 1966.
G. M. HIDY and J. R. BROCK	*The Dynamics of Aerocolloidal Systems*, Oxford: Pergamon Press, 1970.
N. A. FUCHS	*Mechanics of Aerosols*, New York: Pergamon Press, 1964.
H. L. GREEN and W. R. LANE	*Particulate Clouds: Dusts, Smokes, and Mists*, E. and F. N. Spon, Belfast: Universities Press, 1964, 2nd ed,
C. N. DAVIES, Ed	*Aerosol Science*, New York: Academic Press, 1966.
B. J. MASON	*The Physics of Clouds*, 2nd ed., Oxford: Clarendon Press, 1971.

QUESTIONS

1. What adverse effects are caused, at least in part, by aerosols?

2. What are the characteristics of giant, large, and Aitken particles?

3. Give some examples of breakup that add aerosols to the air of your community.

4. If water condenses preferentially on larger particles, why does an Aitken counter provide a good estimate for the number density of particles in the sampled air?

5. When considering the diffusion of particles in the air, why is it incorrect to assume that the rate at which they move depends solely upon the average thermal energy of the particle?

6. What distinction is made between agglomeration and coagulation? And what size particles are most affected by these processes?

7. In what ways does a maritime aerosol differ from the composition of a polluted continental aerosol?

8. Explain why the concentration profile of Fig. 11.8 shows the most marked variation with altitude in the lower stratosphere and lower troposphere, when the troposphere is heavily polluted.

9. Explain how the Junge distribution of particle sizes explains in large measure why the prevailing visibility through polluted atmospheres seems to depend only upon the mass concentration of the total aerosol which is present.

10. Why does washout not scavenge particles of all sizes from the lower atmosphere with equal effectiveness?

PROBLEMS

1. Estimate how far a particle falls before reaching its terminal velocity assuming that it is affected only by gravity. What is this distance for a particle of radius 100 microns and one of radius 1 micron? Under what conditions would this distance be considered negligible?

2. What is the terminal velocity for a spherical particle with a 2 g/cm^3 density and a 100-micron radius? What is it for a particle of similar density but of a 10-micron radius? Is the difference between the terminal velocities of the two particles more sensitive to differences in their density or differences in their radius?

3. Equation (7.3) shows that the viscosity of air depends upon the air density, the average speed of air molecules, and the mean free path. If the mean free path is $l = 1/n\pi a^2$, where a is the radius of an air molecule and n is the number density of air molecules, how would you expect the terminal velocity of giant particles to depend upon altitude?

4. Estimate the total mass of Aitken particles in the aerosol which is characterized by the solid curve in Fig. 11.5, as compared with the total mass of large and giant particles. Assume that the density is the same for all of the particles.

5. If a mass of maritime air moves over a continent at an average wind speed of 10 km/hr, how far inland would the mass penetrate before sea salt particles of radius $R = 10$ micron had settled to the ground from an altitude of 0.3 km?

6. During each second, how many times is an aerosol particle of a radius of 0.5 micron struck by an air molecule? Your answer suggests why agglomeration is an important mechanism in particle growth.

7. Assuming an eddy diffusivity $D = 10$ m^2/sec for aerosol particles of a 0.3-micron radius, what concentration profile would account for a steady-state balance between sedimentation and upward diffusion if the ambient concentration of these particles is about 10^3 per cm^3? How is this profile influenced by the ambient temperature?

8. Find an equation for the position of a particle that is accelerated by gravity but also experiences a drag force as given by Eq. (11.3). Can you define a time constant that describes how long it takes for the particle to achieve its terminal velocity? How far has it fallen during this time, if the particle started from rest?

CHAPTER 12

CONTROL OF AIR POLLUTION

What can be done to avoid air pollution? There are many reasons why this question has no simple answer. Practically every industrial process adds some contaminants to the air. Domestic activities such as cooking a meal or heating the interior of a house contribute as well. Use of automobiles releases not only the gaseous and particulate matter, which we have discussed in connection with photochemical smog, but metallic particles and asbestos from brake linings and rubber from tire wear. Practically every activity releases some contaminants into the atmosphere. And for this reason it is generally agreed that air pollution will never be completely eliminated. Thus in a pragmatic sense we should more properly rephrase the question as: What can be done to reduce air pollution to acceptable levels? This chapter deals with a partial answer. We shall examine concepts which have enjoyed success in at least partially meeting the challenge and which show promise of future utility. Our purpose is to supplement the discussion of previous chapters in which we focused on the types of devices and techniques for physically controlling the emission of pollutants. Here we shall be most concerned with concepts which can be implemented as legal statutes or economic incentives to encourage the use of these controls.

Legal actions

There are three general approaches by which legal actions have been employed to place controls over the emission of air pollutants—public nuisance statutes, private litigation, and prescribed emission standards—and we shall deal with each of these in turn.

In most countries a *public nuisance* exists when conditions cause discomfort, inconvenience, damage to property, or injury; and the causing of a public nuisance is prohibited by law. If the person or corporation responsible for the conditions can be identified, public authorities can seek a court injunction to prevent continuation of the nuisance and impose criminal sanctions. This may be successful if a single, clearly identifiable source is responsible, as when an exhaust stack emits great quantities of large particles which quickly fall out in the immediate neighborhood. But the assessment of responsibility is nearly impossible in the case of community smog, where emissions from more than one source may act synergistically. In addition, prosecution under public nuisance statutes occurs after the fact, and so the procedure is evidently unsatisfactory for establishing preventive measures. When health is concerned, absolute proof of physical injury may be hard to come by, as we have seen during our discussions of epidemiologic analysis of pollution effects. Thus control of pollution sources by use of existing laws governing public nuisances generally has not been found satisfactory.

Under the common law in most countries, an individual suffering damage from contaminants released by another may seek by *private litigation* to recover the cost of damages he has sustained and cause an injunction to be issued to compel cessation of the harmful conditions. The burden of proof rests upon the complainant; to be successful, he must clearly link the damage he has sustained to the pollutants emitted by the defendant. The complainant may find this expensive, for he must have technical experts available to support his case. Furthermore, legal tradition stipulates that not only must damage be proven, but also must the actions of the defendant be shown to be unreasonable. In some instances a court will weigh the costs of improving conditions against the anticipated benefits. For these reasons it is difficult for an individual to seek redress by legal suit. And as for use of public nuisance statutes, a court suit after the fact is evidently an unsatisfactory procedure for preventing damage and protecting public health.

Thus *government* at various levels has shouldered the burden of protecting the public interest. This has been approached in a number of ways: through research into the causes and effects of pollution, development of devices for pollution control, establishment of guidelines for air quality, introduction of tax incentives for the installation of control equipment on sources, and perhaps, most importantly, promulgation and enforcement of ordinances for restricting the emission of contaminants, to name but a few. Most pollution control laws rest on the police powers of the state which derive from the right of people through organized government to protect their health and property. For example, the United States enacted a Clean Air Act in 1963, amended in 1970, involving federal authority for abating air pollution which "endangers the health or welfare of any persons." Thus actual injury need not be sustained before the government can act.

With government legislation, has come a shift in the burden of proof. An example is the 1970 Amendments to the Clean Air Act in the United States, which in certain instances require that, before emissions are commenced, those discharging certain pollutants show that their actions will not be harmful. Such legislation, when enforced, greatly eases the burden placed on air pollution control authorities.

12.1 AIR QUALITY CRITERIA

There is a logical and by now conventional method by which a government can attack the problem of air pollution. The first step is to define the levels at which a contaminant is considered a pollutant. By this we mean the adoption of a clear statement of what quality for ambient air would ensure that the public health and public welfare would be protected. Such prescriptions are usually called *air quality criteria*. They are determined on the basis of known or suspected adverse effects: toxic effects in humans or animals, damage to vegetation or materials, and degradation of aesthetic aspects of the environment. An air quality criterion indicates the level above which the presence of a pollutant is considered to have an adverse effect. Thus it may indicate a threshold concentration, dosage, and exposure time for each pollutant or combination of pollutants which act synergistically.

An air quality criterion is merely a technical statement. Standing alone, it has no practical significance, for it prescribes no legal standard nor does it indicate how the criterion is to be met. The *desired* levels of air quality may be different from the criterion and are defined by what we shall call *air quality goals*.

12.2 AIR QUALITY GOALS

Two factors have traditionally motivated the establishment of air quality goals: (1) public opinion, and (2) appreciation by responsible government agencies of demonstrated or suspected adverse effects. The first factor is very much evident today and has been largely responsible for prompting the adoption of criteria in many nations around the world. The second factor is in principle a recognized and continuing activity of government, for it is the responsibility of government to protect the public health and take cognizance of technical aspects in which the public at large has little expertise. However, it is generally recognized that many government bodies will not honor this responsibility without the support and perhaps impetus of public opinion.

In setting goals, there is an inevitable balance of societal values against cost. This occurs for many reasons, but perhaps the prime one is the fact that in the long run, the *economic cost* of pollution and its control will be borne by the citizen, in taxes and in the prices he pays for goods and services.[1,2] The more stringent the goal, the more stringent will be the controls on the sources of pollution and the more costly it will be for maintaining and policing these controls.

In highly polluted regions, there may in fact be a direct economic benefit to the citizen from the first major decrease in pollutant levels, resulting from a reduced toll of material damage and perhaps health expenses. This possibility is illustrated in Fig. 12.1 where we show a hypothetical case in which the economic loss from pollution damage decreases as the ambient pollution level decreases.

On the other hand, the marginal cost for introducing emission controls generally increases as the controls are made more stringent. For example, the cost of reducing the SO_2 emissions by a factor of 10 from a power plant exhaust stack may be partially or completely offset by the revenue gained from the sale of the collected sulfur. But to reduce the emissions by another factor of 10 would be considerably more expensive, not only because the control equipment would be more sophisticated and costly but also because the plant would obtain only one-tenth as much sulfur to sell. Thus in Fig. 12.1 we show the capital and

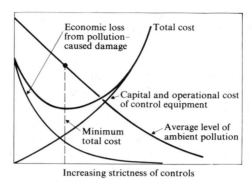

Fig. 12.1   Schematic portrayal of a citizen's share of the cost of air pollution and its control.

operational costs of control equipment increasing as the pollutant level decreases.

It is clear that the total direct and indirect economic cost to the citizen has a minimum value, and to establish an air quality goal on the basis of minimizing the total cost means accepting as that goal the ambient level of pollution at the corresponding point above the minimum of the cost curve. Similarly, to achieve this criterion would require an investment in emission controls sufficient to provide the strictness of controls indicated by the corresponding point on the horizontal axis. Thus a purely economic condition can be used to define an air quality criterion.

The example which we have just given is an oversimplification for practical situations. For example, it has neglected the temporal variations in meteorological conditions which necessitate a more sophisticated means for relating pollution levels to the resulting economic damage. Perhaps it will be found that the yearly average concentration of a pollutant is the important parameter by which pollution levels should be measured, but very likely a more complex measure will prove to be more directly relevant. Our simple model has also neglected the fact that whereas the economic damage from pollution is generally suffered by those in the immediate vicinity of the source, the cost of controls over the particular industry—if met by an increased price in its products—is borne by the consumers, who may represent quite a different group of people. Thus the money for repairing pollution damage and for emission controls need not come from the same pocket. An accurate economic model for predicting the total cost of pollution will involve many considerations we have overlooked; nevertheless, the principle we have illustrated by the simple model is important and is directly relevant to the establishment of air quality criteria.

If an air quality goal is to have significance, it must reflect the values prevalent in society. Perhaps the economic condition of minimal cost just described is not the determining factor. The population in a locality may believe instead that the preservation of *aesthetic factors* such as odor or visibility is a more important goal. Thus a community may decide that the air quality goal for suspended particulate matter should be determined on the basis of what prevailing visibility would be considered adverse. But here again there will be a trade-off between benefit and cost, for maintaining good visibility generally requires much stricter controls over emissions than would be defined by a goal based on minimum cost. It is refreshing to be able to see for a distance of 100 km from the top of a mountain peak, but would this distance be acceptable as the basis for a goal? Meeting it would require, for example, a tenfold reduction in the aerosol content from that level of emission which permitted a prevailing visibility of only 10 km. The cost of meeting the more stringent requirement would be paid in both economic and social terms. Industries would need to filter their exhaust more stringently, farmers could no longer burn refuse in their fields, and homeowners would perhaps not be permitted to burn logs in their home fireplaces.

Another consideration is the fact that aerosols from natural sources frequently reduce the visual range to less than 100 km, and on these occasions there would be little point in keeping the concentration from anthropogenic sources much below the naturally occurring level. And what percentage of a population would ordinarily be at a location where they would have an unobstructed view of more than 10 km on even the clearest days? It would be an unusual community that could justify maintaining a prevailing visibility of 100 km in the face of the economic and social costs. But inhabitants of the arid American southwest who appreciate their vistas very likely may be willing to bear the cost and demand better visibility than, for example, employees of steel mills in the Ruhr Valley of Germany, for whom the same prevailing visibility may be considerably more costly. In Fig. 12.2 we indicate schematically one measure of the economic cost of better visibility, taking cognizance of the economic factors we have discussed previously.

It is generally agreed that an air quality goal ought to be sufficiently strict to safeguard *public health*. This is a straightforward consideration when pronounced adverse effects and their causes are well-established scientific or medical facts. Data summaries such as we have illustrated in Figs. 8.2 and 8.3 and 10.12 can be

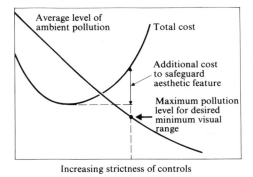

Increasing strictness of controls

Fig. 12.2 An air quality goal based solely on visual range can increase a citizen's share of the cost of air pollution and its control.

helpful in the decision-making process. However, when dealing with marginal effects, such as temporary and reversible changes of the chemical equilibria in human pulmonary tissue, there may be no clear indication as to whether the effect is an adverse one. Another ambiguity is associated with the extension of the results of toxicologic studies on animals to humans, who may be considerably more—or perhaps less—sensitive to pollutants. There is then an element of *medical opinion* that bears on the establishment of air quality goals which are designed to protect health. The degree of conservatism incorporated in a goal is a proper subject for debate, and will be influenced by public preference and expressions of willingness to shoulder added costs for a more certain measure of protection.

An additional philosophical issue with far-reaching implications is whether a goal ought to be sufficiently stringent to safeguard *all* of the members of a population, including the few very sensitive and susceptible members. This issue can be phrased in a perhaps more palatable way by asking what percentage of a population would be encouraged to move away to where the air is cleaner. When sufficient medical data becomes available, this question can be tackled on a more quantitative basis by use of curves for the threshold distributions of a population, such as we have examined in Section 2.1. At present, however, very little information of this nature is available.

A clear example of the influence of value judgments when agencies set air quality goals is the short shrift commonly given to some segments of the agricultural community. Certain plants are considerably more sensitive to a given pollutant than are humans, but few goals now in effect give them the required measure of protection. Lettuce, for example, is easily damaged by ozone, and orchids by ethylene; in fact, aside from reduced visibility, plant damage is often the first sign available to the general public that photochemical smog has invaded a region. The bases for goals have favored the protection of public health but have rarely provided for the protection of all vegetation, on the argument that the additional cost for pollution controls is not justified.

In Fig. 12.3 we have summarized many of the key factors which influence the formulation of air quality goals. This figure emphasizes the fact that decisions on goals involve an interplay of technical considerations and value judgments. And the value judgments include not only those which derive from society, but properly include medical, scientific, and technical opinions for evaluating the significance and applicability of available data. It should be kept in mind that when we say the object of formulating these goals is usually to seek a balance between benefit and cost, we do not mean just economic cost. The cost in social factors is also important. For example, the imposition of a goal, with anticipation that government agencies will subsequently regulate emissions from sources so as to meet the goal, can more seriously affect some segments of a population than others. As a case in point, suppose that an ambient-air-quality goal could be met only by making it illegal to operate an automobile which is more than 10 years old; the resulting hardship would generally be borne by the poorest

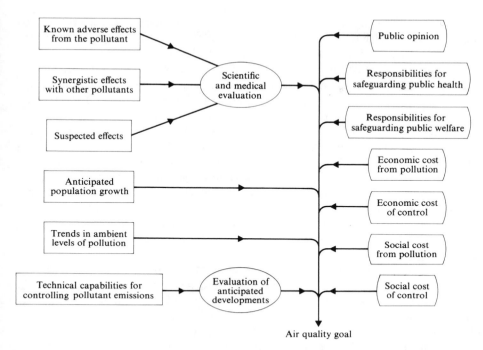

Fig. 12.3 Factors which influence the formulation of an air quality goal or standard for a pollutant.

members of the population. It is not difficult to imagine other examples which discriminate against certain socioeconomic groups. So we emphasize in Fig. 12.3 that the costs of controlling pollution are not just economic costs.

Concepts for Criteria and Goals

Two types of criteria have commonly been employed to define air quality: the maximum ambient concentration X_c (equivalent to a threshold) and the maximum period of exposure to a given average ambient concentration (or dosage D_c). The distinction between the two is important. If X_c is so small that exposure to this maximum level for any anticipated likely duration in time τ produces a dosage considerably less than the maximum approved dosage (that is, if $X_c \tau$ is less than D_c), then only a restriction on the ambient concentration X_c is significant. On the other hand, if a population will experience an ambient concentration X which is generally less than X_c but is endured for a sustained period such that $X \tau$ exceeds D_c, then an air quality criterion must contain a restriction on both X_c and D_c.

Usually insufficient data are available to define a dosage, and in many cases it is not clear that such a simple criterion applies to biological systems, since natural processes eliminate much of an ingested pollutant if X is sufficiently low. Thus it has become popular to define criteria by two parameters: the average concentration $\overline{X}_{\bar{c}}$ and the time τ over which the average is taken.

There may be a hierarchy of criteria for a given pollutant, with thresholds defined on the basis of the different effects it causes. For example, we might require that the ambient concentration of SO_2 be less than 0.5 ppm for 10 minutes to avoid disagreeable odors; but to avoid a deterioration in the health of bronchitis patients and an increased incidence of cardiorespiratory disease, we might also set a daily maximum of $\overline{X}_c$ at a lower average of 0.1 ppm for 24 hours. In view of the importance of synergism, this level might be further lowered if the particulate mass concentration averages over 100 $\mu g/m^3$. More elaborate criteria for SO_2 and other pollutants can of course be devised; however, there is often insufficient epidemiologic or toxicologic data upon which to develop them.

Instead of an absolute limit on the maximum tolerable concentration X_c, or a limit on the average concentration $\overline{X}_c$, a criterion can be formulated that permits occasional violations of the adverse level. This has certain practical advantages, since it takes into account the fact that ambient concentrations fluctuate markedly in accordance with changing rates at which contaminants are emitted and with varying meteorological conditions. A hypothetical but typical example is illustrated in Fig. 12.4, from which it is evident that very high concentrations occur but infrequently. These few occasions may not in fact be harmful or objectionable to a population. The criterion for prevailing visibility, for example, might be relaxed to the extent of permitting lower visibility on a few days each year with little adverse reaction from the population.

The advantage of permitting occasional exceptions is that a simple criterion can be formulated to govern what might more strongly affect people—the more frequent exposure to peak pollution levels of lower magnitude. Thus a criterion could be formulated whereby excursions above a level indicated by X_a in the figure would be considered adverse if they occurred more frequently than, say, 1% of the time, or perhaps three times a month for an hour's duration each. This seeks to control principally the secondary maxima that occur more frequently; it

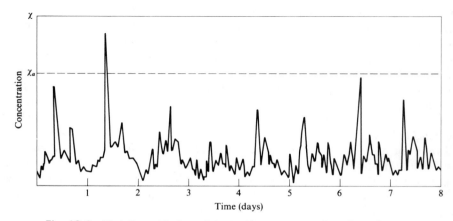

Fig. 12.4   Variation with time of the ambient concentration of a pollutant.

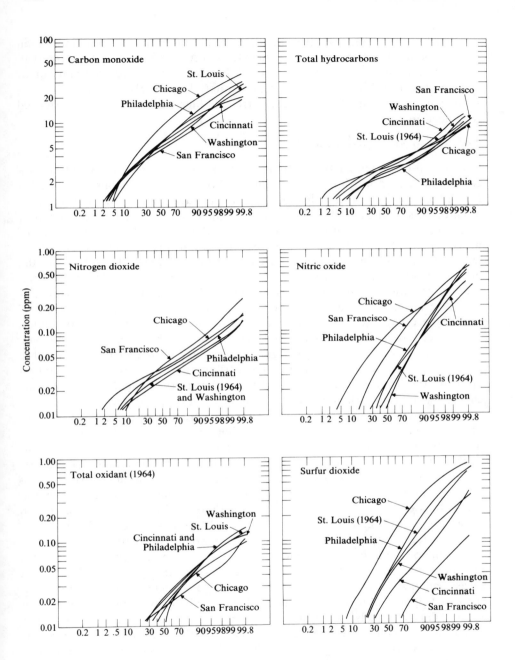

Fig. 12.5 Cumulative frequency distributions of gaseous pollutant data. The horizontal axis gives the percent of measurements equal to or less than the concentration stated on the vertical axis. One-hour average data from 1962 through 1964. (United States Continuous Air Monitoring Program.)

is a hybrid condition, less stringent than an absolute limit X_c, but more than just a condition on the average concentration $\overline{X}_c$ for the same time interval. It is becoming more generally recognized that the average levels of pollution, or perhaps the frequent excursions to moderately high levels, may in fact be as significant in affecting health as the extremely high peaks that are experienced only occasionally.[2]

Frequency of occurrence

It is a well-documented fact that the highest ambient levels of pollution in a community occur only infrequently. This is illustrated by the data summarized in Fig. 12.5 which give examples of one-hour averages of the air pollution at a monitoring site in each of six cities in the United States. The horizontal axis of these charts is a cumulative frequency scale; that is, it gives the percentage of measurements for which the ambient concentration was found to be equal to or less than the concentration indicated on the vertical scale. Thus 99.8% of the measurements in the represented cities yield concentrations equal to or less than the concentration given by the highest point on the respective curve. It is remarkable that in almost all cases, concentrations which are within a factor of 2 of this highest value occurred less than 5% of the time. Thus the commonly experienced pollution levels are considerably lower than the maximum level.

The horizontal scale in these figures is arranged so that a straight line in the figure would correspond to a log-normal distribution with a median concentration given by the level corresponding to 50%. Although the general trends of some curves favor a log-normal distribution, there are substantial deviations.

But now we might ask ourselves how the frequency distribution of concentrations averaged over one hour, as illustrated in Fig. 12.5, corresponds to the frequency distribution for longer averaging times, since in defining air quality criteria long-term effects are important for some pollutants. The answer depends upon many factors, including the specific pollutant, distribution of sources, and meteorology. But measurements in various cities reveal a common trend. It is convenient to illustrate this by a plot of the average concentration versus the averaging time, and we show in Fig. 12.6 an example of this for the SO_2 concentrations monitored in Chicago over a period of six years.

Each solid curve indicates the levels below which ambient concentrations are found to occur for the indicated percentage of the time. Thus 99% of the measurements showed that the one-hour average concentration of SO_2 was less than 0.64 ppm, and for the remainder of the one-hour averages it exceeded this value. (This could also have been deduced from the curves in Fig. 12.5.) The trends of the upper curves in Fig. 12.6 indicate that the level for the 99th percentile decreases for longer averaging intervals. If we consider a daily average, we find that 99% of the measurements showed an SO_2 concentration less than 0.49 ppm. An extrapolation of the trend for monthly and annual averages is given by the dashed lines in the figure, although there would be little statistical significance to the 99th percentile in this region owing to the relatively small number of

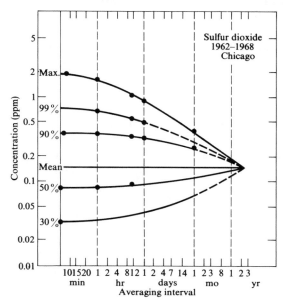

Fig. 12.6 Concentration, cumulative frequency of occurrence, and averaging interval for SO_2 levels in Chicago during the period 1962–1968. The curves indicate the average concentration which is not exceeded by the given percentage of measurements. (United States Continuous Air Monitoring Program; data from R. I. Larsen, Journal of the Air Pollution Control Association, **20**, 214, 1970.)

months and years included in the survey. Nevertheless, a convergence of the curves for this and other percentiles to a 3-year mean of about 0.13 ppm is suggestive of the anticipated feature that annual averages show relatively less variation from year to year to year than do averages over much shorter intervals.

Analyses such as we have illustrated in Fig. 12.6 are useful, because they incorporate the three significant parameters used in the United States to define air quality criteria—concentration, averaging time, and frequency of occurrence. Once established for a given pollutant, they can be used to determine whether various air quality goals based on different adverse effects are equally restrictive or whether one is significantly more restrictive than the others. For example, suppose we wish to establish an SO_2 goal and we have the following two *hypothetical* adverse effects which we wish to avoid:

1. The public in a particular community is known to react unfavorably if the odor recognition threshold of about 0.5 ppm for SO_2 is exceeded more than 10% of the time.

2. In this community there is evidence that plant damage occurs if the daily SO_2 level exceeds 0.1 ppm more than 10% of the time.

If the frequency of occurrence of various levels of pollution in this community were found to be identical to those shown in Fig. 12.6, we could conclude that the existing air quality could meet condition (1), because even with an averaging interval of only 1 minute, the 90% curve falls below 0.5 ppm. However, the existing air quality could not meet condition (2) and, to do so, would require a scaling down of ambient levels by about a factor of 3, all other factors being equal.

If absolute limits are desired for air quality goals, with no exceptions permitted, then reference should be made to the "maximum" curve in Fig. 12.6, for this indicates the maximum concentration observed for the corresponding averaging intervals. Suppose, for example, that a community wished the SO_2 one-hour average never to exceed 0.5 ppm; if conditions in this community were described by the data of Fig. 12.6, we would conclude that the ambient concentration would have to be reduced by about a factor of 4 to meet such an absolute criterion. This perhaps would involve ensuring that emissions from all sources were reduced by the same factor.

We shall later give some specific examples of air quality goals, but first we turn to consider the ways in which ambient concentrations of pollutants are measured, since such measurement plays an important part in the strategy to understand and control air pollution and to set air quality goals.

12.3 MONITORING DEVICES

Formulations of air quality goals have little relevance unless means are available to monitor the existing air quality. Great progress has been made within the past five years in developing monitoring techniques and in placing installations where they will provide the most significant information. There are several reasons for monitoring air pollution:

1. To chart current levels and thus furnish a basis for decisions regarding the course of actions required to establish safe and acceptable conditions.

2. To monitor the progress toward meeting air quality goals.

3. To provide data for refining mathematical models by which the dispersal of pollutants can be predicted.

4. To establish background information for epidemiologic studies which gauge the effects of various pollutants in the region.

Monitoring is an essential component in the effort to control and understand the effects of pollution, even though the measurement of levels of pollution does nothing in itself to make the air any safer.

The difficulty with measuring ambient levels of pollutants is the fact that they ordinarily occur at very low concentrations. Most gaseous pollutants occur at levels under one part per million, corresponding roughly to a mass concentration of less than one milligram per cubic meter of air; and particulate matter is generally found with a mass concentration of much less than one milligram per cubic meter of air. Hence there is not much pollutant to be measured.

The problem is not so great when measuring the amount of a pollutant in industrial *emissions*, for the concentrations in the effluent going to an exhaust stack are generally a factor of 100 or more greater than typical ambient concentrations. Monitoring emission concentrations will play an ever-increasing role in the enforcement of emission controls by government agencies, and the principles of monitoring techniques for emission levels are closely related to those for ambient levels.

Years ago, when analytical techniques were cumbersome, it was popular to take vessels containing samples of air to a central laboratory for the identification and measurement of its contaminants. This "grab sample" technique is still used for spot checking, but more common now are continuous monitoring systems based on instruments in the field, made possible by advances in automated methods of analysis. These instruments incorporate several different principles, including the colorimetric, conductrimetric, amperometric, flame emission, and gas chromatographic methods.

A *colorimetric* or *photometric* technique common in many instruments is illustrated in Fig. 12.7. Air to be analyzed for the presence of a specific pollutant is pumped at a constant rate through a reaction chamber and is then exhausted back into the atmosphere. During its residence in the chamber, the pollutant has time to react with chemicals which are circulated in a closed cycle loop, and the product of the reaction is chosen to have a signature which can be identified

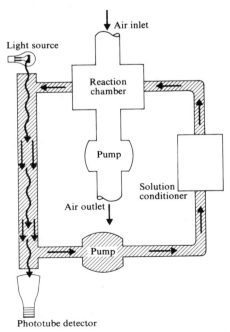

Fig. 12.7   Colorimetric analysis on a continuous basis.

by its light absorption characteristics, perhaps by a change in the color of the solution. The reaction product, whose concentration is determined by the concentration of pollutant in the sampled air, is then pumped into a photometer where the concentration of the light-absorbing substance is indicated by the reduction in light intensity that reaches a phototube from a light source. The solution emerging from the photometer can then be chemically standardized and recirculated through the reaction cell. When properly calibrated, the electrical signal from the phototube provides a measure of the concentration of pollutant in the sampled air. The response time for such a device is defined as the time interval required for its reading to come to 90% of its ultimate steady value after a change has occurred in the concentration of pollutant in the sampled air. The response time is generally limited by the reaction rate of the pollutant in the cell and the circulation rate of the solution, and it determines the maximum rate of change of pollution levels that can be monitored faithfully.

Not all monitoring instruments can be operated in continuous fashion like this one. Some involve batch operations in which a fixed amount of an air sample is analyzed for its content. The response time for batch processing is limited by the interval required for the batch operation and therefore can be much longer than the averaging period during which the air sample was gathered.

Most instruments have been developed to measure specific pollutants, although there are some which can measure several different chemicals. We shall indicate below a representative list of the more common techniques.

Particulate matter

Most devices for monitoring the particulate content of air are merely collectors; and the analysis of the weight or number of particles is done separately. The simplest method for determining the mass concentration of particulate matter depends upon pumping a known volume of air through a fine filter and measuring the increase in weight due to the trapped particles. The filter, possibly of fibrous or granular material, physically blocks the passage of larger particles and forces air to follow a tortuous route, so that inertia causes some of the smaller ones to impact on the material where the direction of air motion changes abruptly. Diffusion also carries particles to the filter surfaces, where very small particles may be trapped. Some glass fibers are moderately effective for trapping particles as small as 0.05 micron. If the filter is designed to trap all aerosol particles greater than 0.1 micron (that is, the large and giant particles), a good approximation to the total aerosol mass can be obtained. Because a large volume of air is monitored by such procedures, the technique is often called a *high volume sampler*.

Another popular method for gauging the atmosphere's particulate burden in high volume sampling is to pass a known volume of air through a piece of filter paper, then remove the paper with its circular smudge and place it in front of a bright light. Measurement of the intensity of light which has passed through the smudge yields a measure of the scattering and absorptive properties of the trapped aerosol, which is expressed on a relative scale known as the *coefficient of haze* (CoH). A CoH value of 4 or more is considered unusually dense pollution, 3–4 is heavy, and 1–3 is light to moderate. Because of the nature of the measurement, this scale is more directly related to visibility reduction than to the mass concentration or number density of the aerosol.

More sophisticated devices allow the measurement of the mass distribution according to particle size. Cascade impactors such as the one depicted in Fig. 12.8 direct the air stream onto successive collection surfaces, and by use of successively smaller jets, and consequently higher air speeds, smaller particles are successively collected in later stages. The amount collected on each surface can be determined by weighing, and a count of the number of particles can be made by observation through a microscope. Because this is an inertial collector, it does not collect very small particles.

The above are devices which generally provide the basis for instruments used in day-to-day monitoring of community air pollution, although other devices are available for particle collection and analysis, including those based on principles mentioned in Section 9.1 where we discussed emission control devices.

Quite promising is the use of optical techniques, especially several that depend upon the light-scattering properties of aerosols. One technique uses a laser to send a pulse of light into the atmosphere, and the light scattered back to a detector is monitored.[3] The delay time for the return pulse and its intensity are a measure of the position and concentration of the aerosol. This so-called *Lidar* method is similar to the principal of microwave radar. Efforts are underway to devise techniques that will provide an unambiguous measure of the size distribution and composition of the light-scattering particles.

Sulfur dioxide

A traditional method for determining the amount of SO_2 in a measured volume of air is the *conductrimetric* technique in which the air is bubbled through a dilute

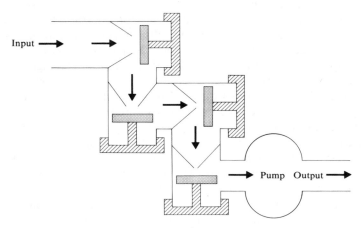

Fig. 12.8   Four-stage cascade impactor.

solution of sulfuric acid and hydrogen peroxide (H_2O_2). Sulfur dioxide contained in the air is oxidized to H_2SO_4, thereby increasing the electrical conductivity of the solution; a measure of the conductivity then gives an indication of the amount of SO_2 in the sample. Detection of SO_2 at levels as low as 0.01 ppm is possible.

Another way to measure SO_2 concentrations is by an electrochemical technique called the *amperometric* method. An air sample is continuously bubbled through an electrolytic solution for which current produced between the electrodes represents a measure of the SO_2 in the sample. A solution commonly used is 0.1 M of KBr in $2N$ sulfuric acid, with an additional concentration of bromine that is automatically maintained by a feedback system. Sulfur dioxide in the air flowing through the electrolyte decreases the bromine concentration according to the two reactions

$$H_2O + SO_2 = SO_3^= + 2H^+$$
$$H_2O + SO_3^= + Br_2 = SO_4^= + 2Br^- + 2H^+.$$

(11.1)

The concentration of Br^-, indicated by the electrode current, is thus related to the SO_2 concentration of the sample. This method permits the continuous monitoring of ambient SO_2 levels.

However, oxidizing agents such as ozone, or reducing agents such as hydrogen sulfide will interfere with an accurate measurement by both the amperometric and conductrimetric techniques. Consequently a filtering system must be used to eliminate those contaminants if they are present in appreciable quantities. For example, a roll of silver gauze heated electrically to a temperature of about 120°C will cause ozone to decompose and hydrogen sulfide to form silver sulfide.

West and Gaeke[4] have introduced a method which has gained wide popularity. Sulfur dioxide from a measured quantity of air is first absorbed in a 0.1 N solution of sodium tetrachloromercurate which prevents further oxidation of SO_2 into SO_3 by the formation of dichlorosulfitomercurate. This is then reacted with formaldehyde and bleached pararosaniline, forming a reddish-purple color, which can be detected by photometric methods. The *West-Gaeke* (or *pararosaniline*) method can be made more sensitive if the photometer light is filtered to pass only wavelengths near 0.56 micron where strong absorption from the solution occurs. If the solutions are cycled, the technique permits continuous monitoring with a response time of about 15 minutes. Ambient concentrations between 0.002 ppm and 5 ppm can be measured, but interference results if there are also high concentrations of nitrogen dioxide and sulfite salts.

Finally we mention a *flame emission* technique whereby the total sulfur content of a volume of air can be determined. The sample of contaminated air is mixed with an excess of hydrogen and burned; under such conditions there is a strong luminescence from the sulfur in a band of wavelengths centered at about 0.394 micron in the near ultraviolet portion of the spectrum. When the intensity of the emitted light close to this central wavelength is monitored with a photomultiplier tube, the electrical signal from the tube is a measure of the total sulfur concentration, including the contributions from SO_2, SO_3, H_2S, and mercaptans.

Carbon monoxide

Many techniques for measuring CO are not specific to this gas; they depend upon its reactions with other agents or its light absorption characteristics, neither of which are unique signatures, at least as monitored by many popular methods. Nevertheless, nonspecific indicators may be useful, since CO usually occurs at much higher concentrations than other atmospheric contaminants, and so contributions from other pollutants by comparison are negligible.

One type of instrument exploits the fact that CO strongly absorbs infrared radiation in two bands of wavelengths centered at 4.6 and 4.7 microns. If a spectrophotometer is employed, in which a scanning procedure enables the absorption of a sample of gases to be measured at each wavelength, the technique is said to be *dispersive*, and a unique identification of the absorbing gases may be possible. On the other hand, if the absorption is simultaneously measured without filtered light for a wide band of wavelengths, the measurement is *nondispersive*. Convenience and economy have made nondispersive, infrared photometers popular instruments for measuring CO concentrations, with a detection limit of about 1 ppm.

A newer technique requires first that CO be converted to methane in a reactor; then the presence of the methane is detected by a *gas chromatograph*. The chromatograph, first employed by W. Ramsey in 1905, utilizes a batch process in which the sample is mixed with an inert carrier gas and is slowly percolated through a column of densely packed material, such as silica gel or fine charcoal. Some chromatographs may have a nonvolatile solvent coating the packing. In either case, the column selectively retards the passage of various gaseous components in the carrier gas; and the length of the column is chosen to ensure that the components form separate bands of high concentration within the carrier. Thus the various constituents appear in a time sequence when emerging from the column. A flame detector or other technique is used to indicate the concentration of carbon emerging at each instant of time after the sample has begun its transit of the column. Thus from a previous calibration of the transit time for various gases, the concentration of each gas in the sample can be deduced. The time for analyzing each batch of sample air is approximately 20 minutes.

Ozone

One of the first techniques for monitoring ozone in the atmosphere was based upon measurements of the depth of cracks which developed in stressed rubber exposed to the oxidant. However, a more reliable method is the *potassium iodide colorimetric analysis*. The air sample is passed through a 1% solution of KI which has a buffering agent to establish neutrality. With ozone present, the following reaction occurs:

$$O_3 + 2\,KI + H_2O = O_2 + KOH + I_2, \tag{11.2}$$

and the solution becomes colored due to the iodine produced. The concentration of iodine can then be indicated by a photometer. Competing reactions reduce the iodine concentration if the solution is basic (for example, pH of 8) or acidic (pH of 6), so buffering is essential for accuracy. Interference may result from NO_2 to the extent that NO_2 yields about 10% of the response of O_3 when each is tested at equal concentrations. Peroxides and PAN also produce positive interference, but the reducing agents SO_2 and H_2S cause negative interference. The response time to changes in ambient concentration of O_3 is 8–10 minutes, and the detection limit is about 0.01 ppm.

Electrochemical methods can also be applied to monitor atmospheric O_3. One amperometric device uses a neutral buffered solution of KI and KBr through which the sample of air is bubbled. The resulting I_2 is reduced at one electrode to yield I^- and causes a current to flow through an external circuit to the other electrode, which is therefore a measure of the ozone concentration in the sample.

Nitrogen dioxide

The most widely used procedure for determining NO_2 levels is one first described by B. E. Saltzman.[5] It is a colorimetric technique utilizing a multicomponent solution through which is bubbled the sample air. The oxidizing effect of NO_2 results in a purplish color, whose presence can be measured by a photometer. The detection limit for NO_2 in the atmosphere is about 0.02 ppm. There is a negative interference from SO_2 and other strong reducing agents.

Application of infrared lasers to monitoring NO, NO_2, and other contaminants shows great promise. Sensitivies of up to a few parts per billion have been achieved by measuring the energy absorbed by a sample of air when the wavelength of the laser's radiation is appropriate for absorption by the pollutant.[6]

Hydrocarbons

The "total hydrocarbon" content of air is most commonly determined by the *flame ionization* method, for the response is rapid and over a wide range of concentrations it is proportional to the number of carbon atoms in the sample. The sample is burned in a flame fed by hydrogen gas (perhaps containing up to 40% nitrogen), which has the effect of breaking up a certain fraction of the hydrocarbon contaminants and forming CH radicals. Through a series of reactions in the flame, CH is oxidized to CHO^+, yielding a free electron; and the CHO^+ then combines with H_2O to form CO and H_3O^+. By maintaining a pair of electrodes in the flame, across which is established an electrical potential difference of several hundred volts, the electrons are collected at the anode and the H_3O^+ at the cathode. The resulting current in the electrode circuit can be monitored and is approximately proportional to the number of carbon atoms present. Total hydrocarbon concentrations can thereby be determined to within about 0.1 ppm. But this method is considerably less sensitive to carbon atoms bound to oxygen atoms and for that reason underestimates the concentration of formaldehyde and other oxygenated hydrocarbons of low molecular weight.

Nondispersive infrared techniques also are widely used for hydrocarbon measurements, especially for monitoring automobile exhaust. Based on a photometric method for measuring the fraction of infrared light absorbed over a band of wavelengths near 3.4 microns, devices of this type have a response to paraffins, other than methane, which is proportional to the number of carbon atoms. However, the response is unfortunately lower by a factor of 3 or more to methane and the reactive hydrocarbons ethylene, propylene, and certain aromatics such as toluene. Furthermore, there is positive interference from any carbon dioxide and water vapor present.

Gas chromatography is a more discriminating method for measuring ambient concentrations of various hydrocarbons, but suffers from its long response time and complexity. Nevertheless, it is a highly sensitive analytical technique.

As we have noted in the above descriptions, many air pollution monitoring devices are sensitive to more than one contaminant, and therefore spurious results from interference effects can be obtained if care is not taken. To maintain a degree of standardization, a statement of an air quality goal is therefore often accompanied by a stipulation about how the pollutant is to be measured.

12.4 AIR QUALITY STANDARDS

It is important to keep in mind that an air quality *criterion* is merely a description; that is, it indicates the concentration and dosage of a pollutant or combination of pollutants above which the atmosphere is considered to have an adverse effect. An air quality *goal* stipulates the maximum acceptable pollution levels. Neither criteria nor goals have legal significance, in the sense that they do not require compliance.

Some governments prescribe legal standards for ambient air quality. In the United States for example they are called *air quality standards*, and there are two types: *Primary standards* are based on the protection of health, and *secondary standards* on the protection of public welfare, including protection against known or anticipated effects of air pollution on property, materials, climate, economic values, and personal comfort. They indicate ambient levels of pollution that cannot legally be exceeded in a specific geographical region; and they may be imposed by national, state, and local governments. National standards have been adopted by the United States Environmental Protection Agency. How these standards are to be met is left to local governments, the states, and their respective air pollution control organizations; and it is anticipated that these local governments when necessary will impose emission restrictions over at least the most important sources of pollution within their jurisdiction.

The formulation of air quality standards proceeds in much the same way as that of air quality goals. The considerations indicated in Fig. 12.3 again apply.[6] Of particular importance are the technical factors listed in the bottom left-hand portion of the chart, because several questions are crucial in the decision-making: Can a standard be met with existing technology? If not, can advance notice of the imposition of a standard be expected to prompt the necessary developments of

TABLE 12.1

Air quality standards, 1971

Pollutant	Averaging time	Federal standards Primary	Secondary	Method
Photochemical oxidants (corrected for NO_2)	1 hr	$160 \, \mu g/m^3$ (0.08 ppm)[a]	Same as primary standard	Chemiluminescent method
Carbon monoxide	12 hr	—	Same as primary standard	Nondispersive infrared spectroscopy
	8 hr	$10 \, mg/m^3$ (9 ppm)[a]		
	1 hr	$40 \, mg/m^3$ (35 ppm)[a]		
Nitrogen dioxide	Annual average	$100 \, \mu g/m^3$ (0.05 ppm)	Same as primary standard	Colorimetric method using NaOH
	1 hr	—		
Sulfur dioxide	Annual average	$80 \, \mu g/m^3$ (0.03 ppm)	$60 \, \mu g/m^3$ (0.02 ppm)	Pararosaniline method
	24 hr	$365 \, \mu g/m^3$ (0.14 ppm)[a]	$260 \, \mu g/m^3$ (0.10 ppm)[a]	
	3 hr	—	$1300 \, \mu g/m^3$ (0.5 ppm)[a]	
	1 hr	—	—	
Suspended particulate matter	Annual geometric mean	$75 \, \mu g/m^3$	$60 \, \mu g/m^3$	High volume sampling
	24 hr	$260 \, \mu g/m^3$ [a]	$150 \, \mu g/m^3$ [a]	
Lead (particulate)	30 day average	—	—	—
Hydrogen sulfide	1 hr	—	—	—
Hydrocarbons (corrected for methane	3 hr (6–9 a.m.)	$160 \, \mu g/m^3$ (0.24 ppm)[a]	Same as primary standard	Flame ionization detection using gas chromatography
Visibility reducing particles	1 observation	—	—	—

[a]Not to be exceeded more than once a year.

California standards		Objectives of standard
Concentration	Method	
0.10 ppm (200 μg/m^3)	Neutral buffered KI	To prevent eye irritation and possible impairment of lung function in persons with chronic pulmonary disease. Also to prevent damage to vegetation
10 ppm (11 mg/m^3)	Nondispersive infrared spectroscopy	To prevent interference with oxygen transport by the blood based on carboxyhemoglobin levels greater than 2%
—		
40 ppm (46 mg/m^3)		
—	Saltzman method	To prevent possible risk to public health, and atmospheric discoloration
0.25 ppm (470 μg/m^3)		
—	Conductimetric method	To prevent possible increase in chronic respiratory disease and damage to vegetation
0.04 ppm (105 μg/m^3)		
—		
0.5 ppm (1310 μg/m^3)		To prevent possible alteration in lung function, and irritating odor
60 μg/m^3	High volume sampling	To improve visibility and prevent acute illness when present with about 0.05 ppm sulfur dioxide
100 μg/m^3		
1.5 μg/m^3	High volume sampling, dithizone method	To protect health
0.03 ppm (42 μg/m^3)	Cadmium hydroxide STRactan method	To prevent offensive odor
—	—	—
In sufficient amount to reduce the prevailing visibility to 10 miles (16 km) when the relative humidity is less than 70%		To improve visibility

emission control devices? Standards make sense only if they can be met in principle. Even for a region now meeting the prescribed levels, this is a matter of concern because future growth of population and the introduction of new industries would mean the need for more stringent controls on the new sources, as well as the existing ones, if possible. These issues are always raised; and others must be faced as well. For some industries, the alternative to an investment in costly control equipment is termination of operations with a resulting loss of jobs. Hence the imposition of air quality standards and their enforcement means that hard economic and political decisions must be made, as well as technological ones.

As one example of a set of air quality standards, we list in Table 12.1 those which have been adopted by the United States in 1971; for comparison, we list the ones then existing in the State of California. The responsibility for devising ways to meet these standards is in the hands of local Air Pollution Control Districts, each comprising a county or a group of counties in an air basin defined by meteorological and geographical features.

Reducing ambient levels

It is often said that a way in which local control districts could reduce ambient levels of pollution is by artificially improving meteorological ventilation. For example, to alleviate the conditions over Los Angeles, giant fans might be installed in tunnels penetrating through the mountains which ring the basin. Unfortunately, a simple calculation will quickly reveal that this would necessitate the expenditure of such great quantities of energy as to be impractical.[7] Thus artificially improving ventilation of this type does not hold much promise for the future. This is not to say that one should not take advantage of natural conditions where ventilation is good. If governments can include such factors when locations for industrial zones are established, the effects of the effluent can be minimized. It is also possible to exploit the buoyant potential of hot exhaust gases to penetrate some inversions if large sources are grouped together, but this may only locally improve convection.

It has also been suggested that industrial activity and automobile traffic should be reduced during episodes of high pollution, to reduce the rate of emission of pollutants. Controls then need to be applied only during conditions of atmospheric stagnation. Once ambient levels approach a predetermined value, law enforcement agencies would ensure compliance by industry and the public. Unfortunately, such an idea does not take cognizance of the fact that industrial emissions cannot always be turned on and off as easily as a home stove; the time lag between notification and response will vary from industry to industry. Furthermore, how is the reduction to be prescribed? Some sources may emit pollutants at a crucial stage in manufacturing and would be more severely hurt economically than others. And who is to say that the shutdown of a power plant may not cause more harm or discomfort when a consequence is that commercial offices can no longer benefit from the air-conditioning? Only a few specialized industries might be able to switch to special processes or utilize special control devices that are ordinarily too expensive to use continuously. For these reasons and others, the curtailment of activity to reduce emissions during pollution episodes has been regarded as a measure of last resort. It is not a practical means to ensure continual compliance with air quality standards.

The problem of pollution control boils down to the problem of devising emission standards which will continually be in force and can be practically achieved.

12.5 EMISSION STANDARDS

A legal restriction on the amount or conditions of release of a pollutant from any source is called an *emission standard*. In principle, a desired level of air quality can be achieved by ensuring that there is a sufficiently low rate of pollutant emissions from all relevant sources.

Setting an emission standard is a relatively simple matter if only one source causes the pollution; a series of measurements of the ambient concentration can be used to deduce by what factor the emissions should be reduced. Then the choice of an emission standard can be decided on the basis of one of several possible strategies, including (1) to respect the desired maximum ambient level regardless of consequences; or (2) to require the use of the most efficient existing pollution control devices. If (1) cannot be achieved by (2), two other strategies could be applied in an attempt to improve on the *status quo*: (3) to require slightly stricter controls than are currently feasible in anticipation that they will encourage technical progress (or, in the very least, that they could be met by the source operating under reduced capacity); or (4) to set up a schedule for future periodic tightening of standards, assuming that research and development will enable them to be met. It is a considerably more complex problem when multiple sources of different types are involved and when secondary pollutants are produced.

The problem of emission controls has often been divided into two aspects, controls over *stationary sources*, such as industries and utilities, and controls over *mobile sources*, such as automobiles, aircraft, trains, and ships. In some countries the responsibility over these various sources is divided between local and national agencies. In the United States, local governments such as counties and states have jurisdiction over stationary sources, whereas the federal government has claimed exclusive rights for controlling emissions from automobiles and aircraft. This was done to avoid a multitude of different standards being imposed on sources traveling interstate. And it was argued that the federal government could more effectively impose emission standards than could one state or county.

In connection with our discussion of photochemical smog in Section 10.5, we examined the standards which have been applied to automobiles. Clearly the setting of standards by a national government cannot—because of the realities of politics—be inclusive of situations peculiar to only a few individual localities. Thus the formulation of standards for automobiles has generally involved tightening by stepwise reductions in somewhat arbitrary amounts, with little regard for

TABLE 12.2

Examples of emission standards for single sources

Concept	Sample standards
Gases	
Maximum concentration	Limit maximum concentration of pollutant: 1) No more than 200 ppm of NO_x. 2) No more than 1% sulfur content in the fuel.
Maximum mass emission rate	Limit rate at which pollutant is released: 1) No more than 0.5 Tonne per hour of NO_x (stationary sources). 2) No more than 1.2 g/km of CO (motor vehicles).
Maximum concentration, adjusted for effective release height	Limit maximum concentration of pollutant, with less stringent limit on plumes having a higher rise or emitted from a taller stack: 1) No more than 200 ppm of NO_x for sources greater than 100 MW; no more than 100 ppm of NO_x for smaller sources. 2) No more than 250 ppm of SO_x for sources whose effective stack height exceeds 200 m under neutral conditions and 10 km/hr wind; 150 ppm all others.
Particles	
Maximum opacity	Limit the fraction of light which is absorbed upon traversing a plume: 1) No more than 20% reduction of the intensity of light (20% opacity).
Coloration	Avoid dense black or gray plumes: 1) Plume grayness not to exceed No. 1 on the Ringleman chart except during starting, but not greater than No. 2 for more than 30 min.
Maximum concentration	Limit maximum concentration of particulates: 1) No more than 0.7 g/m^3.
Maximum concentration adjusted for effective release height	Permit higher concentrates in plume with a greater rise: 1) No more than 0.7 g/m^3 for sources greater than 10 MW; no more than 0.5 g/m^3 for all others.
Maximum concentration, specifying size ranges	More control to avoid effects of light scattering: 1) No more than 0.3 g/m^3 for radius above 5 microns; no more than 0.05 g/m^3 for smaller particles.
Prohibition	1) Prohibition on disposal of refuse by incineration except under prescribed conditions.

the detailed consequences for the resulting pollution trends in individual counties or cities.

Local governments need not take such an arbitrary approach when regulating stationary sources, although they usually do. Local governments have several guidelines: the pollution trends during the past years, the results of diffusion model calculations for area-wide pollution, and an estimate for the anticipated growth of the community.

Yet despite this information, which in principle is available to local governments, and which is invaluable in tailoring emission standards to meet local goals, there is a continuing debate about which level of government can most effectively prescribe and enforce emission controls. It is argued, on the one hand, that overall controls ought to be imposed by the national government to avoid the tendency for industries to move from one place to another seeking a locality that would impose the minimum amount of control, thus perhaps economically penalizing those communities imposing stricter standards. On the other hand, if control is left solely to the national government, local values may not be respected. The inhabitants of some communities may prefer cleaner air than would be possible under a national norm. These people would prefer the option of regulating their industries more tightly. If control is solely in the hands of local authorities, an accompanying disadvantage is that the local industries will be able to exert relatively more political influence than they would if controls were imposed by the national government. Experience indicates that in many cases this is to the detriment of air quality. To maximize the benefits of local and national control, it has been argued that a nation should establish uniformly applicable standards, but that local governments be given the power to impose stricter limits.

Many types of emission standards are possible. In the overwhelming number of cases they have been applied only to single sources; that is, to a single exhaust stack or to a single furnace with multiple stacks. Some examples of standards for single stacks are listed in Table 12.2. The first column lists the concepts upon which standards can be based and is divided into those which apply to gases and those to particulate matter. The second column gives one or more examples of possible standards. The specific numbers cited in the examples are meant to be illustrative and should not be interpreted as recommendations. Ordinances prescribing such standards are often tailored to local conditions of meteorological ventilation and the local goals for air quality.

Several emission concepts illustrated in Table 12.2 warrant detailed examination. For the sake of clarity, we shall denote each by a letter:

a) The *maximum concentration* condition is founded on the desirability of diluting contaminants as much as possible before release. This is most important for sources whose effluent is released near the ground because the general public may experience the polluted exhaust only a short time later, before atmospheric turbulence can dilute the contaminants to inconsequential levels. However, the standard is usually applied to large sources such as power plants as well.

b) A *maximum mass emission rate* places a limit on the total amount of a pollutant that any source can emit during a prescribed time interval, something that the maximum concentration standard does not accomplish. Limiting the *amount* of released pollutant is important in air basins where pollutants may accumulate as a result of terrain and meteorological factors. Furthermore, the diffusion theory discussed in Chapter 7 indicates that the maximum ground-level concentration of a pollutant downwind from an elevated source, such as a tall exhaust stack, is directly proportional to the mass emission rate of the pollutant, not necessarily to its concentration in the effluent. Thus in some respects the mass emission rate is a more relevant parameter for large sources. Unfortunately, few communities have ordinances which limit it.

c) Both minimum dilution and maximum mass emission rates can be qualified by a parameter which is related to the effective release height of the exhaust plume. This is the *effective stack height*, the sum of the geometric stack height and the additional height attained by the rise of the plume due to its buoyancy and initial upward momentum. Higher-rising plumes have more time to become diluted by atmospheric turbulence before they touch ground and therefore, in many cases, need not be as severely restricted. An indirect parameter related to plume buoyancy is the rate at which heat is released by combustion in a furnace, often measured in terms of the megawatts of thermal energy released (MWt).

Similar concepts apply to controlling particulate emissions, in addition to which separate restrictions can be placed on particles of different sizes. Complaints from citizens about the objectionable appearance of dense, white plumes or black smoke have resulted in the adoption of criteria specifically aimed to improve the appearance.

d) The *maximum opacity* is a measure of how transparent a plume should be. Specifically it is a limit on the reduction of light intensity which would result when a beam of light were directed transversely through the center of a plume. The measurement can be made by mounting a laser on one wall of an exhaust stack near the orifice in such a way that the beam traverses the interior of the stack and impinges on a photocell on the opposite wall. The decrease in the beam intensity can be deduced from the intensity measured by the photocell and the known original intensity from the laser. Thus a criterion for 20% maximum opacity requires that the beam intensity be reduced by less than 20% compared with the amount transmitted when the plume is perfectly transparent.

A plume of water droplets can often be distinguished from a plume of particulates by observing the downwind dispersal. Droplet or "steam" plumes usually end abruptly when conditions are appropriate for the water to evaporate; but particulate plumes become progressively fainter as the particles disperse. The opacity of water droplets is not a measure of the accompanying particulate burden of the effluent.

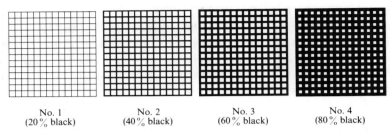

<table>
No. 1 (20% black)   No. 2 (40% black)   No. 3 (60% black)   No. 4 (80% black)
</table>

Fig. 12.9 Ringleman charts which provide a gray scale for judging the blackness of smoke. Ringleman no. 0 is white and no. 5 is black. When the charts are held 2 m from the eye, the lines and squares blur together and the overall shade of gray can be compared with that of a plume.

e) The grayness of a plume can be judged by visual observations which compare the plume's appearance to a standardized gray scale, as first suggested in 1898 by Maximilian Ringleman, a French engineer. Number 5 on the Ringleman scale corresponds to dense, black smoke. Number 0 is perfectly white (as distinct from transparent). Determination of the Ringleman number of a plume can be accomplished by a trained observer who matches one of a series of charts shown in Fig. 12.9 to the appearance of the plume, with the sun behind the observer so as not to unduly affect the comparison. But both opacity and Ringleman standards fail to limit the total output of particulate matter because, as we have previously noted, particles of different size affect visible light in different ways.

f) For completeness, we have included a concept called *prohibition*. This hardly needs explanation; but at the risk of stating the obvious we note that emission control devices may not exist which are sufficient to reduce emissions from certain industries, and these industries therefore might not be permitted to operate. For example, the burning of refuse outdoors, including agricultural burning, releases great numbers of particles that are very effective in scattering light and impairing visibility. Complaints from the public have led in many instances to a complete ban on open burning, affecting not only farmers and construction workers, but owners of domestic backyard incinerators as well.

Occasionally an emission standard is written in a manner that permits larger sources to emit more pollutants than smaller ones. This is implicit for example in maximum concentration standards. But some jurisdictions control emissions of particulate matter by setting the maximum mass emission rate as a fixed percentage of the total hourly weight of all materials introduced into a manufacturing or processing operation, excluding perhaps liquid and gaseous fuels and air. Without additional safeguards to ensure proper dilution of the effluent, such an emission standard cannot be expected to guarantee respect of the ambient air quality standards.

12.6 ENFORCEMENT AND INCENTIVES

Perhaps it is not too great an oversimplification to say that the reason we have air pollution is because a polluter believes it to be in his best interest to pollute. The road to improvement thus lies in changing this attitude. One way is to offer incentives for all of us to reduce the amount of contaminant matter we release into the air and thus make the reduction of pollution an action in our best interest. Emission standards, for example, cannot be effective without the incentive which comes from the threat of legal action against violators. There are many other types of incentives, both rewards and penalties; the more obvious ones include economic gain or loss, the desire for a pleasant environment, the desire for good health, and concern for the well-being of others. Recently we have seen develop still another: a moral attitude that polluting the environment is "bad." It is not our purpose to evaluate the relative importance of these. Surely they influence the decisions of individuals and the policies of industries and governments, but to differing degrees. We shall here briefly consider the incentives which governments have traditionally applied to achieve their goals: legal and economic incentives.

Once a government has prescribed emission standards, the responsibility for enforcing them is usually delegated to air pollution control authorities, a branch of the administration which has police powers. The details of the administrative organization vary from nation to nation, but the same pattern of operation is generally followed. Through monitoring and inspection of sources, compliance with the law can be verified. Violators are cited and brought to appear before a court of law, where guilt or innocence is determined and punishment consists of a *fine*. The charge is generally classified as a misdemeanor, so the fine involved is relatively small. Much public criticism is leveled at the insignificance of the amount involved when industries are at fault. Unless the fine is substantial, there is little incentive for a company to make costly investments in control equipment, and they may choose instead to pay an occasional fine.

One disadvantage in exclusive reliance upon fines as a deterrent to pollution is the lack of incentive to *minimize* the emission of contaminants, and encourage people to reduce emissions to levels closer to air quality goals. *Pollution emission taxes* can provide that incentive. A tax or charge can be assessed on each source according to the type and amount of pollutants released into the air. Thus, for example, coal could be taxed according to its sulfur content, assuming the user has no means of cleaning SO_2 from the effluent. Or a steel mill could be taxed on the basis of the amount of particulate matter it contributes. It is possible by these means to make polluting costly and continually encourage industries to seek means of further reducing their emissions of contaminants. Financial incentives offer an advantage over prescribed emission standards and fines in that they are generally easier to administer and they preserve a measure of decentralization in the decision-making of large corporations.[8] Other economic incentives can be provided by tax laws. These include flexible schedules for the depreciation of capital costs of control equipment and extra write-off allowances or tax credits for funds invested in research and development of control techniques.

It seems a simple matter to assess taxes on the basis of the total amount of a pollutant emitted by a source; but several other issues must be faced if the tax is to reflect the severity of adverse effects which might result from the release of the pollutant. For example, a tax based on average monthly or yearly emissions does not influence emission rates over shorter periods. Further, a tax based on mass emission rate does not necessarily encourage a reduction in the pollution concentration in the effluent, or release from higher exhaust stacks, or construction of new sources in regions of better ventilation. No tax can be completely equitable, because of the many arbitrary decisions involved. Nevertheless, pollution taxes are appealing because they do require polluters to pay for the damage they cause. If the polluter is an industry, the cost is passed along to the consumer, who thus pays an amount closer to the actual "cost" of producing the goods or service.

It has occasionally been suggested that governments might subsidize control operations, and in effect provide economic support to an industry in proportion to the amount of pollutant that is *not* emitted. Unfortunately, such plans are difficult to administer, because they depend on establishing a measure for how much pollutant would have been emitted without the control, a quantity that often cannot be determined. In addition, the intent may be perverted by implementation of methods which intentionally produce large quantities of pollutants, to collect higher subsidies when the pollutants are scavenged from the effluent.

It would appear that an effective way to proceed to solve the air pollution problem would be through a combination of emission taxes which would depend on the amount and mode of pollutant release (to encourage everyone or at least the major sources to minimize emissions) and prescription and enforcement of emission standards (to ensure a minimal air quality).

A useful administrative tool for pollution control authorities is the *permit system*. In one version, a construction permit is required from control authorities before any structure that might be a potential source of pollution can be built or modified. This affords an apportunity for an inspection of the plans by an engineering staff of the control authority to verify that the structure will be able to meet the emission standards. Then a second permit called an "operating permit" is required before the source can begin operations, and periodically thereafter the operating permit must be renewed. The time of issuance and renewal provides a natural opportunity for inspection of the sources themselves. This system gives the control authorities the additional possibility of withholding an operating permit from any violator. The threat of forced closure may be sufficient incentive for small industries to improve conditions. But it is less effective against large industries upon which many jobs, as well as the economic well-being of a community, depend; few governments in a democracy could withstand the wrath that would follow in the wake of massive unemployment.

Strict emission standards often hurt most severely only a few concerns or individuals in any one community. It may be appropriate in such cases to provide *subsidies* to spread the cost to the beneficiaries and ease the burden on the minority. When a smokeless zone is created in the United Kingdom, a subsidy is simultaneously provided to cover part of the cost of equipment modifications to control the emission of particulates. Subsidies need not be direct payments, but could be incorporated into tax laws as we have mentioned above.

12.7 THE FUTURE

In the early Middle Ages it was a mark of distinction to have plumes of smoke continually issuing from every chimney of a chateau, for only the rich could afford the luxury of using fuel wastefully. Then came the industrial revolution and the time when the prosperity of a community was often gauged by the number of exhaust stacks belching black smoke. Now, however, attitudes are changing. Warmth and sustenance are taken for granted; health and aesthetics occupy a higher position in our list of priorities. Befouling the air is no longer praised or acceptable.

And what of the future? In what areas can we expect to make progress in controlling air pollution? Some trends are now appearing. Ordinances prescribing emission standards, once a window dressing for the inactivity of air pollution control districts, are now being enforced. And they are steadily being tightened, partly in response to improving technology and partly to encourage further developments. Furthermore, there is more emphasis to base them on realistic predictions from pollution dispersal models for particular communities, thereby founding them on more substantial estimates of their true effects.

One will find more comprehensive regional planning, with air pollution as one of the factors which will enter decisions on land zoning. Long-range planning will become ever more important to pace emission control efforts with community growth. The true cost of damage from pollution will be better known and will play a more important role in guiding decisions. Health effects, too, will be more precisely identified, and control efforts and the composition of fuels such as gasoline and fuel oil will be altered in view of our better understanding of the chemical mechanisms that govern the evolution of photochemical and sulfurous smog. More emphasis will be placed on modes of transportation which provide an alternative to the widespread use of the internal combustion engine in automobiles. Power plants will be built with higher stacks and greater plume buoyancy, to facilitate plume penetration of low-lying inversion layers. Perhaps "emission rights" will be introduced, in which emission standards are written in terms of mass emission rates for an entire area: Then if a new source moved into the area, existing sources would have to reduce their emissions to ensure overall compliance. All these are logical extensions of present trends.

But continued population growth and ever-increasing consumption of material goods would continue to challenge the effectiveness of air pollution controls, which could only be met by more investments in research, development, and the installation of control equipment. Perhaps even this would not be enough. Many issues are involved in the problem of air pollution, not all of them technical. How much in economic and social costs will be invested to control contaminants is in many cases a direct reflection of the attitudes of the public. As one member of a county air pollution advisory committee has said, just before the committee voted on a proposed tough standard, "It's now no longer a question of the technical trade-offs; it's a question of where the votes lie." This reflects the complex interrelationships between what course of action is obvious from technical considerations and what the public is willing to shoulder. The problem of air pollution will not be solved on scientific facts alone.

SUMMARY

Two governmental approaches to the control of air pollution have gained popularity during the past decade: (1) imposition of emission standards on individual sources, and (2) assessment of emission taxes from individual sources on the basis of the amount and conditions of release of pollutants. In principle both are enacted with the intent of meeting air quality criteria, goals, or standards for ambient air, and thus the establishment of individual source criteria logically should include recognition of the cumulative effects of multiple sources and the local meteorology. Emission standards when intelligently formulated and enforced can offer the guarantee of a minimal air quality, but pollution taxes can provide additional incentive for polluters to do better than simply meeting minimal standards. Depending on desired results, individual source emission standards can take various forms: Maximum concentration, mass emission rate, and opacity are common examples. These may be made more or less stringent to include the conditions of pollution release—whether in a plume with little buoyancy or one emitted from a low exhaust stack. The formulation of air quality standards and the requisite emission standards involve more than technical considerations. Legal standards will impose an uneven burden among different industries and different socioeconomic groups, and the possibility exists that enforcement of emission standards will have economic and sociological consequences. Air pollution is part of a much more general problem of waste disposal; one of the greatest challenges facing society today is the question of how to deal with this broad problem in the most equitable and beneficial way.

NOTES

1. R. G. Ridker, *Economic Costs of Air Pollution*, New York: Praeger Publishers, 1967.

2. L. B. Lave and E. P. Seskin, "Air Pollution and Human Health," *Science* **169**, 723 (1970).

3. W. B. Johnson, "Lidar Applications in Air Pollution Research and Control," *J. Air Poll. Control Assoc.*, **19**, 176 (1969).

4. P. W. West and G. C. Gaeke, *Anal. Chem.*, **28**, 1916 (1956).

5. B. E. Saltzman, *Anal. Chem.*, **26**, 1949 (1954); B. E. Saltzman and A. F. Wartbarg, *Anal. Chem.*, **37**, 1961 (1965).

6. L. B. Kreuzer, N. D. Kenyon, and C. K. N. Patel, "Air Pollution: Sensitive Detection of Ten Pollutant Gases by Carbon Monoxide and Carbon Dioxide Lasers," *Science* **177**, 347 (1972).

7. J. M. Heuss, G. J. Nebel, and J. M. Colucci, "National Air Quality Standards for Automotive Pollutants—A Critical Review," *J. Air Poll. Control Assoc.* **21**, 535 (1971); see also discussion on page 544.

8. M. Neiburger, "Weather Modification and Smog," *Science*, **126**, 637 (1957).

9. R. M. Solow, "The Economist's Approach to Pollution and Its Control," *Science*, **173**, 498 (1971).

FOR FURTHER READING

A. ATKISSON   *Development of Air Quality Standards*, Columbus, Ohio: Charles
and R. S. GAINES   E. Merrill 1970.

G. H. HAGEVIK   *Decision Making in Air Pollution Control*, New York: Praeger Publishers, 1970.

F. P. GRAD,   *Environmental Control. Priorities, Policies, and the Law*, New
G. W. RATHJENS,   York: Columbia University Press, 1971.
and A. J. ROSENTHAL

J. L. SAX   *Defending the Environment*, New York: Alfred A. Knopf, 1971.

R. G. RIDKER   *Economic Costs of Air Pollution*, New York: Praeger Publishers, 1967.

A. V. KNEESE   "Environmental Pollution: Economics and Policy," *Amer. Econ. Rev.* (Papers Proc.), **61**, 153 (1971).

A. J. VAN TASSEL, Ed.   *Environmental Side Effects of Rising Industrial Output*, Lexington, Mass: Heath Lexington, 1970.

QUESTIONS

1. What is the distinction between air quality criteria, air quality goals, air quality standards, and emission standards?

2. Suppose that you are on an advisory committee which must recommend an air quality standard for a toxic gas. What types of standards could conceivably be applicable?

3. Why is the mean SO_2 concentration in Fig. 12.6 independent of the averaging interval?

4. Draw a sketch for the frequency of occurrence of various one-hour average concentrations of SO_2 *versus* the average concentration, using the data available in Fig. 12.6. Is this frequency distribution symmetrical about the median level? What is the relationship between your curve and the SO_2 curve for Chicago shown in Fig. 12.5?

5. You are on a planning commission which has the responsibility for selecting a site which will be zoned for heavy industry. What criteria concerning aspects of air pollution bear on this matter?

6. To reduce SO_x emissions within a city, the local air pollution control authority decides to prohibit the use of coal and fuel oil, with the implication that furnaces will be converted to the use of natural gas. What economic and social consequences could reasonably be expected to follow implementation of such a prohibition?

7. From the data in Fig. 12.5 give a summary of the characteristic features of air pollution in San Francisco. What are the similarities and differences between pollution in that city and the situation in Philadelphia?

8. What percentage of your income would you be willing to pay each year if by doing so you would be assured of no adverse effects from air pollution?

9. Suppose that you suspect that a local copper smelter has emitted sulfuric acid droplets which, upon fallout, have damaged the paint on your automobile. What evidence and testimony do you need to prove this if you decide to seek relief by private litigation against the owners and operators of the smelter? What are the weakest points in your case?

10. What is the difference between the "conductrimetric" and "amperometric" methods for determining the presence of an air pollutant? How do dispersive and nondispersive photometric techniques differ and what are the advantages of each?

11. What pollutants can be detected by the flame ionization technique? Is it equally sensitive for all molecules within the class(es) you mentioned? Explain.

PROBLEMS

1. If it were desired that SO_2 daily average levels in the city of Chicago were to exceed 0.1 ppm on no more than 1% of the days, by how much would ambient concentrations (and emissions) have to be reduced? (Consult Fig. 12.6.)

2. With reference to Fig. 12.6, which of the following two criteria would be more stringent for SO_2 ambient levels in Chicago: (a) the hourly average of SO_2 is not to exceed 0.5 ppm; or (b) the daily average is not to exceed 0.1 ppm on at least 99% of the days?

3. Suppose that 5000 one-hour averages for the SO_2 concentration in St. Louis were taken in 1964. Using the appropriate data from Fig. 12.5, draw a curve giving the expected number of these which would be found at concentrations between 0.01 ppm and 0.60 ppm, using a linear scale for the concentration axis. How is this frequency distribution related to the cumulative frequency distribution in Fig. 12.5?

4. Estimate the opacity of the effluent from a stack, assuming that it consists of transparent gases and small spherical particles of 0.5 micron radius at a concentration of 10^6 per cm^3. Each particle has a dielectric constant of 1.22 (see Fig. C.4 for the necessary light-scattering data). The opacity is measured with green light of wavelength 0.50 micron, and the test beam is sent through a 1-m thick portion of the plume.

5. Suppose that the particles in the plume in the preceding problem have a radius of 0.1 micron. What concentration of particles would give the same opacity as the 0.5 micron particles? Compare the mass concentrations corresponding to these two cases of similar opacity.

APPENDIX A

GLOSSARY AND USEFUL CONSTANTS

GLOSSARY

In most instances the notation used in this book conforms to what is most commonly found in the literature. For this reason the same symbol may represent different quantities in different chapters. Listed below are the definitions which apply to the chapters indicated.

Symbol	Definition	Section
a	An arbitrary constant	2.3
	Acceleration	3.2, C.1
	Albedo	4.2
a_{ce}	Acceleration for circular motion	3.3
a_C	Coriolis acceleration	5.2, F.3
A	Empirical constant	2.3
	Name of an observer	3.3
	Area	3.4
A_λ	Rate at which a square meter of surface absorbs radiation at wavelength λ	D.1
$\mathbf{A}$	Arbitrary vector	F.1
b	Scattering coefficient for I_1 and I_2	2.3
b'	Scattering coefficient for I_0	2.3
B	Name of an observer	3.3
	Turbidity	4.8
$\mathbf{B}$	Arbitrary vector	F.1
c	Speed of light	Introduction to Chap. 4, C.1, D.1, E.2
c_v	Specific heat at constant volume per kilogram of air	5.5, 5.6
C_v	Specific heat at constant volume	5.5
$C(x)$	Perceived contrast of a target	2.3
$\left.\begin{array}{l}C_y\\C_z\end{array}\right\}$	Diffusion parameters	7.5
d	Depth below ground	5.8
D	Molecular diffusivity	7.2
	Eddy diffusivity	7.3
$\left.\begin{array}{l}D_y\\D_z\end{array}\right\}$	Eddy diffusivity in y- and z-directions	7.4, 7.7, 7.8
D_c	Threshold dosage	12.2
D_e	Mean distance between sun and earth	4.1

D_s	Smokestack diameter	7.6
e	Base of natural logarithms	2.3, 4.8, 7.4, 7.5
	Electronic charge	C.1
E	Energy	3.1, 4.1, 4.3, 10.2
	Electric field	C.1, E.2
E_0	Amplitude of electric field	C.1
E_λ	Power radiated per square meter of surface at wavelength λ	D.1
$\left.\begin{array}{l} E_n \\ E_m \end{array}\right\}$	Allowed energy of molecule	E.2
E_p	Energy of photon	E.2
E_r	Rotational energy of molecule	E.4
E_s	Amplitude of electric field of scattered radiation	C.1
E_{bb}	Power radiated by a square meter of black body surface	4.1
E_v	Vibrational energy of molecule	E.4
f	Frequency of turbulent eddy	7.3
F	Force	3.2, F.3
F_C	Coriolis force	5.2, 5.4, F.3
F_{ce}	Centrifugal force	3.3
F_d	Drag force from air viscosity	11.5
F_f	Frictional force	5.4
F_g	Gravitational plus centrifugal forces	3.3, 11.5
F_g^*	Gravitational force	3.3
F_p	Pressure-gradient force	5.4
F_{ce}	Centrifugal force	3.3
g	Acceleration due to F_g	3.2, 3.5, 5.5, 5.6, 11.5
g^*	Acceleration due to F_g^* alone	3.2
g_0^*	Acceleration due to F_g^* at earth's surface	3.2
G	Universal gravitation constant	3.2
h	Planck's constant	D.1, E.2
	Relative humidity	4.6
h_0	Relative humidity at sea level	4.6
h_r	Plume rise above stack	7.6
h_g	Geometrical stack height	7.6
H	Release height of a plume	7.5
I	Intensity of solar radiation	4.8
I_0	$\left\{\begin{array}{l}\text{Intensity of extraneous light}\\\text{Intensity of sunlight in the outer atmosphere}\end{array}\right.$	2.3, C.1 4.8
I_s	Intensity of scattered light	C.1
I_1	Intensity of light from target direction	2.3
I_2	Intensity of light from background	2.3
I_λ	Intensity of radiation at wavelength λ	D.1
k	Absorption coefficient	2.3
k_B	Boltzmann's constant	3.5, D.1
K	Empirical constant	2.3
	Force constant for electron in dielectric	C.1
K_h	Thermal conductivity	5.8

K_s	Scattering area coefficient	C.3
K_λ	Absorption coefficient	D.1
l	Mean free path of air molecule	7.2
L	Latent heat of evaporation per kilogram of water	5.6
	Constant in Wien displacement law	4.1
	Vertical optical path length through atmosphere	4.8
	A length	7.6, 7.7, 7.8
L_v	Meteorological range	2.3
m	Mass of a body	3.2, 3.3, 4.1
	Mass of an air molecule	3.5, 7.2
	Mass of an electron	C.1
	Cosecant (ϕ)	4.8
$\left.\begin{array}{l} m_1 \\ m_2 \end{array}\right\}$	Mass of a body	3.2
M	Aerosol mass concentration	2.3, 11.6
	Mass of an air parcel	5.5, 5.6, F.3
M_a	Molecular weight of air	3.4, 3.5
M_e	Mass of earth	3.2
M_p	Molecular weight of a pollutant	7.4, B.2
n	Number density of molecules	3.4, 7.2, 7.4, 7.8
	Number density of aerosol particles	11.4, 11.6
N	Avogadro's number	3.5
P	Pressure	3.4, 3.5, 4.6, 5.4, 5.5
P_0	Pressure at sea level	3.5, 4.6
P_x	x-directed momentum	7.2
q	Rate of heat flow	7.2
q_x	Rate of diffusion of molecules in x-direction	7.2
Q	Pollutant emission rate	7.4, 7.5, 7.7, 7.8
Q_h	Heat emission rate	5.8, 7.6
Q_M	Pollutant mass emission rate	7.4
Q_V	Pollutant volume emission rate	7.4
ΔQ	Energy given to a parcel of air	5.5
r	Distance between two points	3.2
	Ratio of mass of condensing water vapor to mass of air	5.6
$\mathbf{r}_e$	Position in earth-bound reference system	F.2
$\mathbf{r}_i$	Position in inertial reference system	F.2
R	Particle radius	8.1, 11.5, 11.6, C.1, C.3
R'	Universal gas constant per mole	3.4, 3.5
R_a	Gas constant per kilogram of air	3.4, 5.5, 5.6
R_e	Radius of earth	3.2, 4.1, 4.2, 5.2
R_e	Reynolds number	7.6
R_s	Radius of sun	4.1
s	Distance electron is displaced from equilibrium	C.1
S_0	Solar constant	4.1, 4.2
t	Time interval	3.4, 5.4, C.1, F.2

T	Absolute temperature	3.4, 3.5, 4.1, 5.5, 5.6, 5.8, 7.6, D.1
T_*	Mean surface temperature of earth	4.2, 4.6, 4.7
u	Wind speed	7.3
$\bar{u}$	Average wind speed	7.2, 7.3, 7.4, 7.5, 7.6, 7.7, 7.8
u'	Turbulent component of wind in x-direction	7.3
U	Internal energy of an air parcel	5.5
v	Average speed of molecules	3.4, 7.2
v'	Turbulent component of wind in y-direction	7.3, 7.5
$\left.\begin{array}{c} v_S \\ v_E \end{array}\right\}$	South and east components of wind velocity	5.2
v_g	Wind speed at geostrophic balance	5.4
$\mathbf{v}_e$	Velocity of wind in earthbound reference system	F.2, F.3
$\mathbf{v}_i$	Velocity of wind in inertial reference system	F.2
V	Volume of a quantity of gas	3.4, 5.5
	Exit speed of plume	7.6
	Speed of falling particle	11.5
	Volume of a particle	C.1
V_t	Terminal velocity of falling particle	11.5
V_p	Prevailing visibility	2.3
W	Height of inversion base	7.8, 8.5
ΔW	Work done by an air parcel	5.5
w'	Turbulent component of wind in the z-direction	7.3, 7.5
x	Horizontal distance	2.3, 7.4, 7.5, 7.6, 7.7
y	Horizontal distance perpendicular to plume line	7.4, 7.5
z	Altitude	3.2, 3.5, 4.8, 5.5, 5.6, 7.2
	Height above ground	7.4, 7.5, 7.6, 7.7
α	Wind inclination to horizontal	7.4
	Parameter for particle radius compared to wavelength of light	C.3
α_a	Fraction of light Rayleigh scattered or absorbed by dry air	4.8
α_w	Fraction of light scattered by water vapor	4.8
α_p	Fraction of light scattered by aerosols	4.8
β	Exponent in Junge distribution	11.6
	Angle light is scattered from forward direction	C.1
γ	Angle between electron's acceleration and light scattering direction	C.1
ϵ	Contrast threshold	2.3
Γ	Dry adiabatic lapse rate	5.6
λ	Wavelength of electromagnetic radiation	Introduction to Chap. 4, C.1, C.3, D.1, E.2
λ_m	Wavelength at peak of black body spectrum	4.1
η	Molecular viscosity coefficient of air	7.2
	Eddy viscosity coefficient of air	7.6, 11.5

η_λ	Emission factor at wavelength λ	D.1
v	Frequency of electromagnetic radiation	Introduction to Chap. 4, C.1, E.2
Ω	Angular velocity of earth	3.3, 5.2, 5.3, 5.4, F.2, F.3
$\mathbf{\Omega}$	Angular velocity vector of earth	F.2, F.3
ϕ	Elevation of sun	4.8
	Angle between $\mathbf{A}$ and $\mathbf{B}$	F.1
ρ	Density of air	3.4, 3.5, 5.4, 5.5, 7.2, 7.6
ρ_p	Density of material in a particle	2.3, 11.5
ρ_0	Density of air at sea level	3.5
σ	Black body emissivity	4.1, 4.2
$\left.\begin{array}{c}\sigma_y \\ \sigma_z\end{array}\right\}$	Standard deviations of plume width	7.4, 7.5, 7.6
σ_θ	Standard deviation of horizontal direction of wind	7.5, 7.6
σ_α	Standard deviation of vertical direction of wind	7.5, 7.6
τ	Averaging time or duration of exposure	7.4, 12.2
τ_r	Parameter for Rayleigh scattering	4.8
τ_z	Parameter for absorption by ozone	4.8
θ	Latitude	3.3, 4.8, 5.2, 5.4
	Angle between instantaneous wind and average wind direction in horizontal plane	7.5
Θ_-	Potential temperature	5.6
ϕ	Plume width	7.4
	Azimuth of sun	4.8
	Angle included between vectors A and B	F.1
X	Concentration	7.2, 7.3, 7.4, 7.5, 7.6, 7.7
X_c	Threshold concentration	12.2

USEFUL CONSTANTS

Universal constants

Speed of light	$c = 2.998 \times 10^8$ m/sec
Charge of an electron	$e = -1.602 \times 10^{-19}$ coulomb
Planck's constant	$h = 6.626 \times 10^{-34}$ J-sec
Boltzmann's constant	$k_B = 1.381 \times 10^{-23}$ J/molecule-deg
Avogadro's number	$N = 6.022 \times 10^{23}$ molecules/mole
Gravitational constant	$G = 6.673 \times 10^{-11}$ m^3/kg-sec^2
Black body emissivity	$\sigma = 5.670 \times 10^{-8}$ W/m^2-deg^4
Universal gas constant	$R' = 8.314$ J/mole-deg
Base of natural logarithms	$e = 2.718$

Solar constants

Mass	$M_s = 1.97 \times 10^{30}$ kg
Radius (visible disc)	$R_s = 6.95 \times 10^8$ m

Terrestrial constants

Mass of earth	$M_e = 5.983 \times 10^{24}$ kg
Average radius	$R_e = 6.378 \times 10^6$ m
Acceleration due to gravity (45° latitude at sea level)	$g = 9.806$ m/sec^2
Angular velocity	$\Omega = 7.292 \times 10^{-5}$ rad/sec
Distance to sun	$D_e = 1.49 \times 10^{11}$ m
Solar constant	$S_0 = 1.34 \times 10^3$ W/m^2

Atmospheric constants

Mass of atmosphere	$M_{at} = 5.3 \times 10^{18}$ kg
Mean pressure at sea level	$P_0 = 1.013 \times 10^5$ N/m^2
Mean temperature at sea level	$T_* = 288°$K
Mean humidity at sea level	$h_0 = 0.77$
Dry adiabatic lapse rate	$-\Delta T/\Delta z = 0.00986$ °C/m

Dry air constants

Molecular weight	$M_a = 28.97 \times 10^{-3}$ kg
Gas constant	$R_a = 287.6$ J/kg-deg
Specific heat at constant volume	$c_V = 1.01 \times 10^3$ J/kg-deg
Specific heat at constant pressure	$c_P = 1.42 \times 10^3$ J/kg-deg
Density at 1000 mb and 0°C	$\rho = 1.276$ kg/m^3
Molecular viscosity at 1000 mb and 18°C	$\eta = 1.8 \times 10^{-5}$ kg/m-sec
Molecular thermal conductivity	$K_h = 2.5 \times 10^{-2}$ J/m-sec-°K
Number density at 1000 mb	$n = 2.56 \times 10^{25}$ m^{-3}
Molecular speed	$v = 459$ m/sec
Mean free path at 1000 mb	$l = 6.63 \times 10^{-8}$ m

Water constants

Latent heat of evaporation at 20°C	$L = 2.43 \times 10^6$ J/kg
Latent heat of sublimation at 0°C	$L_s = 2.82 \times 10^6$ J/kg

APPENDIX B

CONVERSION FACTORS

B.1 MULTIPLES AND SUBMULTIPLES OF UNITS

The metric system employs a base unit of measure for each dimensional quantity (for example, gram, meter, and second), and larger or smaller units of measure in multiples of ten of these are denoted by an appropriate prefix. The prefixes and the factors they represent are listed below.

Multiples and submultiples of units

Factor by which unit is multiplied	Prefix	Symbol
10^{12}	tera	T
10^{9}	giga	G
10^{6}	mega	M
10^{3}	kilo	k
10^{2}	hecto	h
10	deka	da
10^{-1}	deci	d
10^{-2}	centi	c
10^{-3}	milli	m
10^{-6}	micro	μ
10^{-9}	nano	n
10^{-12}	pico	p
10^{-15}	femto	f

B.2 CONCENTRATIONS

We have usually expressed concentrations of gaseous pollutants in the units of parts per million (ppm), referring to the number of pollutant molecules per million molecules of air. This is called the *concentration by volume*, since it gives the proportion in which the volumes of gaseous pollutant and air (at the same temperature and pressure) must be mixed to achieve the specified concentration. However, another convention is also commonly found in the literature by which the *mass concentration* of the pollutant is given, equivalent to our use of mass concentration for describing the amount of an aerosol present. Thus the level of a gaseous contaminant might be expressed, for example, as a certain number of micrograms of pollutant per cubic meter. We shall indicate here the formula which relates the mass concentration by volume and mass concentration of a gaseous pollutant at sea level.

We note that the ideal gas law in Eq. (3.8) indicates that one mole of air at 25°C and 1 atmosphere pressure would occupy a volume of 24.5 liters. If M_p is the molecular weight in grams of one mole of a gaseous pollutant, an ambient mass concentration of 1 mg/m^3 would correspond to a concentration by volume of $(24.5/M_p) \times 10^{-6}$. Thus:

$$1 \text{ mg/m}^3 = \left(\frac{24.5}{M_p}\right) \text{ ppm.}$$

For SO$_2$ with $M_p = 64$, we find that a mass concentration of 1 mg/m^3 corresponds to an ambient concentration of about 0.38 ppm at 25°C.

When the ambient temperature or pressure are significantly different from 25°C and 1 atmosphere, respectively, the above formula is not valid. Thus for high altitudes, the correct formula would have the number 24.5 replaced by the number of liters actually occupied by one mole of air.

B.3 CONVERSION BETWEEN DIFFERENT UNITS

We have gathered below quantities of the English and engineering systems of units that are commonly found in the literature on air pollution. Our intention is to list them in such a way that their equivalent in the MKS system of units can be found quickly. Quantities which are listed in each horizontal line are equivalent. The quantity in the middle column indicates the simplest definition or a useful equivalent of the respective quantity in the first column.

Unit (symbol)	Definition or CGS equivalent	MKS equivalent
1 acre	1/640 mi^2	4.047×10^3 m^2
1 Ångstrom (Å)	10^{-8} cm	10^{-10} m
1 atmosphere (atm)	1.013×10^6 dyn/cm^2	1.013×10^5 N/m^2
1 bar (b)	10^6 dyn/cm^2	10^5 N/m^2
1 barrel (bbl)	42 gal, U.S.A.	0.159 m^3
1 boiler horsepower	3.35×10^4 BTU/hour	9.810×10^3 W
1 British Thermal Unit (BTU)	252 cal	1.054×10^3 J
1 BTU/hour	1.93×10^6 erg/sec	0.293 W
1 calorie (cal)	4.184×10^{-7} erg	4.184 J
1 centimeter of mercury (cm Hg)	1.333×10^4 dyn/cm^2	1.333×10^3 N/m^2
1 cubic foot, U.S.A. (cu ft)	2.832×10^4 cm^3	2.832×10^{-2} m^3
1 dyne (dyn)	1 g-cm/sec^2	10^{-5} N
1 erg	1 g-cm^2/sec^2	10^{-7} J
1 foot, U.S.A. (ft)	30.48 cm	0.3048 m
1 foot per minute (ft/min)	1.829×10^{-2} km/hr	5.080×10^{-3} m/sec
1 gallon, U.S.A. (gal)	3.785×10^3 cm^3	3.785×10^{-3} m^3

1 grain	1/7,000 lb (avoirdupois)	6.480×10^{-5} kg
1 grain per cubic foot	2.288 g/m^3	2.288×10^{-3} kg/m^3
1 hectare (ha)	10^8 cm^2	10^4 m^2
1 horsepower, U.S.A. (h.p.)	7.46×10^9 erg/sec	746 W
1 horsepower, metric	7.35×10^9 erg/sec	735 W
1 hour (hr)	3.60×10^3 sec	3.60×10^3 sec
1 inch (in)	2.54 cm	2.54×10^{-2} m
1 joule (J)	10^{-7} erg	1 kg-m^2/sec^2
1 knot (kt)	1.853 km/hr	0.5144 m/sec
1 langley (ly)	1 cal/cm^2	$4.184 \times 10^{+4}$ J/m^2
1 liter (l)	10^3 cm^3	10^{-3} m^3
1 micron (μm)	10^{-4} cm	10^{-6} m
1 mile, U.S.A. (mi)	5,280 ft	1.609×10^3 m
1 mile per hour (mph)	44.70 cm/sec	0.447 m/sec
1 minute (min)	60 sec	60 sec
1 newton (N)	10^{-5} dyn	1 kg-m/sec^2
1 ounce, avoirdupois (oz)	28.35 g	28.35×10^{-3} kg
1 ounce, U.S.A. fluid (oz)	29.57 cm^3	29.57×10^{-6} m^3
1 part per million (ppm)	10^{-6}	10^{-6}
1 part per hundred million (pphm)	10^{-8}	10^{-8}
1 part per billion (ppb)	10^{-9}	10^{-9}
1 pint, U.S.A. dry (pt)	550.6 cm^3	5.506×10^{-4} m^3
1 pint, U.S.A. liquid (pt)	473.2 cm^3	4.732×10^{-4} m^3
1 pound, avoirdupois (lb)	weight on earth of 453.6 g	weight on earth of 0.4536 kg
1 pound per hour (lb/hr)	0.1260 g/sec	1.260×10^{-4} kg/sec
1 poise (P)	1 dyn-sec/cm^2	10^{-1} N-sec/m^2
1 quart, U.S.A. dry (qt)	1.101×10^3 cm^3	1.101×10^{-3} m^3
1 quart, U.S.A. liquid (qt)	946.4 cm^3	0.9464×10^{-3} m^3
1 square foot (ft^2)	929.0 cm^2	929.0×10^{-4} m^2
1 square mile (mi^2)	2.590×10^{10} cm^2	2.590×10^6 m^2
1 square yard (yd^2)	8361 cm^2	0.8361 m^2
1 standard cubic foor per minute (scfm)	472.0 cm^3/sec (at 21°C and 1000 mb)	4.720×10^{-4} m^3/sec (at 21°C and 1000 mb)
1 ton, U.S.A. short (t)	2000 lb avoirdupois	weight of 907.2 kg
1 ton per day (t/d)	83.3 lb/hr	1.05×10^{-2} kg/sec
Tonne, metric (T)	1×10^6 g	10^3 kg
1 torr	1 mm Hg	133.3 N/m^2
watt (W)	10^7 erg/sec	1 J/sec
yard (yd)	91.44 cm	0.9144 m

APPENDIX C

LIGHT SCATTERING BY SMALL PARTICLES

Particulate matter in the atmosphere affects the transmission of light. The index of refraction of the substances which constitute particles differs in general from the index of refraction of air and the resulting nonuniform index causes a portion of the light to be deflected from its original direction. This effect is known as *light scattering*. In heavily polluted air, it is primarily responsible for reducing visibility; and it can affect the earth's energy balance between incoming solar radiation and outgoing thermal radiation. The scientific study of light scattering commenced with J. Tyndall's experiments on aerosols in 1869. Within two years appeared a series of theoretical papers on the subject by Lord Rayleigh, whose name is now associated with the special case of light scattering by very small particles. The characteristics of light scattering are extremely sensitive to the size of the scattering particles. In this appendix we shall investigate the size dependence to determine why aerosols of a certain size are so much more effective as light scatterers than others.

The length by which we gauge the relative size of a particle is the wavelength λ of the incident light beam. Thus a natural parameter for characterizing the size of a particle is the ratio R/λ, where R is the particle's radius. In our investigation, we shall focus our attention on how a single particle affects the incident light beam and will neglect the complications which arise when light scattered from one particle is rescattered from another, a process known as multiple scattering.

We shall first consider the case $R/\lambda \ll 1$ corresponding to Rayleigh scattering, a process that causes the blue of the sky.[1] Then we shall examine the opposite extreme of large particles, $R/\lambda \gg 1$, where scattering can be viewed in terms of the concepts of geometrical optics. With these two cases clarified, what happens with intermediate size particles will become obvious, and we shall find that their features of light scattering are very sensitive to the ratio R/λ.

C.1 SMALL PARTICLES

To appreciate how small particles scatter light we must recall that light, as electromagnetic radiation, has both an electric and magnetic field. Then the process is easily understood by the following line of reasoning: The force of the electric field accelerates electrons bound to the molecules constituting the particle. These accelerated electrons themselves radiate light; and this radiated light is emitted not only in the direction of travel of the incident beam (called the "forward direction") but to the side as well. It is this radiated light, emitted away from the forward direction, that is the scattered light.

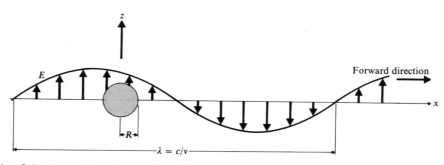

Fig. C.1 A particle, whose radius R is much smaller than the wavelength λ of an incident light beam, experiences a nearly uniform electric field E.

Now let us view this process in more detail. In Fig. C.1 we show the situation at an instant of time when the particle experiences an electric field of the light wave directed along what we shall call the z-axis. The wave itself is traveling toward the right along the x-axis. For simplicity, we show light which is polarized so that the electric field is always parallel to the z-axis. Later we shall generalize our result for the more common instance of unpolarized light, such as that which the sun radiates.

As the electromagnetic wave moves along the x-direction with a speed c, the electric field E experienced by the electrons and the nuclei constituting the particle varies periodically with a frequency $v = c/\lambda$. At the instant portrayed in Fig. C.1, this field exerts a downward force (toward the negative z-direction) on the electrons and an upward force on the nuclei. Being considerably more massive, the nuclei have only a slight acceleration, so small in fact that it can be neglected. On the other hand, the electrons' small mass permits them to respond readily and acquire a high acceleration. Now it is important to recognize that the accelerating electrons radiate energy in the form of an electromagnetic wave having an electric field with a magnitude proportional to the amplitude of the acceleration. This radiated wave is the scattered light.

Suppose that the electric field of the incident light at the site of the particle has the time variation as given by

$$E = E_0 \sin (2\pi v t), \tag{C.1}$$

where E_0 is the amplitude of the field and t the time. We assume that the particle consists of a dielectric material so that the field can penetrate practically without attenuation. The problem is greatly simplified by our assumption that $R/\lambda \ll 1$, for then we can consider the field to be essentially uniform throughout the particle. Thus each electron, with charge e, experiences a force eE. But the electrons also experience forces from their neighboring electrons and nuclei. For our purposes, the details of these other forces need not be considered. All of the forces can be approximated by a net force proportional to the displacement s of the electron from its equilibrium position; so we shall let $-Ks$ be the force that tends to restore each electron to its equilibrium position, where K is an appropriate constant. Newton's second law for the electron's acceleration a is then given in terms of the two forces:

$$ma = eE_0 \sin (2\pi v t) - Ks, \tag{C.2}$$

where m is the electron's mass. This equation has a solution which can be obtained easily from calculus. The result is

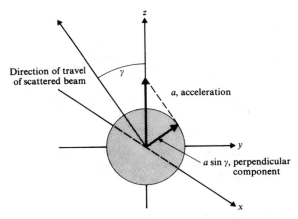

Fig. C.2   The magnitude of the electric field of scattered light is proportional to the component of the electron's acceleration that is perpendicular to the direction of travel of the scattered light. This component has the magnitude $a \sin \gamma$, as illustrated.

$$a = \frac{-v^2}{(K/2\pi m)^2 - v^2} \left(\frac{eE_0}{m} \right) \sin (2\pi v t). \tag{C.3}$$

In most aerosols an electron is so tightly bound to the equilibrium position in its molecule that the factor K/m is much larger than the term v^2 in the denominator of this equation; thus Eq. (C.3) can be approximated by

$$a = \frac{(2\pi)^2 m e v^2 E_0}{K^2} \sin (2\pi v t). \tag{C.4}$$

At long distances from the particle, the amplitude of the electric field of the scattered light is proportional to the amplitude of that component of the acceleration which is *perpendicular* to the direction of travel of the scattered light. According to Fig. C.2, the magnitude of this component of the acceleration is $a \sin \gamma$, where a is given by Eq. (C.4) and γ is the angle between the direction of the scattered light and the electric field of the incident light. The electric field is also proportional to the number of electrons, or the volume V of the particle. Thus the scattered light has an electric field whose amplitude E_s is proportional:

$$E_s \approx \frac{(2\pi)^2 m e v^2 V E_0}{K^2} \sin \gamma. \tag{C.5}$$

We note that no light is scattered along the z-direction ($\gamma = 0$) because no component of the acceleration is perpendicular to this direction. We shall later return to consider the angular dependence of the scattered light. But now the feature of the scattered light we seek is the *intensity*, the rate at which radiant energy crosses a unit area of hypothetical surface normal to its direction of travel. The intensity of light is proportional to the square of the amplitude of the electric field. From Eq. (C.5) we therefore find that the intensity I_s is proportional:

$$I_s \approx v^4 \, V^2 I_0 \sin^2 \gamma, \tag{C.6}$$

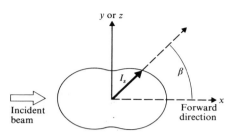

Fig. C.3 Angular dependence of the intensity I_s of light scattered by a small particle for an unpolarized incident beam. The intensity outbound along a direction β is proportional to the length of a radius vector I_s from the origin to the solid curve.

where I_0 is the intensity of the incident wave which, of course, is proportional to E_0^2.

From this relation, several important deductions can be formulated:

1. The fraction of the incident light energy scattered by a small particle is proportional to v^4 or λ^{-4}. Therefore, a larger fraction of blue light ($\lambda = 0.4$ micron) than red light ($\lambda = 0.7$ micron) is scattered.

2. The loss of intensity from the incident beam due to *scattering* is proportional to V^2. This contrasts with the volume dependence for the loss which would arise from *absorption* of a portion of the incident beam, for the absorptive loss is proportional simply to the number of absorbing molecules in the particle, and thus to the volume V. As a result, for sufficiently small particles, absorption will be relatively more important than scattering for reducing the intensity of a light beam in the forward direction.

3. The intensity of scattered light is not the same in all directions. This is clear from the dependence on γ exhibited in Eq. (C.6), but even more relevant is the anisotropy of the scattered light when the incident beam is unpolarized.

For the unpolarized case, we can imagine contributions of the form given by Eq. (C.6) with the difference that they are averaged for all possible orientations of the incident electric field in the plane perpendicular to the x-direction. When this is done, the intensity of scattered light has the angular dependence as illustrated by the polar diagram in Fig. C.3. The intensity depends only upon the angle β through which the light was scattered from the forward direction. The dependence of the scattered intensity on v^4 and V^2 is identical for polarized and unpolarized incident light.

A remarkable feature of Fig. C.3 is that as much energy is scattered backward ($\pi/2 < \beta < \pi$) as forward ($0 < \beta < \pi/2$). However, the intensity of the scattered light is not isotropic, being somewhat less for light scattered perpendicular to the forward direction. We shall see that, despite this, the scattering by small particles is considerably less direction dependent than that from large ones.

C.2 LARGE PARTICLES

If the radius of the particle is more than 20 times the wavelength of light, the situation can be described by elementary optics. In this view, the incident beam is affected by three distinct processes: reflection, refraction, and diffraction. All three processes have the effect of scattering light, for they change its direction of propagation. In Fig. 2.7 we have shown a portion of the incident beam striking the surface of a particle where part is reflected and part is refracted. In addition, some absorption of the refracted light can occur within the particle. When the remaining refracted light strikes the far surface, a portion of the light is reflected internally and a portion is again refracted. The emerging beam will not in general be parallel to the incident one, unless by coincidence the particle has the proper shape. Thus refraction generally contributes to light scattering. In the absence of a wavelength dependence for the absorption and index of refraction, the scattered light will have about the same color as the incident light.

A feature that distinguishes scattering by large particles from that by small ones is the fact that the direction of the refracted and diffracted light differs little from the forward direction. Therefore most of the scattered light goes preferentially near the forward direction. Only a relatively small portion of the incident light is reflected. For spherical particles of radius R, it can be shown that the amount of *diffracted* light is equal to the amount of light that would be *reflected* if the particle were considered to be perfectly reflecting. Since the cross-sectional area of the particle that intercepts the beam is πR^2, the effective cross-sectional area for diffraction is also πR^2. Thus the effective cross-sectional area for the total light scattering from large dielectric particles is $2\pi R^2$.

C.3 INTERMEDIATE SIZES

The scattering process is more complicated to treat when the wavelength of light is comparable to the particle size. The small particle approximation is not adequate, because the electric field is not uniform throughout the particle, and electrons in various locations may be accelerating in different directions. On the other hand, geometrical optics is not applicable either because diffraction cannot be distinguished from the scattering that occurs within the bulk of the sample. In fact an appreciable fraction is scattered by large angles with respect to the forward direction. Hence a more exact solution must be sought to obtain the characteristics of the scattered light. This scattering problem was first solved exactly by G. Mie in 1908 for spherical particles, and solutions now exist for particles of more irregular shape.[2]

The total amount of light that is scattered depends upon R/λ and the index of refraction of the dielectric material of the particle. It is convenient to characterize the total scattered intensity by a factor K_s called the *scattering area coefficient*. The amount of scattered light is considered to be the amount intercepted by a particle of cross-sectional area $K_s \pi R^2$. Thus the dependence of the total amount of scattered light on the ratio R/λ can be characterized by the dependence of K_s on R/λ. In view of our discussion concerning large particles, we would anticipate that K_s would be equal to 2 for very large values of R/λ.

The calculated values of K_s for a spherical particle are shown in Fig. C.4 for the special case where the index of refraction of the particle is 1.33, the same as the index for water. The value the curve approaches for large particles confirms our expectations. For small particles, K_s increases as R^4, in accordance with the factor of V^2 which occurs in Eq. (C.6). Thus there is a sharp rise in scattering effectiveness with increasing particle size until R is approximately equal to λ. For still larger particles, the effectiveness decreases because the electric field within the particle is oriented in one direction in some regions and is oriented in the opposite direction elsewhere. The complex pattern of field directions results in a complex pattern for the acceleration of electrons in different

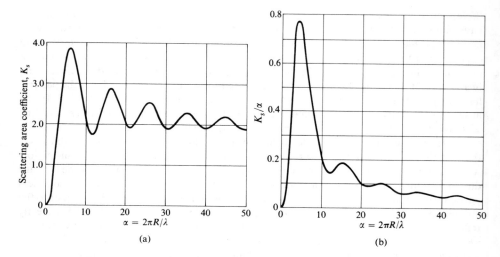

Fig. C.4 (a) Dependence of the scattering area coefficient on R/λ for a spherical particle whose index of refraction is 1.33. (b) Dependence on R/λ for the fraction of incident light scattered by an aerosol of identical spherical particles with the same index of refraction, for a given mass concentration. The fraction scattered per meter of travel for a light beam of 1 m^2 cross section is obtained by multiplying K_s/α by $3\pi M/2\rho_p\lambda$.

locations within the particle. The net effect is destructive interference in the scattering of light from different portions of the particle. This situation gives rise to an oscillatory dependence of K_s on R/λ, since the condition for maximum interference is met periodically for particles of succeedingly larger sizes.

Figure C.4 illustrates an interesting fact about the color of scattered light. For small values of R/λ, Rayleigh scattering is more effective on short wavelengths than long, so blue light is preferentially scattered. Conversely, if an aerosol consists of particles of the same size with a radius corresponding to the decreasing portion of the curve in Fig. C.4 (to the right of the primary scattering maximum), then red light is preferentially scattered. As a consequence, when the air contains the proper size aerosol, the sun may appear green or blue, as has been reported on occasions following a forest fire or volcanic eruption.

In gauging the effects of air pollution, it is often more appropriate not to ask for the amount of light scattered per particle, but to ask for the amount scattered by a given *mass concentration* of particles. The fraction of radiant energy scattered for each meter of optical path length can be calculated from the data in Fig. C.4(a) if for simplicity we assume that the particles in the path of the beam all have the same radius. A beam of 1 m^2 cross-sectional area encounters $3M/4\pi R^3\rho_p$ particles per meter of travel, where M is the aerosol mass concentration and ρ_p is the density of material in the particle. Multiplication of this number density by $\pi R^2 K_s$ gives the fraction of the incident beam which is scattered. Thus the dependence of this fraction on particle size is given by the dependence of K_s/R (or equivalently K_s/α) on particle size. Figure C.4(b) shows how K_s/α depends on α. From this curve it is clear that for a given mass concentration, particles are overwhelmingly more effective in scattering light if R is approximately equal to λ.

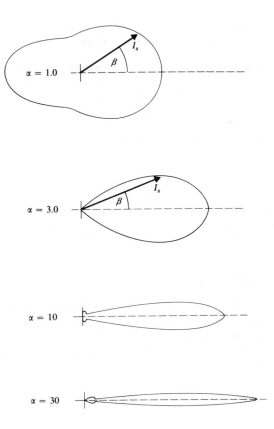

Fig. C.5 Angular dependence of the intensity of light scattered by a spherical particle of index of refraction 1.20, shown for different values of $\alpha = 2\pi R/\lambda$. The intensity is arbitrarily normalized in each case. A complex structure of maxima and minima is not shown. (From L. E. Ashley and C. M. Cobb, *Journal of the Optical Society of America*, **48**, 261, 1958.)

Another factor further reduces the relative importance of scattering from large particles, for a given mass concentration: the dependence on particle size of the *angular distribution* of the scattered radiation. The reason lies in the fact that most of the diffracted and refracted light from large particles is propagating very nearly parallel to the forward direction. Thus, although strictly speaking it is classified as scattered light, this diffracted and refracted light does not markedly remove energy from the direction of travel of the incident beam. More significant is the large angle scattering of light accomplished by smaller particles.

The progression with increasing particle size from predominantly large-angle scattering to small-angle scattering is illustrated in Fig. C.5. These polar diagrams are similar to the one in Fig. C.3, except that they represent the angular dependence of the scattered light intensity I_s from particles with larger values of the parameter $\alpha = 2\pi R/\lambda$. Each diagram has been normalized to keep it to a convenient size; otherwise the patterns in the lower diagrams would be considerably expanded relative to the upper ones. There is a pronounced shift of the scattered light toward the forward direction with increasing particle size. This is true especially in a comparison of the patterns for $\alpha = 1.0$ and $\alpha = 3.0$.

In Fig. C.5 we have illustrated that smaller particles, although scattering a smaller fraction of the incident light, scatter more of it by large angles. For a given concentration of aerosol particles, the intensity of a light beam will be diminished most effectively in the forward or near-forward direction by particles whose sizes correspond approximately to the radius for the highest peak in the scattering area coefficient in Fig. C.4. For this reason, particles between 0.1 and 1.0 micron size are said to be the most effective scatterers.

REFERENCES

1. S. Chandrasekhar and D. D. Elbert, *Trans. Am. Phil. Soc.*, *44*, 643 (1954).

2. See for example M. Kerker, *The Scattering of Light and Other Electromagnetic Radiation*, New York: Academic Press, 1969.

FOR FURTHER READING

M. MINNAERT   *The Nature of Light and Color*, New York: Dover Publications, 1954.

M. KERKER   *The Scattering of Light and Other Electromagnetic Radiation*, New York: Academic Press, 1969.

H. C. VAN DE HULST *Light Scattering by Small Particles*, New York: John Wiley and Sons, 1957.

APPENDIX D

BLACK BODY RADIATION

One of the most important physical processes influencing ambient temperatures and meteorological patterns of ventilation is the transfer of energy from one place to another by radiation. It is, of course, the key process which establishes the mean temperature of the earth through an energy balance between incoming solar radiation and outgoing terrestrial radiation; but it also directly affects local patterns of air circulation: the subsidence inversions of the semipermanent anticyclones, sea and land breezes, nocturnal or radiational inversions, and the mountain-valley breezes—to name but a few. The importance of radiation in meteorology justifies a close examination of this phenomenon.

In this appendix, we shall investigate why the emission spectrum of thermal radiation from natural objects—such as the surface of the oceans, the ground, and the air—is often found to be very similar to the spectrum known as *black body radiation*. The fact that this is so permits us to characterize the spectrum from natural objects by a black body spectrum for the same temperature, and thus from the known features of the latter we can estimate such important quantities as the rate at which energy is emitted and the wavelength for the peak of the spectrum.

However, before we go into this, let us ask a simple question: Why do objects at, say, room temperature emit radiation? The answer lies in the fact that any accelerating charge emits electromagnetic radiation. And so long as there is a thermal agitation of molecules and atoms which constitute an object, the charged electrons will occasionally be accelerated. Because the characteristics of the radiation depend on the degree of thermal agitation and thus on the temperature, it is often called *thermal radiation*. The detailed processes by which radiation occurs are complicated; but the thermal origin of the agitation permits us to neglect such details for a special class of objects called *black bodies*. For these we can regard only the thermodynamic or statistical characteristics. As a result, we find that the characteristics of the radiation from these objects depends only upon their temperature. Natural objects, having characteristics nearly those of black bodies, also emit radiation which depends upon the temperature. A familiar example of this is the color of the radiation given off by a bar of metal when heated. The emission spectrum depends upon the temperature, changing from "red hot" to "white hot" and perhaps even "blue hot" as the bar is heated to ever higher temperatures.

D.1 KIRCHHOFF'S LAW

We shall begin our discussion of black body radiation by considering a very simple experiment. Imagine that we have an object suspended inside a closed box whose walls are maintained at a uniform temperature. The intervening space is evacuated and we suppose that no energy can be conducted to or from the object via its suspension. Only

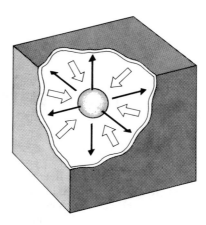

Fig. D.1 An object enclosed by a box whose walls are at a uniform temperature will eventually reach the same temperature as its enclosure, through the radiative exchange of energy.

by thermal radiation can the object and its enclosure exchange energy as depicted in Fig. D.1. If we wait for a sufficiently long time, the exchange of radiant energy between the object and its enclosure will establish an equilibrium; that is, the object will come to the same temperature as the box. Of course it continues to emit and absorb radiation; but when its temperature ceases to vary, it maintains an energy balance.

Let us denote the rate at which radiant energy at a given wavelength λ is emitted from a square meter of the object's surface by the symbol E_λ. If the object is at equilibrium, the emitted radiation can depend only upon the characteristics of the object, for example its chemical composition and temperature. Thus we can equate E_λ with a quantity $\eta_\lambda(T)$ that represents all of the relevant properties of the object:

$$E_\lambda = \eta_\lambda(T). \tag{D.1}$$

The quantity η_λ is called the *emission factor* for the surface of the object. But the surface of the object is also bathed in radiation from the inside walls of the box. Suppose that we let A_λ be the rate at which a square meter of the object's surface absorbs energy; then A_λ must equal the fraction K_λ of the incident radiation intensity I_λ that is absorbed:

$$A_\lambda = K_\lambda I_\lambda. \tag{D.2}$$

The subscript λ again indicates that each of the quantities may depend upon the wavelength of the radiation. The radiant intensity I_λ is the rate at which energy is incident from all directions on to a square meter of surface, and the parameter K_λ is called the *absorption coefficient*. We can now obtain a significant prediction if we note that at equilibrium the object has neither a net gain nor loss of energy, so $E_\lambda = A_\lambda$. From Eqs. (D.1) and (D.2) we thus obtain a condition known as *Kirchhoff's law*:

$$\frac{\eta_\lambda(T)}{K_\lambda} = I_\lambda(T). \tag{D.3}$$

Since the left-hand side of the equation depends upon T, so must the right, and we have so indicated this fact. Kirchhoff's law states that the ratio of the emission factor to the absorption coefficient depends solely on the energy spectrum of the radiation in the enclosure. If the absorption coefficient K_λ is not unity, a certain fraction $(1 - K_\lambda)$ of the incident energy will be reflected without absorption. However, this has no bearing on the validity of Eq. (D.3). The special case of a body whose surface absorbs all incident radiation $(K_\lambda = 1)$ has the interesting and significant feature that the spectrum of its emitted radiation is identical to the spectrum of radiation in the enclosure. Such a body we have called a *black body*.

Kirchhoff's law does not determine the wavelength dependence of the radiation intensity in the enclosure $I_\lambda(T)$. The fact that the radiation within the enclosure must also be in its own equilibrium condition does uniquely determine it. The equilibrium distribution of radiation energy among the possible wavelengths can be predicted by quantum statistics, and this was first achieved by Max Planck in 1901. His success in describing features of the thermal radiation provided the first evidence that energy is quantized in units of energy called photons, and that light has radiated therefore a particle-like aspect. We shall not recount the derivation here, since it is primarily a statistical development and such an explanation would not further our understanding of the radiation process. Planck's result leads to the following formula for $I_\lambda(T)$:

$$I_\lambda(T) = \frac{2\pi hc^2}{\lambda^5 \left(e^{hc/\lambda k_B T} - 1\right)} \tag{D.4}$$

where h is called Planck's constant, k_B is Boltzmann's constant, and c is the speed of light. Comparison of this formula with the observed emission spectrum of bodies with temperatures from $113°K$ to about $2100°K$ has yielded excellent agreement provided that a suitable value for the constant h is chosen. The currently accepted value is $h = 6.626 \times 10^{-34}$ J-sec. The shape of this spectrum for several different temperatures is given in Fig. D.2.

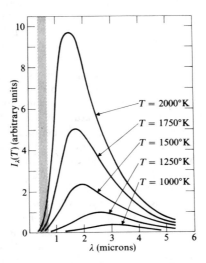

Fig. D.2 Spectrum of radiation from a black body at several temperatures. The shaded band indicates the visible region of the electromagnetic spectrum. (From R. M. Eisberg, *Fundamentals of Modern Physics*, New York: John Wiley and Sons, 1961.)

D.2 BLACK BODIES

From our previous discussion, we conclude that the emission spectrum from a black body is identical to the Planck distribution illustrated in Fig. D.2. And from Eq. (D.4) it is possible to show that the emitted radiation therefore has the characteristics we have summarized in Chapter 4; *viz.* the rate at which energy is radiated from the body's surface increases as the fourth power of the absolute temperature [Eq. (4.2)], and the peak in the spectrum is given by the Wien displacement law [Eq. (4.4)].

An important conclusion we derive from Eq. (D.1) is the fact that the emitted radiation is independent of the incident radiation, so long as the temperature of the body is uniform and very nearly constant—in other words, so long as it is at equilibrium. Under these conditions, a black body will emit a Planck spectrum whether or not it is in a radiative equilibrium with its surroundings. This feature permits us to apply the black body concepts to the analysis of the radiative energy exchange between the earth and its surroundings, even though there is not a true radiative equilibrium such as we supposed to have existed within the enclosure shown in Fig. D.1.

D.3 GRAY BODIES

Most objects, however, do not absorb all of the radiation incident upon their surface and consequently are not ideal black bodies. A surface coated with a layer of carbon black perhaps most closely approximates a black body, at least for its absorption in the visible portion of the spectrum. The data in Tables 4.1 and 4.2 indicate that many natural surfaces of terrestrial objects do not have true black body characteristics, although the differences are slight.

In seeking a means to describe the radiative properties of natural surfaces, it is reasonable to consider a particularly simple case of an object which absorbs a constant fraction of the incident energy independent of wavelength. Then the absorption coefficient K_λ is independent of λ but is less than unity. Such objects are called *gray bodies*. The terminology has no connection with the color of the object as perceived by a human. The characteristics of natural surfaces are more akin to those of gray bodies than black bodies.

However, gray bodies and others with more complicated absorption behavior can be made to approximate more nearly a black body by roughening the surface. The surface cavities which result tend to reflect radiation more diffusely and assist in establishing a radiative equilibrium at the temperature of the surface. An extreme example would be the effect of a cavity almost completely enclosed by the body, with just a small peephole to enable light to enter and exit. Any radiation entering the hole would have little chance of leaving before being absorbed by the interior walls. Thus the absorption coefficient of the hole is unity. The radiation in the cavity would be black body radiation if the walls were at a uniform temperature; thus the radiation which leaves has the Planck spectrum. The oriface of such a cavity has the necessary characteristics of a black body. For similar reasons, the surface of a porous material is more nearly like that of a black body than a smooth surface composed of a material of the same chemical composition. This is one major factor why natural surfaces such as those formed by soil or grasses have emission spectra characteristic of gray bodies with a very high absorption coefficient.

MOLECULAR ABSORPTION

The greenhouse effect is caused by the selective absorption of radiation in certain spectral regions, primarily by water and carbon dioxide molecules and to a lesser extent by ozone molecules. In this appendix, we shall investigate why these relatively minor constituents of the atmosphere are optically active and play the dominant role, whereas the major constituents—nitrogen and oxygen—have essentially no effect. Many pollutants have unique absorption spectra, and this fact provides the basis for several sensitive optical techniques for measuring their concentration in polluted air.

When a gas molecule absorbs radiation, conservation of energy requires that the energy removed from the electromagnetic wave must appear as a property of the behavior or configuration of the molecule. There are three principal ways this can appear: as the energy associated with (1) *rotational motion*, (2) *vibrational motion*, or (3) the *electrons' orbits*. The three possibilities for a diatomic molecule are depicted in Fig. E.1. A molecule which has absorbed radiation is said to be *excited*, because its behavior according to these schematic illustrations appears to be more agitated, either through a more rapid rotational motion, a vibration of greater amplitude, or an electronic orbital configuration characterized by a higher energy. Let us now see how molecules can be excited by electromagnetic

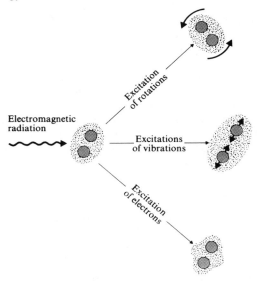

Fig. E.1 Modes by which atmospheric gas molecules absorb electromagnetic radiation.

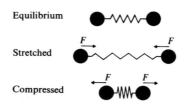

Fig. E.2   Vibrating diatomic molecule schematically indicated as two nuclei joined by a spring. The restoring force on each nucleus provided by the spring is shown.

radiation, for this will tell us under what conditions a gas consisting of these molecules will be absorptive. We turn first to consider the excitation of vibrations, for this is largely responsible for atmospheric absorption in the infrared portion of the spectrum.

E.1 VIBRATING MOLECULES

The electrons which orbit about the two nuclei of a diatomic molecule bind the atoms together and maintain an equilibrium distance of separation between the nuclei. If by some means the nuclei are caused to separate by a slightly greater amount and are then released, the restoring force provided by the electric interactions between electrons and nuclei will cause the molecule to oscillate, with the nuclei moving alternately toward and away from each other as is illustrated in Fig. E.2. The restoring force might be imagined as acting much like a spring joining the two nuclei; consequently there is a well-defined frequency for the oscillatory motion, called the *natural frequency*. For molecular vibrations, the natural frequencies are high, about 5×10^{13} Hz (cycles per second).

Let us now consider how an electromagnetic wave can cause water molecules to vibrate. To do this, we first note the important fact that the water molecule is electrically polarized. In Fig. E.3(a) the equilibrium positions of the oxygen and hydrogen nuclei are shown forming a triangle with an angular separation between the hydrogens of about 105°. As is well known to chemists, electrons are preferentially attracted to the oxygen atom, forming an approximately complete shell of electrons, and thus the hydrogens are left with a net positive charge.

One consequence of charge polarization is the strong attraction that water molecules have for each other in the liquid phase. When oriented so that the net positive charge of one is close to the negative charge of the other there is a strong electrostatic attraction (a feature known as *hydrogen bonding*). This attraction is manifest in an unusually large latent heat of evaporation, which is why the evaporation and subsequent condensation of water is an effective means by which the atmosphere can transport energy from one place to another.

E.2 DIPOLE MOMENT

The charge distribution on a single water molecule forms what is known as an *electric dipole*, a net negative charge at one end of the molecule and a net positive charge at the other, as depicted in Fig. E.3(b). For any polarized molecule, the net charge at one end, when multiplied by the distance of separation between the opposite charges, defines a parameter called the *electric dipole moment*. The larger the net charge on one end of the molecule and the greater the distance between charges, the greater the dipole moment.

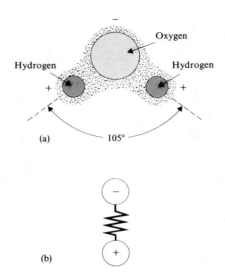

Fig. E.3 (a) Polarized electric charge in a water molecule; (b) simplified model consisting of a positive and negative charge joined by a spring.

Molecules with a dipole moment interact strongly with electromagnetic radiation. The reason for this is illustrated by the second diagram in the sequence shown in Fig. E.4. Here we illustrate an instant of time when the molecule experiences the electric field E of the radiation, which we suppose for simplicity is oriented parallel to the direction of polarization of the molecule. The electric field causes oppositely directed forces on the polarized ends of the molecule as shown, so that the oxygen nucleus and the pair of hydrogen nuclei accelerate toward each other. In addition, the nuclei experience the restoring force from the surrounding electrons which tends to cause them to oscillate back and forth at the natural vibrational frequency of the molecule. If the frequency of the electric field v is equal to the natural frequency of the molecule, the two will remain in synchronism and energy will thereby be absorbed by the molecule from the electromagnetic wave. This condition for synchronism is also known as *resonance*. Only in the resonant condition does the electric force of the wave always have the same direction as the velocity of the end of the molecule upon which it acts; this is shown by the sequence in Fig. E.4. Thus the force always accelerates the nuclei rather than decelerating them.

Our description of the absorption process is useful, for it provides us with a graphic picture of what the resonant condition signifies. However, it is not entirely accurate, and one should instead rely on the more fundamental description provided by the quantum theory of physics developed early in this century. One fundamental concept is the notion introduced by Max Planck and Albert Einstein that electromagnetic waves consist of discrete packets of energy called *photons*, and that the energy of a photon E_p is given by its frequency v according to the relation

$$E_p = hv, \qquad\qquad (E.1)$$

where h is called Planek's constant: $h = 6.63 \; 10^{-34}$ J-sec.

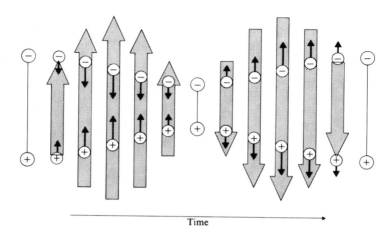

Time

Fig. E.4 Electric dipole of a molecule influenced by an electric field whose frequency coincides with the natural vibration frequency of the dipole. The broad arrow indicates the instantaneous direction and magnitude of the electric field for a sequence of equally spaced time intervals, and the smaller arrows show the direction and magnitude of the forces on the positive and negative charges of the dipole.

Equation (E.1) can also be expressed as an equivalent relationship between the wavelength of a photon and its energy:

$$E_p = hc/\lambda. \tag{E.2}$$

The amount of energy associated with a photon is very small, and that is why in our everyday activities, where illumination involves the production and detection of uncountable numbers of photons, the quantum nature of light is not generally appreciable. Only on the atomic scale does it become an important consideration. The energy of a visible photon with $\lambda = 0.55$ micron is only 3.6×10^{-19} J.

We recall from quantum theory that not only is the energy of electromagnetic radiation quantized in discrete values, but also the energy associated with any periodic motion of a molecule. Thus the energy corresponding to the vibratory motion of the water molecule cannot have any arbitrary value but only discrete values, which we could label as E_1, E_2, E_3, etc. The numerical values of these energies will depend upon characteristics peculiar to a given molecule such as its mass, electronic structure, etc. Consequently the values generally will be different for different types of molecules. The important point is that since energy is conserved, as a photon is absorbed by the molecule, and the molecule is excited to a higher energy, we must have

$$E_m - E_n = h\nu, \tag{E.3}$$

where E_n and E_m are the molecule's initial and final energies, respectively. Thus only radiation which satisfies this condition can be absorbed. The frequencies at which a molecule absorbs are thus a "fingerprint" which indicates the energy differences between its permitted energies, and consequently can be used for its identification. This fact is the basis for many instruments now being employed to monitor ambient concentrations of air pollutants. Typically the energy differences between allowed vibrational

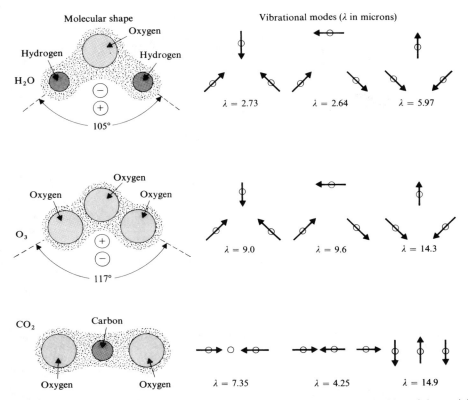

Fig. E.5 Vibrational modes for H_2O, O_3, and CO_2. Arrows indicate the velocities of the nuclei when vibrating, at the instant they pass by their equilibrium positions. The respective values of λ indicate the wavelength of a photon which can excite each mode.

energies of molecules correspond by the quantum condition [Eq. (E.3)] to the absorption of radiation whose wavelength is in the infrared portion of the spectrum.

In Fig. E.5 we show the normal modes by which a molecule can most simply vibrate, for the water molecule and the ozone molecule, whose shape is similar. Each molecule has three modes, which in the figure are distinguished by the arrangement of arrows indicating the velocities of the three nuclei at the moment they are simultaneously passing by their equilibrium positions. Each mode has a natural frequency, and a corresponding difference between successive allowed energies of vibration; the wavelength of radiation whose absorption excites each respective mode is indicated in the figure. By exciting several modes into simultaneous vibration, radiation can cause the molecules to execute complex motions. We see that water should strongly absorb at wavelengths of about 2.6–2.7 microns and 6 microns. On the other hand, ozone should absorb at 9.0–9.6 and at about 14 microns. These do indeed correspond to the observed absorption spectrum of the atmosphere, as given in Fig. 4.5. However, the observed spectrum does not display a single sharp peak at the precise wavelength, but instead has broad absorption bands. The reason for this will be deferred until we have first examined the special question as to why carbon dioxide strongly absorbs radiation.

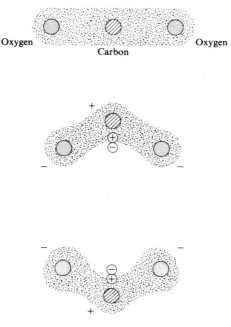

Fig. E.6   Carbon dioxide has an ocillatory dipole moment when it vibrates in a transverse mode as illustrated.

E.3 OSCILLATING DIPOLES

Carbon dioxide provides a different example, because it has a linear structure. The symmetry of the molecular shape precludes a net dipole moment, for both ends of the molecule are identical. However, the molecule can be bent easily, and once the central atom is displaced transversely relative to an imaginary line joining its neighbors, the molecule has a dipole moment as illustrated in Fig. E.6. A restoring force will then cause the molecule to vibrate in a transverse mode and form a dipole moment that varies with time. Such an *oscillating dipole moment* can interact strongly with radiation. The energy required for exciting this transverse vibrational mode is so low that it can be obtained simply through collisions with neighboring molecules in a gas. The further excitation of this mode through absorption of infrared radiation is of particular importance to the energy balance of the earth because it is responsible for strong atmospheric absorption by CO_2 at wavelengths near 15 microns, which is approximately the wavelength where the surface of the earth has the peak in its emission spectrum of thermal radiation.

E.4 ROTATING MOLECULES

Molecules possessing dipole moments can also be excited into rotational motion through the absorption of radiation. Figure E.7 shows a time sequence illustrating how an electric field E oscillating at the same frequency as the rotational motion supplies a torque which

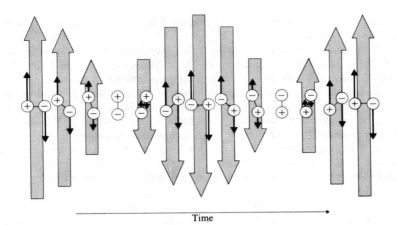

Fig. E.7 Rotational motion of a molecule caused by an electric field whose frequency coincides with the molecule's rotational frequency. The broad arrow indicates the instantaneous direction and magnitude of the electric field, and the smaller arrows show the direction and magnitude of the forces on the positive and negative charges of the dipole.

tends to accentuate the motion. Optical activity is enhanced by a large dipole moment because the torque is similarly enhanced.

Rotation is a periodic motion, so the energy of rotational motion must be quantized as was the energy of vibrational motion. But the kinetic energy of a rotating molecule is quite small, and thus the corresponding allowed energies E_{r1}, E_{r2}, E_{r3}, etc., and their differences are also small. For excitation of rotational motion, a typical molecule would absorb a photon with a wavelength of about 6000 microns, corresponding to a frequency of only 5×10^{10} Hz in the microwave portion of the spectrum.

The reason that rotational excitations are significant is because, in combination with vibrational excitations, they explain the existence of absorption *bands* such as are found in the absorption spectrum of the atmosphere (Fig. 4.5). This is due to the fact that a molecule may absorb a photon and share the energy between both vibrational and rotational motion. It is required that energy be conserved; therefore if ΔE_v is the energy for the vibrational excitation and ΔE_r for rotational, we must have

$$\Delta E_v + \Delta E_r = hc/\lambda. \tag{E.4}$$

Now ΔE_v is considerably greater than the smallest value of ΔE_r which can excite rotational motion, so clustered around the wavelength λ' that satisfies $\lambda' = hc/\Delta E_v$ will be many wavelengths slightly shorter than λ' that will satisfy Eq. (E.4). If the absorption measuring instrument does not have sufficient wavelength resolution to identify absorption when each quantum condition is met, the spectrum will not appear as a series of sharp peaks but will be smeared out into what appear to be bands.

There may be absorption at wavelengths exceeding λ' as well, for the molecule may have been rotating before it absorbed a photon and would thus require less energy to, say, cease rotating and only vibrate. The bands of an absorption spectrum may spread to wavelengths both shorter and longer than the wavelength that satisfies the conditions for vibration excitation alone. For wavelengths exceeding 20 microns, the strong atmospheric absorption is, in fact, caused by excitation of purely rotational motion of water molecules.

Now we can understand why nitrogen and oxygen do not contribute significantly to absorption in the infrared. Because of their symmetrical diatomic arrangement, neither has a dipole moment or an oscillatory moment. Since an electric field would influence both ends of the molecule equally, it cannot cause either vibration or rotation.

Only for very short wavelength photons is absorption by N_2 and O_2 significant, and then it is through the excitation of electrons in the molecules. We shall not say much about this except to note that many molecules can be excited in this way only if the wavelength of the radiation is 1 micron or less. For O_2 it must be in the ultraviolet, as mentioned in Chapter 4. In fact electrons can absorb a photon having so high an energy that they can no longer bind a molecule together, and the molecule dissociates. This process occurs for oxygen molecules at a wavelength of 0.24 micron or less and for nitrogen at 0.12 micron or less and, as we saw in Chapter 4, this photodissociation is an important process in the upper atmosphere. If a molecule is ionized or dissociated, the end products do not execute periodic motion, and consequently their energies are not quantized. A photon of any energy above the minimal amount will suffice with the result that the absorption spectrum is truly continuous. This mechanism is responsible for the complete absorption of solar radiation at very short wavelengths, in the far ultraviolet. However, relatively little solar energy is emitted in this spectral region, so electronic excitation does not play a significant direct role in the greenhouse effect.

THE CORIOLIS FORCE

Macroscale patterns of atmospheric motion over the earth are strongly influenced by an inertial effect known as the Coriolis force. In this appendix we give a completely general derivation of the inertial forces which affect matter when we wish to describe the kinematics with respect to fixed reference points on the earth's surface. The fact that the earth rotates, and therefore that the reference system is not an inertial one, necessitates the *ad hoc* introduction of these inertial forces to express the tendency for many objects to continue their motion in a straight line with a constant velocity. Thus when applying Newton's second law to predict the motion of air masses over a continent, we must include not only real forces such as those caused by pressure differences, but also inertial forces. The general form of the inertial forces can quickly be derived by employing vector notation.

F.1 VECTOR PRODUCT

We shall first review one pertinent definition of vector algebra, which is called the *cross product* or *vector product*. We suppose that we have at hand two vectors **A** and **B** which for the moment we allow to be perfectly arbitrary vectors. The vector product of **A** and **B** is a vector denoted by **A** × **B** which is perpendicular to the plane in

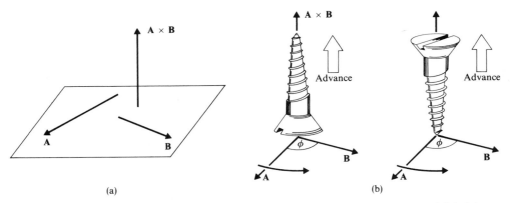

(a)   (b)

Fig. F.1   (a) The vector **A** × **B** is perpendicular to the plane in which **A** and **B** lie. (b) A right-threaded screw in either orientation perpendicular to the plane would advance in the same direction if its head were rotated with **A** when swung through the angle ϕ toward **B**. The direction of advance defines the direction of **A** × **B**.

which **A** and **B** lie. Thus **A** × **B** is perpendicular to both **A** and **B**. This is illustrated in Fig. F.1.

   Determining the sense of **A** × **B**—that is, whether **A** × **B** should point up or down in Fig. F.1(a)—is often a source of confusion. Perhaps the simplest rule is to proceed as follows: Translate **A** toward **B** without rotating **A** so that the tails of the vectors touch. Then imagine that **A** is rotated toward **B** in the plane through the smaller included angle ϕ. The direction of **A** × **B** is the direction that a right-threaded screw would advance if its axis were perpendicular to **A** and to **B** and if its head were rotated in the same sense as **A** is rotated. Figure F.1(b) illustrates the two possible orientations for the screw, showing that either choice yields the identical result. The definition of the vector product also stipulates that the magnitude of the vector **A** × **B** is equal to $|\mathbf{A}||\mathbf{B}|\sin\phi$ where $|\mathbf{A}|$ denotes the magnitude of **A** and $|\mathbf{B}|$ denotes the magnitude of **B**. This can be interpreted as meaning that the magnitude of **A** × **B** is equal to the magnitude of **A** multiplied by the magnitude of that component of **B** perpendicular to **A**.

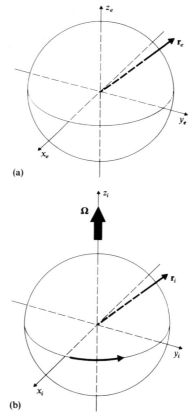

Fig. F.2 (a) Earth-bound reference system in which the z_e-axis is along the polar axis and the x_e and y_e-axis are in the equatorial plane. (b) Inertial reference system in which the earth is seen to make one complete rotation each day. The earth's angular velocity in this system is represented by the vector $\boldsymbol{\Omega}$ which points from the South Pole toward the north, parallel to the polar axis as shown.

F.2 DERIVATION

We are seeking a valid means of describing the movement of objects with respect to a reference system fixed to the earth; we illustrate such a system in Fig. F.2(a). Position, velocity, and acceleration vectors of any object whose components are measured with respect to the coordinate axes (x_e, y_e, z_e) of this earth-bound system will be denoted by a subscript e. For example, the position of an object slightly above the surface of the earth might be denoted by the vector $\mathbf{r}_e$ as shown in the figure. If the object is stationary with respect to the earth's surface, it has no velocity or acceleration in this reference system.

Since the earth rotates, the earth-bound system is not an inertial reference system. Let us therefore introduce an inertial system which is stationary with respect to the fixed stars, and denote vectors whose components are measured with respect to this system by the subscript i. Without loss of generality, we may consider the origins of the earth-bound and inertial systems to coincide.

To fix in our minds what is implied by introducing two reference systems, let us consider a simple example. Suppose that $\mathbf{r}_e$ points to an object with rests on the surface of the earth. Therefore this vector does not vary with time in the earth-bound system. But the same position vector when given in terms of its components in the inertial system, $\mathbf{r}_i$, is not stationary; the head of the vector $\mathbf{r}_i$ will describe a circle as the earth makes a complete revolution. We can therefore anticipate that the position, velocity, and acceleration of an object will be quite different when viewed with respect to the coordinate axes of one system as compared with the other.

For convenience we can denote the angular velocity of the earth's rotation as a vector quantity $\boldsymbol{\Omega}$ in the inertial system. It has a magnitude equal to the angular velocity Ω of the earth's rotation, and is directed along the polar axis as shown in Fig. F.2(b). Suppose now that at an instant of time t an air mass is located at a position denoted by the vector $\mathbf{r}_i(t)$, as shown in Fig. F.3. Should the mass be moving, a short time later Δt it will be located at a new position denoted by $\mathbf{r}_i(t + \Delta t)$. During this increment of time, however, the earth-bound system has rotated eastward by an angle $\Omega \Delta t$, so that the change in position of the mass in this system is not the same as in the inertial system, but differs by the amount $\mathbf{r}_{\text{rot}} = (\boldsymbol{\Omega} \times \mathbf{r}_i)\Delta t$. The geometry indicated in Fig. F.3 thus shows

$$\Delta \mathbf{r}_i = \Delta \mathbf{r}_e + (\boldsymbol{\Omega} \times \mathbf{r}_i)\, \Delta t. \tag{F.1}$$

Our ultimate goal is to determine how the *acceleration* of the air mass in the earth-bound system is related to the acceleration in the inertial system and hence discover how the apparent forces are related. Thus we first divide Eq. (F.1) by Δt to find how the velocities are related:

$$\left(\frac{\Delta \mathbf{r}}{\Delta t}\right)_i = \left(\frac{\Delta \mathbf{r}}{\Delta t}\right)_e + \Omega \times \mathbf{r}_i \tag{F.2}$$

or

$$\mathbf{v}_i = \mathbf{v}_e + \Omega \times \mathbf{r}. \tag{F.3}$$

Thus the velocities differ by the quantity $\Omega \times \mathbf{r}$. We have omitted the subscript on this quantity, for it can of course be expressed in terms of its components in either the earth-bound or inertial reference systems.

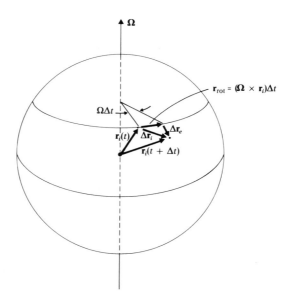

Fig. F.3 During a time interval Δt an air mass moving by an amount $\Delta \mathbf{r}_i$ will appear in the earth-bound system to have moved only an amount $\Delta \mathbf{r}_e$ because the earth has rotated by an angle $\Omega \Delta t$.

A little reflection should convince the reader that Eq. (F.2) is true for *any* vector **A**, provided that the second term on the right is expressed in its most general form $\Omega \times \mathbf{A}$. This equation is a prescription for determining the relationship between rates of change of a given vector **A** as viewed in the two systems. Explicitly stated, we have

$$\left(\frac{\Delta \mathbf{A}}{\Delta t}\right)_i = \left(\frac{\Delta \mathbf{A}}{\Delta t}\right)_e + \Omega \times \mathbf{A}. \tag{F.4}$$

To find the relationship between the acceleration of an air mass as measured in the two systems, we first apply this equation to the vector $\mathbf{v}_i$:

$$\left(\frac{\Delta \mathbf{v}_i}{\Delta t}\right)_i = \left(\frac{\Delta \mathbf{v}_i}{\Delta t}\right)_e + \Omega \times \mathbf{v}_i.$$

Then we use Eq. (F.3) to eliminate $\mathbf{v}_i$ from the right-hand side and obtain our desired result:

$$\left(\frac{\Delta \mathbf{v}_i}{\Delta t}\right)_i = \left(\frac{\Delta \mathbf{v}_e}{\Delta t}\right)_e + 2(\Omega \times \mathbf{v}_e) + \Omega \times (\Omega \times \mathbf{r}). \tag{F.5}$$

The left-hand side of this equation is the acceleration of an air mass or other object as measured with respect to an inertial reference system. The right-hand side contains only quantities measured with respect to the earth-bound system, as well as the angular velocity of the earth-bound system. The positions of the parentheses in the last term are not arbitrary. The cross product of the vectors within the parentheses should be calculated first, and then the cross product of Ω with ($\Omega \times \mathbf{r}$).

F.3 INERTIAL FORCES

We can most simply make sense of Eq. (F.5) by considering what happens when no forces are acting when viewed from the inertial system. This means that we somehow turn off gravity and eliminate all mechanical forces. According to Newton's second law, there will then be no acceleration of an object in the inertial system and therefore the left-hand side of Eq. (F.5) is zero. But the remaining part of the equation says that nevertheless there will be an acceleration when viewed from the earth-bound system. The right-hand side of Eq. (F.5) predicts an acceleration given by

$$\left(\frac{\Delta \mathbf{v}_e}{\Delta t}\right)_e = 2(\mathbf{v}_e \times \mathbf{\Omega}) + \mathbf{\Omega} \times (\mathbf{r} \times \mathbf{\Omega}). \tag{F.6}$$

(Some negative signs have been eliminated from the original expression by reversing the order of the vectors in the cross products.)

If both sides of Eq. (F.6) are multiplied by the mass of the observed object, the equation has the form of Newton's second law in which the object appears to be affected by two forces. The last term on the right is then the centrifugal force, as will be made evident by a careful comparison with Eq. (3.7). The first term is the Coriolis force. It appears only when an object is *moving* on the earth, so that $\mathbf{v}_e$ is not zero. For the motion of an air mass with respect to the surface, $\mathbf{v}_e$ is simply the wind velocity. If in addition to these inertial forces there are mechanical forces, e.g. due to gravity or pressure variations, they must also be added to the right-hand side of the equation.

Let us consider the magnitudes of the two terms in Eq. (F.6). For most meteorological effects, the wind speed $|\mathbf{v}_e|$ is much smaller than the tangential speed $|\mathbf{r} \times \mathbf{\Omega}|$ with which a stationary air mass on the surface of the earth hurtles through space due to the earth's rotation. Therefore the centrifugal force is considerably larger than the Coriolis force (except at the poles!). The dominance of the centrifugal force does not mean that the Coriolis force is negligible, for the two forces act in different directions. In Section 3.3 we noted that the surface of the earth has actually deformed under the influence of the centrifugal force, so that the net centrifugal and gravitational force near the surface is everywhere directed vertically downward. Only the Coriolis force has a horizontal component and can therefore deflect the winds.

Equation (F.6) shows that the Coriolis acceleration depends only upon the magnitude of the wind velocity $v_e = |\mathbf{v}_e|$, the rate of angular rotation of the earth $\Omega = |\mathbf{\Omega}|$, and the angle between the velocity and the axis of the earth's rotation. It will be easier to visualize the situation if we assume that the wind velocity is horizontal. This is an excellent approximation for all types of macroscale motion. The definition of the vector cross product then indicates that only the component of $\mathbf{\Omega}$ which is *perpendicular* to $\mathbf{v}_e$ contributes to the Coriolis force. Therefore the magnitude of the Coriolis acceleration, which we call a_C, is given by the product of the wind speed, v_e, and the magnitude of the vertical component of $\mathbf{\Omega}$:

$$a_C = 2v_e\Omega \sin \theta, \tag{F.7}$$

where θ is the latitude of the air mass. If M is the mass of a parcel of air whose position and velocity are measured with respect to features on the earth's surface, the Coriolis force F_C affecting the parcel is given by

$$F_C = 2Mv_e\Omega \sin \theta. \tag{F.8}$$

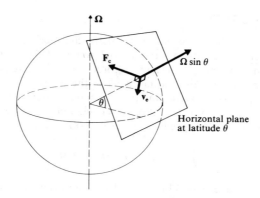

Fig. F.4   Wind velocity $\mathbf{v}_e$, vertical component of the earth's angular velocity $\Omega \sin \theta$, and the Coriolis force F_C for a wind at latitude θ.

According to Eq. (F.6), the direction of the acceleration and Coriolis force is perpendicular to the wind velocity, as illustrated in Fig. F.4. It is in such a sense that an observer with his back to the wind will see the wind accelerate toward the right (in the Northern Hemisphere) or to the left (in the Southern Hemisphere). A remarkable feature of the Coriolis force is that at a given latitude, its magnitude is independent of the direction of the wind. There is no Coriolis force at the equator, where the vertical component of the earth's angular velocity is zero.

BUILDING BLOCKS OF ORGANIC CHEMISTRY

Organic compounds are compounds containing at least carbon and hydrogen. The name derives from the fact that the first known members had been synthesized by plants and animals. In this appendix are summarized the essential elements of organic chemistry which are needed to understand the important contributions of organic molecules to photochemical smog. The main emphasis here is in relating the nomenclature to structural formulas. To a lesser degree, these structures are analyzed in terms of their chemical reactivity.

In Appendix E we discussed the importance of electric forces and electrons in binding together hydrogen and oxygen atoms to form the water molecule. Although the electrons tended to concentrate near the oxygen nucleus, thereby polarizing the charge distribution, the transfer was not complete. One pair of electrons was shared between a hydrogen atom and oxygen atom. A bond formed between two atoms by sharing a pair of electrons is called a *covalent bond*. It contrasts with *ionic bonds* in which there is a complete transfer of one or more electrons from one atom to the other. The classic example of ionic bonding is the salt crystal (NaCl) in which one electron from the sodium atom is lost to a neighboring chlorine. In other compounds bonding is possible with various degrees of covalency or ionicity. Covalent bonding between carbon and other atoms is the basis of organic chemistry.

Carbon has four valence electrons, and can share bonds with anywhere between one and four neighboring atoms including other carbon atoms. For this reason carbon chemistry can have a high degree of complexity. Carbon hydrides or *hydrocarbons* are molecules which consist solely of hydrogen and carbon. Electrons forming the bond between these dissimilar atoms are so equally shared that the bond shows but little ionic character.

The nomenclature of organic chemistry may at first appearance seem confusing because portions of several systems are in common use. However in many cases the terminology follows a systematic convention established by the International Union of Pure and Applied Chemistry. Unfortunately, other systems are also in common use for some of the simpler compounds. In structural formulas, covalent bonds may explicitly be denoted by one or more straight lines connecting the bound atoms. The number of lines indicates the number of *pairs* of shared electrons.

G.1 ALKYLS

The simplest building block is the carbon-hydrogen bond (C—H). This can be only a single bond corresponding to the one electron which hydrogen can contribute to the pair. But carbon, with four valence electrons, can also form single bonds with three other atoms. From these possible arrangements for bonds are developed several groups of atoms, known collectively as the *alkyls*. The first five alkyls are listed below, together with their chemical and structural formulas:

Name	Chemical formula	Structural formula
hydrogen	H·	H—
methyl	CH_3·	$\begin{matrix} & H & \\ & | & \\ H- & C & - \\ & | & \\ & H & \end{matrix}$
ethyl	C_2H_5·	$\begin{matrix} H & H & \\ | & | & \\ H-C & -C & - \\ | & | & \\ H & H & \end{matrix}$
n-propyl	C_3H_7·	$\begin{matrix} H & H & H \\ | & | & | \\ H-C & -C & -C- \\ | & | & | \\ H & H & H \end{matrix}$
n-butyl	C_4H_9·	$\begin{matrix} H & H & H & H \\ | & | & | & | \\ H-C & -C & -C & -C- \\ | & | & | & | \\ H & H & H & H \end{matrix}$

For completeness we have included atomic hydrogen as a member of the alkyl series although of course it is not a hydrocarbon. The names of the other alkyls beyond butyl are derived from the Greek numerals corresponding to the number of carbon atoms, for example, octyl, C_8H_{17}.

A generic notation which can represent the chemical formula for any group in the alkyl series is the letter R. The alkyls are reactive because each has an unshared valence electron. We have represented the unshared electron by the dot in the chemical formula (for example, R·) and by a dangling bond in the structural formula. Such atomic groups with unpaired electrons are called *free radicals*. A specific example would be CH_3·, the methyl radical. Free radicals have an affinity for bonding to other atoms or molecules at the site of the unpaired electron.

The two members of the alkyl group listed at the end deserve special comment. The structural formula shown in the right column is not uniquely determined by the respective chemical formula in the center column. These alkyls can also be represented by a second possible structure:

Isopropyl	C_3H_7·	$\begin{matrix} & & H & \\ & & | & \\ & HH-C & -H \\ & | & | & \\ H-C & -C & - \\ & | & | & \\ & H & H & \end{matrix}$

$$
\begin{array}{ccccc}
& & & \overset{\displaystyle H}{\underset{\displaystyle |}{}} & \\
& & & HH\!-\!C\!-\!HH & \\
& & & |\qquad|\qquad| & \\
\text{Isobutyl} \qquad C_4H_9\cdot \qquad & & H-\!&C-C-C-& \\
& & & |\qquad|\qquad| & \\
& & & H\quad H\quad H & \\
\end{array}
$$

Molecules or radicals with the same chemical formula but different structures are called *isomers*. *Continuous chain* molecules or radicals have each of their carbons bonded to no more than two other carbons. The prefix *n* (normal-) designates this type of structure, as can be seen from the names and structural formulas of the above pairs of isomers. A molecule or radical in which a carbon is joined to more than two other carbons, such as the isobutyl radical is said to be a *branched chain*. The isopropyl alkyl forms a branched molecule only if the unshared electron on the central carbon is attached to another carbon. Note also that the isopropyl structure may be viewed as consisting of a carbon to which is attached one hydrogen and two methyl groups. Should a molecule be a linear continuous chain except for two methyl groups (or other alkyls) at one end, the prefix *iso-* is used. If three methyl groups (or other alkyls) are at one end, the prefix *neo-* is used.

G.2 ALKANES

An alkyl group may acquire another hydrogen atom in order to share its unpaired electron and it then becomes a stable molecule. All of the carbon bonds are shared with hydrogen atoms, except for the minimum number which are necessary for carbon-carbon bonding. Such a molecule is said to be *saturated*. A consequence of saturation is that such molecules are quite inert. These molecules of the general chemical formula C_nH_{2n+2} are called *alkanes*. The suffix of the name of each alkane (except hydrogen) is derived from the corresponding alkyl by an obvious rhyme: methane (CH_4), ethane (C_2H_6), propane (C_3H_8), *n*-butane (C_4H_{10}), and so on. Of the alkanes, only the preceding four are gases at normal ambient temperatures and pressures. Butane can be liquefied at high pressures and is commonly used as "bottled gas." Alkanes with more than 17 carbons are solids. The alkanes are also known as *paraffins* (from the Latin *parum affinis*, not enough affinity) because they are less reactive with acids and oxidizing agents than are corresponding unsaturated molecules.

Paraffins generally react with other atoms or molecules by substitution. One of the hydrogens is replaced by the new atom or molecule. For example, should the new atom be a halogen the resulting molecule has the generic name alkyl halide. The order of the words supposes that the halide is the important functional part of the molecule, so it is distinguished by being the base name. In this case the organic part is viewed as an attached alkyl group. A specific example is ethyl chloride C_2H_5Cl. There is no ambiguity in the structural formula for ethyl chloride since all hydrogen sites on the two carbon atoms are equivalent. Therefore whether the single chlorine is on one or the other carbons is irrelevant.

However, isomerism is possible whenever *two* substituents are introduced into the ethane molecule, for they may be located on either the same carbon or on different ones. The distinction between the two structures is made by use of numerical prefixes which cite the position of each carbon atom to which a chlorine is joined:

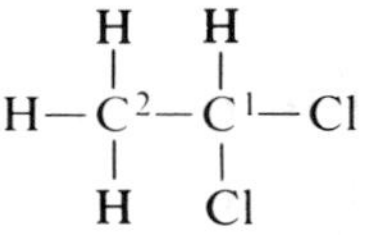

$$
\begin{array}{cc}
\quad H \quad H \\
\quad | \quad\; | \\
H-C^2-C^1-Cl \\
\quad | \quad\; | \\
\quad H \quad Cl
\end{array}
\qquad\qquad
\begin{array}{cc}
\quad H \quad H \\
\quad | \quad\; | \\
H-C^2-C^1-Cl \\
\quad | \quad\; | \\
\quad Cl \quad H
\end{array}
$$

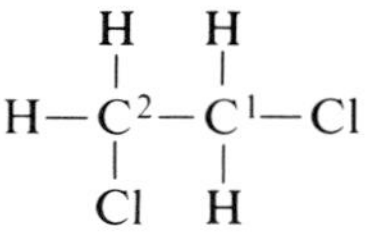

$$CH_3CHCl \qquad\qquad\qquad ClCH_2CH_2Cl$$

1,1 dichloroethane   1,2 dichloroethane

In these examples the number 1 indicates the right-hand carbon atom, and 2 its neighbor. It is conventional to begin counting from the end of the carbon chain which keeps the sum of the prefix numbers as small as possible. Note also that the base word upon which the name is developed is now the name of the alkane. This is usually the case for all but the simplest of molecules.

In general, for molecules which are more complicated than substituted *n*-hexane or the higher *n*-alkanes, the name of the longest continuous chain of carbon atoms is taken as the base name. It is modified by the names of the substituted alkyl, halogen, or other groups, with their positions indicated by prefix numbers. As an illustration consider 2,3 dimethyl pentane:

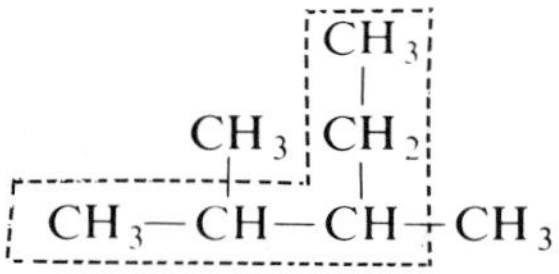

$$
\begin{array}{c}
CH_3 \\
| \\
CH_3 \quad CH_2 \\
| \\
CH_3-CH-CH-CH_3
\end{array}
$$

The longest chain of carbons is indicated by the dotted lines. It contains five carbons, so the base name is "pentane." To fix the conventions of notation more firmly in mind, consider a case in which two halide groups are substituted on the same carbon atom:

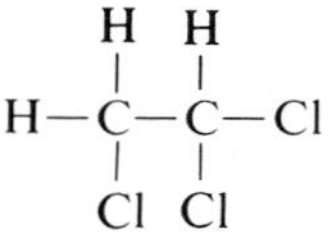

$$
\begin{array}{c}
H \quad H \\
| \quad\; | \\
H-C-C-Cl \\
| \quad\; | \\
Cl \quad Cl
\end{array}
$$

$$ClCH_2CHCl_2$$

1,1,2 trichloroethane

If a complicated substituent group is attached to the base chain, the carbons of of the group are numbered commencing with the one attached to the base:

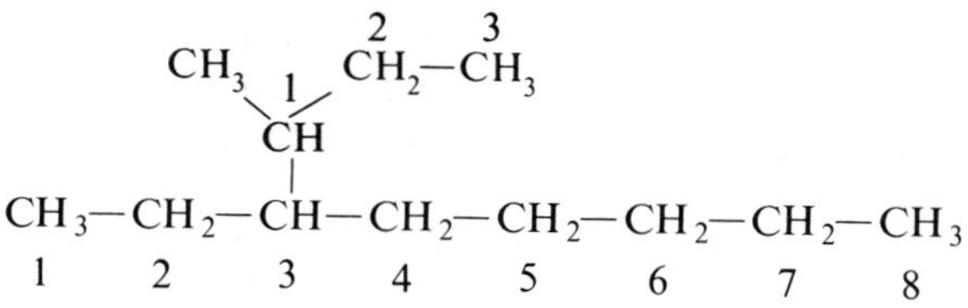

$$
\begin{array}{c}
\quad\quad\quad 2 \quad\;\; 3 \\
CH_3 \quad CH_2-CH_3 \\
\;\;\searrow\; 1 \;\swarrow \\
CH \\
| \\
CH_3-CH_2-CH-CH_2-CH_2-CH_2-CH_2-CH_3 \\
1 \quad\;\; 2 \quad\;\; 3 \quad\;\; 4 \quad\;\; 5 \quad\;\; 6 \quad\;\; 7 \quad\;\; 8
\end{array}
$$

3-(1 methylpropyl)-octane

In cases involving two or more different substituents, an arrangement of the names in alphabetical order may be preferred:

$$CH_3-CH_2 \quad CH_3$$
$$CH_3-CH_2-CH_2-CH-CH-CH_2-CH_3$$

4-ethyl-3 methylheptane

Alkanes with both alkyl and halogen substitution groups are commonly named as haloalkylalkanes (not as alkylhaloalkanes).

G.3 ALKENES

A series of molecules which are more reactive than the paraffins are the *alkenes*, whose members are described by the general formula C_nH_{2n}. In these molecules two neighboring carbons each contribute two electrons to be shared mutually, forming a double bond. The position of the first of the two unsaturated carbons is indicated by a number usually included at the end of the name of the molecule, but occasionally found after the names of all substituent groups. The suffix of the name of the corresponding member of the alkyl group undergoes a change to rhyme again with the generic name (alkene). A common departure from the systematic nomenclature sees the simpler alkenes named *alkylenes*.

Systematic	Common		
ethene	ethylene	C_2H_2	$CH_2 = CH_2$
propene	propylene	C_3H_6	$CH_3CH = CH_2$
n-butene-1 (or 1-n-butene)	n-butylene-1	C_4H_8	$CH_3CH_2CH = CH_2$
n-butene-2 (or 2-n-butene)	n-butylene-2	C_4H_8	$CH_3CH = CH\, CH_3$

A logical extension of this terminology applies to the other alkenes, but other systems of notation are commonly found in the literature.

In contrast with the rather inert alkanes, alkenes are very reactive. The double bond can be converted easily into a single bond between the respective carbons, with each of the carbons acquiring a new side group. Consider the reaction between dimethylethylene and chlorine:

$$
\begin{array}{ccc}
CH_3 & H & \\
\diagdown & \diagup & \\
C=C & + Cl_2 \longrightarrow & CH_3-\overset{\displaystyle Cl}{\underset{\displaystyle CH_3}{\overset{|}{\underset{|}{C}}}}-\overset{\displaystyle Cl}{\underset{\displaystyle H}{\overset{|}{\underset{|}{C}}}}-H \\
\diagup & \diagdown & \\
CH_3 & H &
\end{array}
$$

This process of addition also contrasts with the process of substitution which is more common to the alkanes.

Members of a series of compounds with similar chemical structures and smoothly varying physical properties (melting point, boiling point, viscosity, etc.) and which differ from one another only by the number of atoms in their structural backbone are said to be *homologous*. Therefore the *n*-alkenes are homologous, just as the *n*-alkanes form their own homologous series. However, branched chain molecules generally do not exhibit such regular variations of physical properties.

The oily feel of certain liquid dihalides of alkenes, compounds produced by the reaction of gaseous low molecular weight alkenes, led to a description of the initial hydrocarbon as an *olefin* (oil-forming substance). The preceding reaction is an example of an olefin being converted into a dihalide alkene. In the petroleum industry the term "olefin" usually denotes an alkene with a low boiling point. Alkenes with five carbons or less are in fact gases at normal temperatures and atmospheric pressure. However, organic chemists commonly apply the term to *any* hydrocarbon that has one or more double bonds. We shall adopt this more general meaning. Because of their reactivity, olefins play a central role in photochemical smog.

A greater degree of unsaturation with respect to the alkenes can be achieved by inclusion of a triple bond between two carbons. This group of molecules is known as *alkynes*. We shall not deal with this group further except to note that an example of one such *alkyne* is the linear molecule acetylene ($H—C \equiv C—H$).

We previously mentioned that an alkane-based compound with more than one carbon chain (a branched chain compound) is assigned a basic name corresponding to the name of the longest carbon chain. If a carbon-carbon double bond is present, the longest continuous chain containing the double bond forms the base name, with the systematic name of the respective alkene. For example, consider the following molecule based on *n*-butene-1:

$$\begin{array}{c} CH_3 \\ | \\ CH_3CH_2C{=}CH_2 \end{array}$$

2-methyl-butene-1

If an alkene is converted to a free radical by loss of a hydrogen atom, the resulting atomic group has the generic name *alkenyl*. Such radicals are substitutional groups in their own right, analogous to the alkyls, which is why the generic name ends with -*yl*.

Compounds with two or more double bonds are known as *alkadienes*, *alkatrienes*, *alkatetraenes*, etc., with the suffix indicating the number of double bonds. Numbers indicate the positions of these double bonds. An example is 1,3 butadiene:

$$CH_2{=}CH—CH{=}CH_2.$$

G.4 CYCLOALKANES

An important homologous series of *cyclic hydrocarbons* are the *cycloalkanes*. As the names imply, the structure of these molecules is based on a closed ring of carbon atoms. The chemical properties of cycloalkanes with small rings are intermediate between those of the saturated alkanes and the unsaturated alkenes. The name of each molecule is obtained by adding the prefix *cyclo-* to the name of the corresponding alkane having the same number of carbon atoms as in the ring. The chemical formulas are represented by the general form C_nH_{2n}. The first few are listed at the top of the next page.

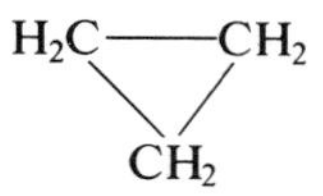

$$H_2C\text{———}CH_2$$
$$\diagdown\diagup$$
$$CH_2$$

Cyclopropane

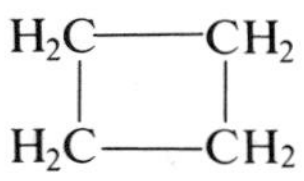

$$H_2C\text{———}CH_2$$
$$|\qquad\qquad|$$
$$H_2C\text{———}CH_2$$

Cyclobutane

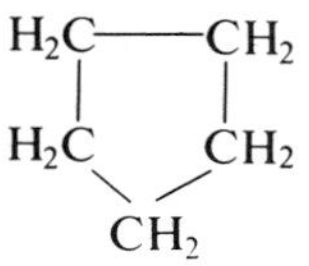

$$H_2C\text{———}CH_2$$
$$|\qquad\qquad|$$
$$H_2C\qquad CH_2$$
$$\diagdown\qquad\diagup$$
$$CH_2$$

Cyclopentane

$$CH_2$$
$$H_2C\qquad\qquad CH_2$$
$$|\qquad\qquad\qquad|$$
$$H_2C\qquad\qquad CH_2$$
$$CH_2$$

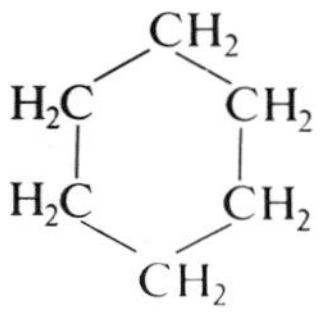

Cyclohexane

Polyalkyl derivatives of cyclopentane and cyclohexane, those in which one or more hydrogens is replaced by an alkyl, are known in petroleum technology as naphthenes. The positions of substituting alkyls are indicated by numbers, in such a way as to minimize their numerical sum. For instance, 1-ethyl-3 methylcyclopentane:

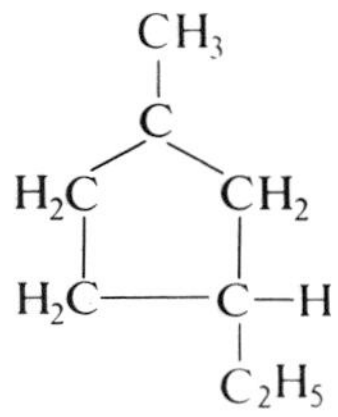

$$CH_3$$
$$|$$
$$C$$
$$H_2C\qquad CH_2$$
$$|\qquad\qquad|$$
$$H_2C\text{———}C\text{—}H$$
$$|$$
$$C_2H_5$$

The relationship between alkanes and alkyls carries through for cycloalkanes and cycloalkyls when a cycloalkane is made into a free radical (or substitutional group) by removal of a hydrogen atom. An example of a cycloalkyl halide is cyclopentyl chloride:

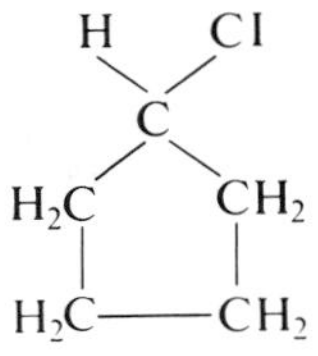

$$H\qquad Cl$$
$$\diagdown\diagup$$
$$C$$
$$H_2C\qquad CH_2$$
$$|\qquad\qquad|$$
$$H_2C\text{———}CH_2$$

G.5 ARENES

A second type of cyclic hydrocarbon has a key part in the production of photochemical smog—the *arenes*. These molecules are also known as *aromatics*. The latter name derives from the fragrant odor associated with many of these compounds. The simplest type from which the others derive is benzene. Its formula C_6H_6 is suggestive of a highly unsaturated hydrocarbon. Yet substitution rather than addition is its characteristic reaction. The resolution of this apparent contradiction came with A. Kekulé's suggestion in 1865 that benzene has a cyclic structure with three double bonds:

An important feature of the structure of this molecule becomes apparent if two of the hydrogen atoms are replaced by methyl groups or other alkyls:

(a)   (b)

The bond arrangement within the closed ring in (a) does not in fact correspond to a different molecule from that shown in (b). Neither arrangement of bonds, or electron orbits, represents what happens in the actual molecule. The configuration which corresponds to the lowest energy level is a hybrid of both, with the electrons resonating between the two. As a result, benzene is more stable than would be indicated by the bond energies of just one arrangement. The resonance aspect is a pure quantum phenomenon and has no analog in our everyday world in which we see only the behavior of large objects. The molecular structure of an arene may be denoted by either configuration of bonds so long as this hybrid structure is implicitly understood. A notation which is appearing more frequently in the literature avoids the arbitrary assignment of double bonds by replacing them with a circle; another shorthand is to omit hydrogen atoms if they are side groups and carbon atoms which compose the ring. The molecular structure is then indicated solely by the arrangement of bonds and positions of the non-hydrogenic side groups. In this notation, benzene is indicated by

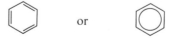

                         or

Disubstitution and trisubstitution of identical alkyl groups may each be accomplished in any of three possible arrangements, as enumerated below

ortho- (*o*-)   meta- (*m*-)   para- (*p*-)

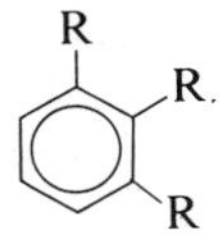

vicinal (*vic-*)

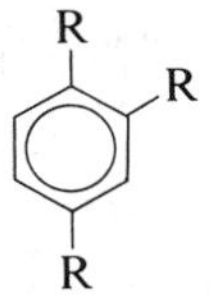

asymmetrical (*as-*)

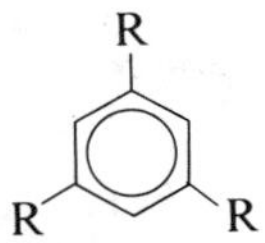

symmetrical (*s-*)

In a chemical formula the particular arrangement is indicated by use of the appropriate prefix given above in parentheses. One well-known compound which is an example of monosubstitution of the methyl group is toluene $C_6H_5CH_3$, which would be known as methylbenzene in the systematic nomenclature. Another is styrene $C_6H_5CH = CH_2$. Further substitution of toluene by a halogen, say bromine, could produce three possible molecules:

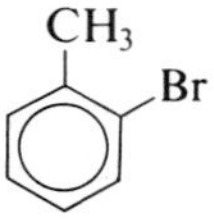

o-bromotoluene 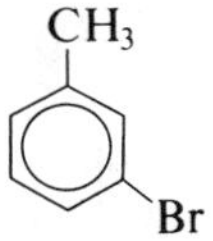   *m*-bromotoluene

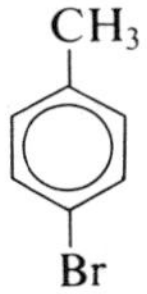

p-bromotoluene

In an atmosphere exposed to light, substitution occurs preferentially by a mechanism that first involves forming a radical on the side chain; in the presence of chlorine, we have:

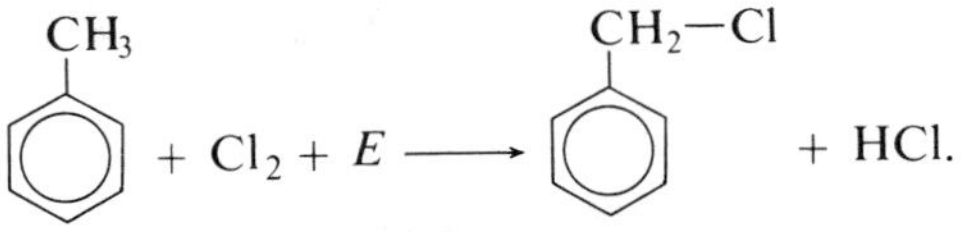

The symbol E represents the radiant energy that is absorbed by the toluene molecule. Participation by the methyl group makes toluene much more reactive in photochemical smog than benzene.

Another common arene is xylene, which has two substitutional methyl groups. The three possible isomers together with their common and systematic names are indicated below:

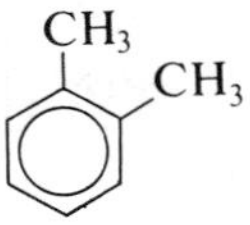

o-xylene   *m*-xylene   *p*-xylene
(1,2 dimethylbenzene)   (1,3 dimethylbenzene)   (1,4 dimethylbenzene)

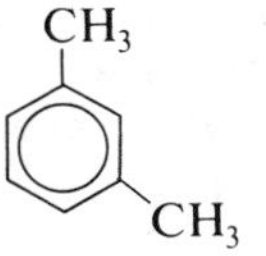

Tri- and tetra-alkyl benzenes are very reactive components of photochemical smog. Ortho- and para-dialkyl benzenes are somewhat less reactive, and benzene itself is not at all reactive.

Numerous polycyclic aromatic compounds are known which consist of benzene rings joined with pairs of carbons sharing a common side. These are known as *polycyclic* or *polynuclear aromatic hydrocarbons*. The three simplest are naphthalene, anthracene, and phenanthrene. The first is a common constituent of moth balls.

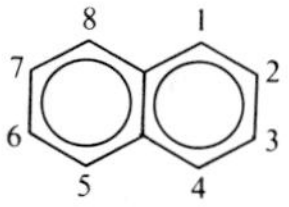

Naphthalene 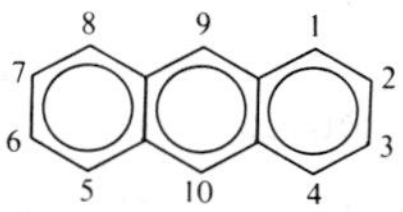   Anthracene 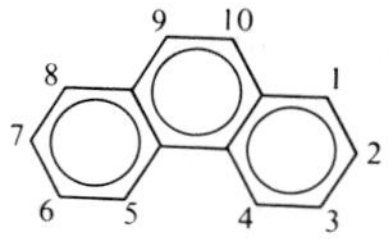   Phenanthrene

The numbers indicate the accepted convention for indicating the positions of substitutional groups. As an illustration, consider 2 methylnaphthalene.

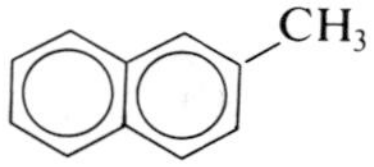

Some polynuclear aromatics in sufficient quantities are known to produce cancer in the skin of animals. The most powerful animal carcinogen found in polluted community air is benzo(a)pyrene or benzo-3,4 pyrene:

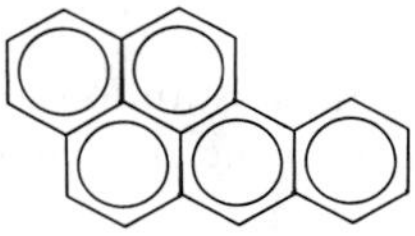

The base for this molecule is the four-ring structure pyrene:

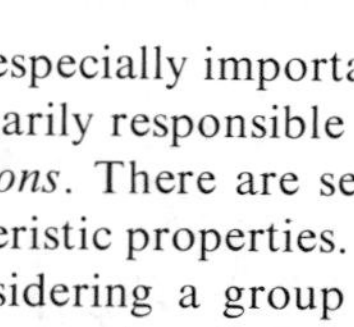

G.6 OXYGENATED HYDROCARBONS

One class of organic compounds is especially important in forming photochemical smog. Members of the class are also primarily responsible for eye irritation and plant damage. These are the *oxygenated hydrocarbons*. There are several distinct groups of oxygenated hydrocarbons, each with its characteristic properties. Perhaps one can most easily become acquainted with them by first considering a group that appears to have no important role in smog, the *alcohols*. Alcohols are not an important ingredient, because they are not emitted in comparatively great quantity into the air.

The simplest alcohols are composed of an alkyl which is bound with a hydroxyl (OH) and therefore have the generic formula ROH. Their names are derived from the respective alkane in which the hydroxyl is considered to have been substituted for a hydrogen. Examples include methanol, ethanol, propanol-l (or *n*-propyl alcohol), and propanol-2 (or isopropyl alcohol). If we wished to be completely systematic, we might include water as the first member of the series. More generally, alcohols comprise the alkane derivatives which have one or more hydrogens replaced by hydroxyls. Just as occurs in water, the hydroxyl group is electrically polarized, so the simpler molecules of the alcohols in the liquid state tend to attract each other. The outermost hydrogen of one hydroxyl aligns itself with the oxygen of another. Alcohol molecules are similarly attracted to water molecules. The three simplest alkyl alcohols are consequently miscible with water in all proportions, although the higher ones are less so.

G.7 ALDEHYDES AND KETONES

Two groups of oxygenated hydrocarbons which are reactive in photochemical smog are the *aldehydes* and *ketones*. These molecules are based on the arrangement in which carbon and oxygen are joined by a double bond, unlike the alcohols where the hydroxyl is joined by a single bond. The basic unit is therefore the *carbonyl*, or carbon-oxygen double bond:

$$\diagdown \atop \diagup \; C{=}O$$

The carbonyl group occurs in contexts which do not involve hydrogen atoms. One common example is carbon dioxide, which has another oxygen joined to the carbon by a double bond ($O{=}C{=}O$). The molecule is linear (as contrasted with water), because the double bond restricts the molecule's ability to bend away from its configuration of highest symmetry.

However our main interest now is on the molecules which can be formed by attaching hydrocarbon groups to the carbon. If one bond is shared with a hydrogen atom and the other with a hydrocarbon group such as an alkyl, R, the possible molecules are known collectively as the *aldehydes*:

$$\begin{array}{c} R \\ \diagdown \\ \diagup \; C{=}O \\ H \end{array}$$

These are commonly found in the exhaust of motor vehicles as a result of incomplete combustion of hydrocarbons. The simplest example of an aldehyde is formaldehyde (CH_2O), where R is a hydrogen atom.

If both bonds of the carbonyl are connected to hydrocarbon groups, the molecules are known as *ketones*. Acetone is an example of a ketone formed by bonding to two methyl groups.

Aldehydes can be produced by oxidation of the alcohols. By this we mean that the ratio of the number of oxygens to the number of hydrogens of the molecule is increased. This can be accomplished in solution for example by reaction with chromic acid. A further oxidation of the aldehyde converts it into a *carboxylic acid*. The sequence for ethyl alcohol is indicated schematically as

$$\underset{\text{Ethyl alcohol}}{H-\underset{\underset{H}{|}}{\overset{\overset{H}{|}}{C}}-\underset{\underset{H}{|}}{\overset{\overset{H}{|}}{C}}-OH} \longrightarrow \underset{\text{Acetaldehyde}}{H-\underset{\underset{H}{|}}{\overset{\overset{H}{|}}{C}}-\underset{\underset{H}{|}}{C}{=}O} \longrightarrow \underset{\text{Acetic acid}}{H-\underset{\underset{H}{|}}{\overset{\overset{H}{|}}{C}}-\underset{\underset{OH}{|}}{C}{=}O}$$

The name of each aldehyde is derived from the name of the corresponding carboxylic acid. We will come back to the question of names shortly. First we wish to point out an important feature of the above sequence of oxidations.

Methane and ethane as paraffins are resistive to oxidation. The fact that the corresponding alcohol does not resist oxidation indicates that the presence of oxygen in a molecule confers a degree of susceptibility to further oxidation at the oxygen site. Thus the first oxidation step leads to formation of a double bond rather than a substitution of oxygen onto another carbon atom. It is empirically determined that an oxidized carbon is the one most susceptible to further oxidation. This also holds for the second step of the reaction. The oxygen atom is added to the carbon which already has an oxygen (actually in this example an OH group substitutes for a hydrogen). The susceptibility of oxygenated hydrocarbons for further oxidation makes them, in gaseous form, active components of the oxidation reactions within photochemical smog.

Let us now return to the system by which the aldehydes are named. We have noted that they are named after the following carboxylic acids into which they are converted by oxidation:

$$\text{Formic acid} \qquad \text{H}-\overset{\displaystyle \overset{O}{\|}}{C}-\text{OH}$$

$$\text{Acetic acid} \qquad \text{CH}_3-\overset{\displaystyle \overset{O}{\|}}{C}-\text{OH}$$

$$\text{Propionic acid} \qquad \text{CH}_3\text{CH}_2-\overset{\displaystyle \overset{O}{\|}}{C}-\text{OH}$$

$$n\text{-butyric acid} \qquad \text{CH}_3\text{CH}_2\text{CH}_2-\overset{\displaystyle \overset{O}{\|}}{C}-\text{OH}$$

The names of propionic and higher acids correspond to the systematic nomenclature we have been using consistently, in which the name is derived from the number of carbons in the longest chain containing the carbonyl. However, an exception is made for the two simplest members, both of which have common names. Thus the corresponding names of the first few aldehydes are

$$\text{Formaldehyde} \qquad \text{H}-\overset{\displaystyle \overset{O}{\|}}{C}-\text{H}$$

$$\text{Acetaldehyde} \qquad \text{CH}_3-\overset{\displaystyle \overset{O}{\|}}{C}-\text{H}$$

$$\text{Propionaldehyde} \qquad \text{CH}_3\text{CH}_2-\overset{\displaystyle \overset{O}{\|}}{C}-\text{H}$$

A more complex aldehyde found in photochemical smog is the alkene-based molecule acrolein:

Acrolein   (2-propenal)

$$CH_2{=}CH{-}\overset{\displaystyle O}{\overset{\|}{C}}{-}H$$

Both formaldehyde and acrolein are eye-irritating components of photochemical smog.

As mentioned, substitution of a hydroxyl for the hydrogen in an aldehyde produces a carboxylic acid. Why is this type of molecule an acid? The reason that it partially dissociates in water solution lies in the fact that by ridding itself of the end hydrogen which clings to the oxygen, it lowers its energy by resonating between two configurations. We can indicate the two conditions by the following schematic representation:

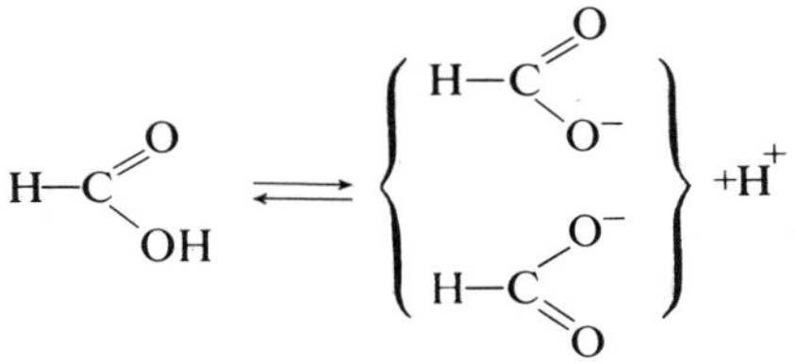

Therefore the driving influence for dissociation is the subsequent resonance. The resulting ionic configuration is stable only in polar solvents such as water, where the energy of the solution is lowered through the mutual attraction of oppositely charged ends of the polar molecules of solvent and solute. Organic acids do not dissociate in nonpolar solvents such as benzene.

Numerous other molecules based on the carbonyl unit are possible. Carboxylic acids may be halogenated to replace the hydroxyl by a halide. The product is an acid (*acyl*) halide, such as formyl chloride

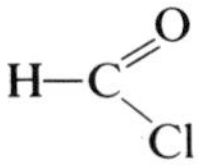

Other derivatives of carboxylic acids also take their base names from the respective acid. Some involve substitution of another group for the hydroxyl. Treatment with nitric acid (HNO_3) will result in NO_3 substituting for OH, yielding the *acyl nitrates*. An example is acetyl nitrate

$$CH_3{-}\overset{\displaystyle O}{\overset{\|}{C}}{-}OH + HNO_3 \longrightarrow CH_3{-}\overset{\displaystyle O}{\overset{\|}{C}}{-}O{-}NO_2 + H_2O.$$

When further oxidized in photochemical smog, the acyl nitrates produce chemicals which are very potent oxidizing agents.

This appendix has not introduced all of the homologous groups of organic chemistry and has barely touched upon the complexities of the chemistry of the more complicated molecules. It was our purpose to summarize only the essentials for a general acquaintance-ship with the organic molecules which are important as air pollutants. For further details the reader should consult a standard text in organic chemistry, such as the ones listed below.

FOR FURTHER READING

JOHN D. ROBERTS and MARJORIE C. CASERIO   *Basic Principles of Organic Chemistry*, New York: W. A. Benjamin, 1965.

LOUIS F. FIESER and MARY FIESER   *Introduction to Organic Chemistry*, Boston: D.C. Heath, 1957.

SUBJECT INDEX